SPECTROSCOPIC DATA

Volume 2

Homonuclear Diatomic Molecules

SPECTROSCOPIC DATA

Volume 1 — Heteronuclear Diatomic Molecules
  Edited by S. N. Suchard • 1975

Volume 2 — Homonuclear Diatomic Molecules
  Edited by S. N. Suchard and J. E. Melzer • 1976

SPECTROSCOPIC DATA

Volume 2

Homonuclear Diatomic Molecules

Edited by

S. N. Suchard and J. E. Melzer

The Aerospace Corporation
Los Angeles, California

SPRINGER SCIENCE+BUSINESS MEDIA, LLC

Library of Congress Cataloging in Publication Data

Suchard, S   N
  Spectroscopic data.

  Vol. 2 edited by S. N. Suchard and J. E. Melzer. Vol. 1: chiefly tables.
  Includes bibliographical references.
  CONTENTS: v. 1. Heteronuclear diatomic molecules—v. 2 Homonuclear diatom-
ic molecules.
  1. Spectrum analysis—Tables, etc. I. Melzer, J. E. II. Title.
QC453.s85   535'.84'0212   74-34288
ISBN 978-1-4757-1387-9   ISBN 978-1-4757-1385-5 (eBook)
DOI 10.1007/978-1-4757-1385-5

Published in 1976 by Springer Science+Business Media New York
Originally published by IFI/Plenum Data Company in 1976
Softcover reprint of the hardcover 1st edition 1976

Preface

During the preparation of this compilation, many people contributed; the compilers wish to thank all of them. In particular they appreciate the efforts of V. Gilbertson, the manuscript typist, and those of K. C. Bregand, J. A. Kiley, and W. H. McPherson, who gave editorial assistance. They would like to thank Dr. J. R. Schwartz for his cooperation and encouragement. In addition, they extend their gratitude to Dr. L. Wilson of the Air Force Weapons Laboratory, who gave the initial impetus to this project.

Contents

I. INTRODUCTION

In recent years, the need for a complete collection of information relevant to diatomic molecules has become evident. Several excellent collections of this type of information have been available for many years (Refs. 1-3); however, the state of our collective knowledge has been considerably expanded since their publication. To this end, a comprehensive compendium of spectroscopic information relevant to selected heteronuclear diatomic molecules was compiled and disseminated by us during FY 74 (Ref. 4). At present, however, if recent information concerning a specific homonuclear species is desired, it is still necessary to institute a time-consuming library search to collect the required information from the large number of sources in which it may have been presented. If information concerning an entire isoelectric molecular series is required, the time involved in collecting the information can be considerable.

This compilation was assembled in the hopes of solving this time problem. We have attempted to gather a complete collection of spectroscopic information relevant to homonuclear diatomic systems. Unlike the compendium for heteronuclear molecules, where the decision was made to restrict our collection to molecules that, from thermodynamic reasoning, could be produced in an electronically excited state by a chemical reaction between a ground state atom and a ground state molecule, this collection includes information relevant to all homonuclear diatomic molecules.

The organization of the material has been patterned after our heteronuclear compendium (Ref. 4). We believe that this form of presentation displays the spectroscopic information in a manner that is amenable to efficient retrieval. The material itself was located with the help of several excellent earlier compendia (Refs. 1-3), a recent bibliography of spectroscopic data (Ref. 5), and from the continously updated Berkeley Newsletters collected by J. G. Phillips and S. P. Davis of the University of California at Berkeley (Ref. 6).

II. ORGANIZATION OF THE SPECTROSCOPIC TABLE

The information that we are presenting deals primarily with the electronic spectra of homonuclear diatomic systems. In general, the spectroscopic constants have been derived from the interpretation of electronic spectra; however, when the amount of data was insufficient, the information was taken from alternative sources.

For simplicity, the molecules are presented in alphabetical order. The information for most of the molecules is broken into five separate sections, with several of the sections broken further into several subsections. Information is presented according to the following format.

METHODS OF PRODUCTION AND EXPERIMENTAL TECHNIQUE

The most favorable sources for the production of the molecule of interest and important experimental techniques used for studying the molecule.

BAND SYSTEMS

This section is broken into two subsections. In the first subsection, a general description of the molecular transition of each system or group is presented; the description is divided into eight headings.

1. System numbers or common designations for the system, or both (e.g., Swan bands of C_2).

2. Transition. Conventional or quantum designation for the states involved. The signs →, ←, and ⇌ refer to systems observed in emission, in absorption, or in both emission and absorption.

3. Sources. The most favorable sources for producing the particular system.

4. Wavelength Limits. Spectral range of the system (Å).

5. Degrading. Direction of band head shading (R ≡ red, V ≡ violet).

6. Band Head $\nu_{0,0}$ or Characteristic Bands λ. Characteristic spectral bands free from overlapping bands or with sharp heads.

7. Remarks. Additional information that is useful in characterizing the particular system.

8. Bibliography. Listing of references that concern themselves with the particular system.

The second subsystem presents a more detailed analysis of the system. Wavelengths (Å) of band heads or origins, intensities, and vibrational classifications are presented where available. Other available and relevant information is presented to characterize the system.

SPECTROSCOPIC CONSTANTS

The molecular constants that totally define the electronic states of the molecule are presented under ten headings. If not specifically mentioned, these constants refer to molecules made of the most abundant isotopes.

1. State. Quantum specification of the electronic state
2. T_e. Electronic energy above ground state (cm^{-1})
3. ω_e. Vibrational spacing (cm^{-1})
4. $x_e \omega_e$. Anharmonic correction to vibrational spacing (cm^{-1})
5. B_e. Rigid rotator rotational spacing (cm^{-1})
6. α_e. Nonrigid rotator correction to B_e (cm^{-1})
7. D_e. Anharmonic correction to rotational spacing (cm^{-1})
8. r_e. Equilibrium internuclear distance (Å)
9. Remarks. Additional relevant information
10. Bibliography. Important references

Other molecular constants or information, if known, are given as footnotes. The dissociation energy D_o^o or D_{298}^o is given in cm^{-1}, kcal/mole, and eV. The greater majority of the values for the molecular dissociation energies have been adopted from the book, Dissociation Energies and Spectra of Diatomic Molecules, by Gaydon (Ref. 7). When molecular values that are more recent than Gaydon's for the dissociation energy exist, the appropriate reference is cited. Dissociation energies taken directly from Gaydon, it should be noted, are not followed by a reference.

PERTURBATIONS AND GENERAL INFORMATION

This section encompasses all other information that would be useful for a complete understanding of the specific molecule. The information, where available, includes predissociations, perturbations, dipole moments, ionization potentials, potential energy curves, Franck-Condon factors, spontaneous lifetimes, rates of production, and deactivation and branching ratios. We hope that this section, in conjunction with the other information set forth, presents a complete description of the physical parameters associated with each molecular species.

BIBLIOGRAPHY

Following the format of our heteronuclear compendium (Ref. 4), in addition to presenting the references used to gather information, a short description of the important points of the paper is given. The bibliography takes into account most papers published through 1974; however, during final preparation of the manuscript, numerous omissions were noted. It is our intention to update this report at suitable intervals; consequently, the omissions will be rectified in a supplement to this report.

Our referencing system is similar to that employed previously in that the references are presented in terms of two numbers. The first number refers to the year of publication, and the second number refers to a running count of the references for each specific molecule.

III. NOTATION AND NOTATIONAL CONVERSION FORMULAS

The total energy of a given state of a diatomic molecule is given by the formula

$$T = T_e + G + F \qquad (1)$$

where T_e is electronic energy, G is vibrational energy, and F is rotational energy. Further breaking down these different forms of energy, the electronic energy T_e is given by

$$T_e = T_o + A\Lambda\Sigma \qquad (2)$$

where T_o is the electronic energy if spin is neglected, A is spin-orbit coupling, Λ is the electronic orbital angular momentum quantum number about the internuclear axis, and Σ is the component of the resulting spin. The vibrational energy G is given by

$$G = \omega_e(v + 1/2) - x_e\omega_e(v + 1/2)^2 + y_e\omega_e(v + 1/2)^3 + \cdots \qquad (3)$$

where v is the vibrational quantum number, ω_e is the harmonic oscillator vibrational spacing, $x_e\omega_e$ is the first anharmonic correction to the vibrational spacing, and $y_e\omega_e$ is the second anharmonic correction. The rotational energy F is given by

$$F = B_v J(J + 1) - D_v J^2(J + 1)^2 + H_v J^3(J + 1)^3 + \cdots \qquad (4)$$

where J is the rotational quantum number, B_v is the rigid rotator rotational spacing, D_v is the first anharmonic correction to the rotational spacing, and H_v is the second anharmonic correction. In addition, there are nonrigid rotator corrections to both B_v and D_v. These corrections are given by

$$B_v = B_e - \alpha_e(v + 1/2) + \gamma_e(v + 1/2)^2 + \cdots \tag{5}$$

and

$$D_v = D_e + \beta_e(v + 1/2)^2 + \cdots \tag{6}$$

where B_e is $\hbar^2/2\mu r_e^2$, μ is the reduced mass, r_e is the equilibrium inter-nuclear distance, α_e and β_e are the first anharmonic corrections, and γ_e is the second anharmonic correction.

Using these formulas, a transition from state 1 at energy T_1 to state 2 at energy T_2 will be at an energy (cm^{-1}) of

$$\upsilon = T_1 - T_2 = (T_{e_1} - T_{e_2}) + (G_1 - G_2) + (F_1 - F_2) \tag{7}$$

Since, in general, the rotational energy changes are much smaller than either the vibrational or electronic changes, neglecting rotation,

$$\upsilon_{v', v''} = \upsilon_e + \omega_e'(v + 1/2) - x_e'\omega_e'(v + 1/2)^2 + y_e'\omega_e'(v + 1/2)^3 + \cdots \tag{8}$$
$$- [\omega_e''(v + 1/2) - x_e''\omega_e''(v + 1/2)^2 + y_e''\omega_e''(v + 1/2)^3 + \cdots]$$

Assuming, as is often the case in absorption, $v' = v'' = 0$, and substituting in Eq. (8),

$$\upsilon_{v', v''} = \upsilon_{0,0} + \omega_o'v' - x_o'\omega_o'v'^2 + y_o'\omega_o'v'^3 + \cdots \tag{9}$$
$$- (\omega_o''v'' - x_o''\omega_o''v''^2 + y_o''\omega_o''v''^2 + \cdots)$$

6

where

$$\omega_o = \omega_e - x_e \omega_e + 3/4\, y_e \omega_e + \cdots$$

$$x_o \omega_o = x_e \omega_e - 3/2\, y_e \omega_e + \cdots$$

$$y_o \omega_o = y_e \omega_e + \cdots$$

A final quantity that is also reported is $\Delta G_{1/2}$. This quantity corresponds to the energy difference between vibrational levels $v = 0$ and $v = 1$, neglecting $y_o \omega_o$, and is represented by

$$\Delta G_{1/2} = \omega_o - x_o \omega_o = \omega_e - 2\, x_e \omega_e$$

Several other molecular constants that are reported for several molecules are represented in the following list:

f = oscillator strength (f-value)

λ, γ = spin-coupling constants for multiplet Σ states (cm^{-1})

q, p = Λ doubling constants (cm^{-1})

μ = electronic dipole moment (D)

R_b^o = reactive branching ratio

IV. CONCLUSIONS ON THE AVAILABILITY OF
SPECTROSCOPIC INFORMATION

It was our hope, when this search was initiated, to find sufficient information in the literature from which to draw definite conclusions regarding the feasibility of the production of an electronic transition chemical laser. As can be seen in the following table, our hopes were not fulfilled. At the present time, there does not appear to be sufficient information about any homonuclear diatomic molecule that would lead one to believe that molecule would definitely produce a chemically pumped electronic transition laser.

We have charted how much is known about the molecular systems we have researched. In general, very little information is available for any given system. The charts immediately precede the detailed information for the collected systems in each volume.

Sufficient information is available for the identification of the emitting species from potential laser systems. If, however, the experiment is to measure the partitioning of energy between the various accessible electronic levels of the molecule, molecular lifetimes and Franck-Condon factors must be known in order to interpret the data. To ascertain the feasibility of a specific molecule for a laser entails knowing not only the reactive branching ratio, but also the lifetime, pumping rates, and deactivation rates. We do not yet have all of this information for a single molecule.

Knowing that, at present, there is almost a complete lack of the necessary information for producing a chemically pumped electronic transition laser, we realize that we are in the same situation as experimenters who were trying to produce vibrational lasers during the early 1960s. At that time, pumping rates, vibrational distributions, and deactivation rates were all unknown, as is now the case for electronic transition lasers. Also in analogy with vibrational lasers, however, the ultimate demonstration of a chemically pumped electronic transition laser should be possible.

The question then arises: What information is required and is there any ordering in the importance of the information? The answer becomes obvious when you remember that an inversion must be created before laser action can occur. Therefore, a knowledge of the reactive branching ratio is absolutely necessary before predictions can be made as to the suitability of a molecule as a potential laser. Even if the lifetimes and pumping rates appear suitable, an inversion must exist before the system can lase. As previously mentioned, the interpretation of spectral intensities to determine the branching ratio for a particular transition requires a knowledge of both the Franck-Condon factors and radiative lifetime for the transition. The Franck-Condon factors, however, cannot be accurately calculated without accurate knowledge of the spectroscopic constants defining the two electronic states between which the transition takes place. So, at least from the experimental point of view, it is imperative that accurate spectroscopic constants and radiative lifetimes be known for the determination of reactive branching ratios. From a theoretical viewpoint, it may only be necessary to have accurate spectroscopic constants.

Whereas a sufficiently large reactive branching ratio may be a necessary condition for a system to reach threshold, it is not sufficient. Even with a sufficiently large reactive branching ratio, if the electronic state of interest is not produced in a time that is fast as compared to the rate at which it is being quenched, either by spontaneous emission or de-excitation, the required inversion cannot be produced. Consequently, once a particular set of reactants have been shown to produce a molecule with a sufficiently large reactive branching ratio, either by experiment, theory, or spin conservation rules, deactivation studies of the molecule should be actively pursued as well as similar studies on other molecules of the same family (e.g., Na_2, K_2, Rb_2, and Cs_2, or O_2, S_2, Se_2, and Te_2). Once a suitable reactive branching ratio is found for a system and the important quenching rates measured, kinetic calculations can be made to determine proper operating conditions for the production of an inversion and laser action.

9

The above approach for determining the feasibility of chemically pumped electronic transition is sound. Whereas the actual production of the laser may prove to be quite difficult, the probability of finding one approaches unity. If one were to assign priorities to the information needed to assist the researcher, they would be as follows:

1. Reactive branching ratios. To ascertain the maximum possible inversion

2. Vibrational level distributions of the product molecules. A total inversion may not exist between the entire upper and lower electronic levels, but it still can exist between specific vibration-rotation levels of the two states.

3. Radiative lifetimes. Necessary for the experimental determination of reactive branching ratios and also for the dictation of minimum reaction rates

4. Spectroscopic constants and dissociation energies. Needed to calculate Franck-Condon factors

5. Pumping and quenching rates. Even if the inversion exists, it must do so on the proper time scale.

Lastly, while the straightforward research approach to the production of a laser may be aesthetically pleasing, the intuitive "shotgun" approach should not be discounted. The largest boost that could accelerate the discovery of new lasers is the discovery and understanding to the first chemically pumped electronic transition laser system.

Since the beginning of the keeping of sports statistics, no human being had run a four-minute mile. Once this record had been broken, however, it was not long before many had done so. Possibly this analogy will hold here.

REFERENCES

1. G. Herzberg, Molecular Spectra and Molecular Structure. I. Spectra of Diatomic Molecules, D. Van Nostrand Co., Inc., Princeton (1950).

2. R. W. B. Pearse and A. G. Gaydon, The Identification of Molecular Spectra, Chapman and Hall Ltd., London (1965).

3. B. Rosen, ed., Selected Constants – Spectroscopic Data Relative to Diatomic Molecules, Pergamon Press, Oxford (1970).

4. S. N. Suchard, ed., Spectroscopic Data, Volume I – Heteronuclear Diatomic Molecules, Parts A and B, IFI/Plenum, New York (1975).

5. R. F. Barrow, Ed., Diatomic Molecules – A Critical Bibliography of Spectroscopic Data, Vols. I-II, National Center for Scientific Research, Paris (1973).

6. J. G. Phillips and S. P. Davis, Berkeley Newsletters, University of California, Berkeley (1960-present).

7. A. G. Gaydon, Dissociation Energies and Spectra of Diatomic Molecules, Chapman and Hall Ltd., London (1968).

SPECTROSCOPIC INFORMATION SUMMARY

MOLECULE	VIBRA-TIONAL CONSTANTS	ROTA-TIONAL CONSTANTS	VIBRA-TIONAL LEVEL DISTRIBU-TIONS	DISSO-CIATION ENERGY	LIFE-TIMES	FRANCK-CONDON FACTORS	BRANCH-ING RATIOS	QUENCH-ING	LASER ACTION OBSERVED	
									VIBRA-TIONAL	ELEC-TRONIC
Ac_2										
Ag_2	X			X						
Al_2	X	X		X						
Am_2										
Ar_2	P	P		X						X
As_2	X	X		X		P		P		
At_2										
Au_2	X	X		X		P				
B_2	X	X		X						
Ba_2										
Be_2				P						
Bi_2	X	P		X						
Bk_2										
Br_2	X	X	P	X	P	X		P		
C_2	X	X	P	X	X					
Ca_2	P	P		X						
Cd_2										
Ce_2				X						
Cf_2										
Cl_2	X	P		X		P		P		
Cm_2										
Co_2				X						
Cr_2				X						
Cs_2	X			X	P					

X=SUBSTANTIAL INFORMATION; P=SKETCHY INFORMATION; NO NOTATION=NO INFORMATION

MOLECULE	VIBRA-TIONAL CONSTANTS	ROTA-TIONAL CONSTANTS	VIBRA-TIONAL LEVEL DISTRIBU-TIONS	DISSO-CIATION ENERGY	LIFE-TIMES	FRANCK-CONDON FACTORS	BRANCH-ING RATIOS	QUENCH-ING	LASER ACTION OBSERVED	
									VIBRA-TIONAL	ELEC-TRONIC
Cu_2	X	P		X		P				
Dy_2				X						
Er_2				X						
Es_2										
Eu_2				X						
F_2	P	P		X						
Fe_2				X						
Fm_2										
Fr_2										
Ga_2				X						
Gd_2				X						
Ge_2				X						
H_2	X	X	P	X	X	X		X		X
He_2	X	X		X						X
Hf_2										
Hg_2				X						
Ho_2										
I_2	X	X	P	X	P	X		X		X
In_2	X			P						
Ir_2										
K_2	X	P		X	P					
Kr_2	X	P		X						X
La_2	P			X						
Li_2	X	X		X	P					

X=SUBSTANTIAL INFORMATION; P=SKETCHY INFORMATION; NO NOTATION=NO INFORMATION

MOLECULE	VIBRA-TIONAL CONSTANTS	ROTA-TIONAL CONSTANTS	VIBRA-TIONAL LEVEL DISTRIBU-TIONS	DISSO-CIATION ENERGY	LIFE-TIMES	FRANCK-CONDON FACTORS	BRANCH-ING RATIOS	QUENCH-ING	LASER ACTION OBSERVED	
									VIBRA-TIONAL	ELEC-TRONIC
Lu_2										
Md_2										
Mg_2	X	X		X		X				
Mn_2				X						
Mo_2										
N_2	X	X	P	X	X	X		X		X
Na_2	X	X		X	P	P				
Nb_2										
Nd_2				X						
Ne_2	P	P		X	P					
Ni_2				X						
No_2										
Np_2										
O_2	X	X	P	X	X	X		X		
Os_2										
P_2	X	P		X				P		
Pa_2										
Pb_2	X			X						
Pd_2				X						
Pm_2										
Po_2	X			X						
Pr_2				X						
Pt_2										
Pu_2										

X=SUBSTANTIAL INFORMATION; P=SKETCHY INFORMATION; NO NOTATION=NO INFORMATION

MOLECULE	VIBRA-TIONAL CONSTANTS	ROTA-TIONAL CONSTANTS	VIBRA-TIONAL LEVEL DISTRIBU-TIONS	DISSO-CIATION ENERGY	LIFE-TIMES	FRANCK-CONDON FACTORS	BRANCH-ING RATIOS	QUENCH-ING	LASER ACTION OBSERVED	
									VIBRA-TIONAL	ELEC-TRONIC
Ra_2										
Rb_2	X			X	P			P		
Re_2										
Rh_2										
Rn_2										
Ru_2										
S_2	X	X		X	P			P		
Sb_2	X	P		X				P		
Sc_2				X						
Se_2	X	X		X				P		
Si_2	X	X		X				P		
Sm_2				X						
Sn_2				X						
Sr_2										
Ta_2										
Tb_2				X						
Tc_2										
Te_2	X	X		X		X				
Th_2				X						
Ti_2				X						
Tl_2				X						
Tm_2				X						
U_2				X						
V_2				X						

X=SUBSTANTIAL INFORMATION; P=SKETCHY INFORMATION; NO NOTATION=NO INFORMATION

MOLECULE	VIBRA-TIONAL CONSTANTS	ROTA-TIONAL CONSTANTS	VIBRA-TIONAL LEVEL DISTRIBU-TIONS	DISSO-CIATION ENERGY	LIFE-TIMES	FRANCK-CONDON FACTORS	BRANCH-ING RATIOS	QUENCH-ING	LASER ACTION OBSERVED	
									VIBRA-TIONAL	ELEC-TRONIC
W_2										
Xe_2	P			X	P			P		X
Y_2				X						
Yb_2				X						
Zn_2				P						
Zr_2										

X=SUBSTANTIAL INFORMATION; P=SKETCHY INFORMATION; NO NOTATION=NO INFORMATION

$$Ag_2$$

Methods of Production and Experimental Technique

Absorption at 1700-1800°C

Emission at 2000°C

$\Big\}$ from a King furnace.

Absorption and emission in quartz discharge tube.

BAND SYSTEMS

	System	Transition	Sources	Wavelength Limits	Degrading	Band Head, $\nu_{0,0}$	Remarks	Bibliography
	I	$A \rightleftharpoons X^1\Sigma_g^+$	Absorption and Emission	5050-4000	R	22984		(66.6, 59.2, 55.1)
	II		Absorption and Emission	3630-3150				(66.6)
	III	$B \rightleftharpoons X^1\Sigma_g^+$	Absorption and Emission	2880-2750	R	35631		(66.6, 59.2)
	IV	$C \rightleftharpoons X^1\Sigma_g^+$	Absorption and Emission	2710-2640	R	37618		(66.6, 63.4, 59.2)
	V	$D \rightleftharpoons X^1\Sigma_g^+$	Absorption and Emission	2620-2560	R	38995.5		(66.6, 59.2)
	VI	$E \rightleftharpoons X^1\Sigma_g^+$	Absorption and Emission	2560-2460	R	40137		(66.6, 63.4, 59.2)

Ag_2

I. $A \rightleftarrows X^1\Sigma_g^+$ System

Intense bands in absorption, λ (55.1):

v', v''	0	1	2
0	4350.86	4387.34	4424.16
1	4322.08	4357.98	
2	4293.79		4365.30
3	4265.89		

II. Two strong bands and patches of weak continua have been seen at 3280.7 and 3382.9Å. They probably arise from a transition between an unstable upper state and an unstable lower state of Ag_2 (66.6).

III. $B \rightleftarrows X^1\Sigma_g^+$ System

Intense band heads in absorption, λ (intensity) (66.6, 59.2):

v', v''	0	1	2	3	4	5
0	2806.5(8)	2822.5(8)	2839.8(4)	2856.4(1)	2873.3(1)	
1	2792.6(10)	2809.6(7)	2825.4(8)	2842.7(5)		
2	2780.1(8)	2796.0(5)	2812.5(5)	2828.6(5)	2845.6(4)	2861.9(2)
3	2768.4(4)					2849.3(2)
4		2771.7(3)			2756.8(1)	
5		2761.4(2)				

IV. $C \rightleftarrows X^1\Sigma_g^+$ System

Intense sequence with R and Q heads. Intense bands in absorption, λ (66.6, 59.2):

(v', v'')		(v', v'')	
(0, 0)	2657.49(Q)	(3, 3)	2661.7
	2656.79(R)	(4, 4)	2663.2
(1, 0)	2645.4	(5, 5)	2665.2
(1, 1)	2658.97(Q)	(6, 6)	2666.8
	2658.78(R)	(7, 7)	2668.7
(2, 2)	2660.45(Q)	(10, 10)	2674.9
	2660.03(R)		

V. $D \rightleftarrows X^1\Sigma_g^+$ System

$\Delta v = 0$ sequence with intense Q and R heads.   Intense bands in absorption, λ (66.6, 59.2):

(v', v'')		(v', v'')	
(0, 0)	2563.63(Q)	(2, 2)	2567.00(Q)
	2562.64(R)		2566.25(R)
(1, 1)	2565.24(Q)	(3, 3)	2568.0
	2564.38(R)	(4, 4)	2570.0

VI. $E \rightleftarrows X^1\Sigma_g^+$ System

Intense band heads in absorption, λ (66.6, 59.2):

v', v''	0	1	2	3
0	2490.72	2502.80	2514.61	
1	2481.9			2517.53
2		2485.2		
3			2488.1	

Ag$_2$

SPECTROSCOPIC CONSTANTS

State	T_e	ω_e	$x_e\omega_e$	B_e	$\alpha_e \times 10^3$	$D_e \times 10^6$	r_e	Remarks	Bibliography
E	40159.1	146.08	1.54						(63.4, 59.2)
D($^1\Pi_u$)[b]	39023.7	166.7	1.134						(59.2)
C($^1\Pi_u$)[b]	37626.9	172.9	1.07						(63.4, 59.2)
B	35827.3	151.3	0.70						(59.2)
A	22996.4	154.6	0.587					$y_e\omega_e=0.0022$[a]	(59.2, 55.1)
X$^1\Sigma_g^+$	0	207.0	0.643	0.496	0.19		2.5	$y_e\omega_e=0.0003$[a]	(66.6, 59.2, 55.1)

[a] Constants for ^{107}Ag^{109}Ag.

[b] $\Omega = 1_u$.

Dissociation energy = 1.68 ± 0.10 eV, 38.7 kcal/mole, 13550 cm^{-1} (60.3).

22

Perturbations and General Information
====================================

Perturbations and predissociation have been observed in the v = 4, 5, 6, and 7 levels of the ground state (66.6).

Ag_2

BIBLIOGRAPHY

(55. 1)   A - X System,
Vibrational Analysis,
B. Kleman and S. Lindkivist,
Arkiv Fysik 9, 385-390

(59. 2)   B, C, D, E ← X Systems,
Vibrational Analysis,
J. Ruamps,
Ann. Physique 4, 1111-1157

(60. 3)   Dissociation Energy,
M. Ackerman, F. E. Stafford, and J. Drowart,
J. Chem. Phys. 33, 1784-9

(63. 4)   Observation of C, E ← X Systems,
R. C. Maheshwari,
Indian J. Phys. 37, 368-374

(65. 5)   M. D. Dolgushin,
"Construction of Potential Curves for Diatomic Molecules of Copper, Silver, and Gold,"
Opt. and Spect. 19, 519-23

(66. 6)   Systems in Emission,
C. Shin-Piaw, W Loong-Seng, and L. Yoke-Seng,
"Emission Band Systems of Ag_2 Produced in Discharge,"
Nature 209, 1300-1302

Al$_2$

Methods of Production and Experimental Technique

Thermal emission from a King furnace.

Knudsen cell mass spectrometric method.

BAND SYSTEMS

	System	Transition	Sources	Wavelength Limits	Degrading	Band Head, $\nu_{0,0}$	Remarks	Bibliography
	I	$A^3\Sigma_u^- \rightarrow X^3\Sigma_g^-$	Thermal Emission	6600-5650	R	17231.3		(65.4, 64.3, 54.1)

Al_2

I. $\underline{A^3\Sigma_u^- \rightarrow X^3\Sigma_g^- \text{ System}}$

Most intense bands, λ (54.1):

v', v''	0	1	2	3
0	5801	5709		
1	5920		5732	
2	6042	5943		5755
3		6064	5965	
4		6190	6086	5986

SPECTROSCOPIC CONSTANTS

Al_2

State	T_e	ω_e	$x_e\omega_e$	B_e	$\alpha_e \times 10^3$	$D_o \times 10^6$	r_e	Remarks	Bibliography
$A\,^3\Sigma_u^-$	17269.3	278.80	0.831	0.1907	1.3	0.38	2.560	-0.010^a	(65.4, 64.3)
$X\,^3\Sigma_g^-$	0	350.01	2.022	0.2053	1.2	0.30	2.466	-0.010^a	(73.5, 65.4, 64.3)

$^a y_e\omega_e$

Dissociation energy = 2.05 eV, 47.3 kcal/mole, 16544 cm^{-1} (73.5).

27

BIOGRAPHY

(54. 1) P. B. Zeeman,
Can. J. Phys. 32, 9-15

(58. 2) W. A. Chupka, J. Berkowitz, C. F. Giese, and M. G. Inghram,
J. Phys. Chem. 62, 611-4

(64. 3) D. S. Ginter, M. L. Ginter, and K. K. Innes,
Astrophys. J. 139, 365-78

(65. 4) D. E. S. Ginter,
"Electronic Spectra of the Homonuclear Molecules of the Group III
Metals,"
Thesis, Vanderbilt Univ.

(73. 5) C. A. Stearns and F. J. Kohl,
"Mass Spectrometric Determination of the Dissociation Energies of
Gaseous Al$_2$, AlSi, and AlSiO,"
High Temp. Sci. 5, 113-27

Ar$_2$

Methods of Production and Experimental Technique

Solid Ne at 6°K.

Discharge in helium continua.

Electron bombardment of high pressure Ar.

BAND SYSTEMS

	System	Transition	Sources	Wavelength Limits	Degrading	Band Head, ν	Remarks	Bibliography
	I	$^3\Sigma_u^+ \rightarrow X^1\Sigma_g^+$	Discharge	1083 - 1073	V	93124.2		(70.10)
	II	$^1\Sigma_u^+ \rightarrow X^1\Sigma_g^+$	Discharge	1073 - 1065	R	93703.9		(70.10)
	III		Discharge	1052 - 1047	V	95050.7		(70.10)
	IV		Discharge	944 - 941	R	106213.5		(70.10)
	V		Discharge	932 - 928	R	107667.7		(70.10)
	VI		Discharge	922 - 918	R	108884.8		(70.10)
	VII		Discharge	903 - 895	R	111722.4		(70.10)
	VIII		Discharge	893 - 883		113050.9		(70.10)
	IX		Discharge	851 - 846	R	118183		(70.10)
	X		Discharge	7000 - 2900		32258		(66.4)

I. $^3\Sigma_u^+ \rightarrow X\,^1\Sigma_g^+$ System

Bands not classified completely, λ (Intensity) (70.10):

v' | v''	0	1	2	3	4
v-7	1082.2(1)	1082.6(2)	1082.8(2)	1083.0(1)	
v-6	1080.7(2)	1081.0(3)	1081.3(2)	1081.4(1)	1081.6(1)
v-5	1079.2(4)	1079.5(4)	1079.8(3)	1080.0(1)	1080.1(1)
v-4	1077.9(6)	1078.2(4)	1078.4(3)	1078.6(1)	
v-3	1076.6(8)	1076.9(4)	1077.1(3)		
v-2	1075.5(9)	1075.8(3)	1076.0(2)		
v-1	1074.6(9)	1074.9(3)			
v	1073.8(10)	1074.1(5)			

II. $^1\Sigma_u^+ \rightarrow X\,^1\Sigma_g^+$ System

Bands not classified completely, λ (Intensity) (70.10):

v' | v''	0	1	2	3	4
v-3	1072.4(2)	1072.7(3)	1073.0(3)	1073.1(3)	1073.3(2)
v-2	1070.8(4)	1071.1(4)	1071.4(5)	1071.6(3)	1071.7(2)
v-1	1069.4(6)	1069.7(5)	1070.0(4)	1070.1(4)	1070.3(2)
v	1068.2(8)	1068.5(6)	1068.7(4)	1068.9(3)	1069.0(1)
v+1	1067.1(10)	1067.4(3)			
v+2	1066.3(3)	1066.5(6)			

Five other bands have been seen and not identified:

1065.1(3)   1065.2(3)   1065.3(3)   1065.4(3)   1065.4(3).

Probably all five originate with some upper level which is located on the potential hump and thus affected by dissociation through the hump (tunneling effect) (70.10).

III.   System

Band heads, λ (Intensity) (70.10):

v' | v''	0	1	2	3	4
0	1052.0(10)	1052.3(7)	1052.5(9)	1052.7(2)	1052.8(1)
1	1051.4(9)	1051.7(3)	1051.9(4)		
2	1050.8(7)	1051.4(8)			
3	1050.3(5)	1050.6(5)	1050.8(7)	1051.0(5)	1051.1(8)
4	1049.9(4)	1050.2(4)	1050.5(3)	1050.6(5)	
5	1049.7(3)	1049.9(4)	1050.2(3)	1050.4(4)	
6	1049.5(2)	1049.7(2)	1049.9(4)	1050.1(2)	1050.2(2)
7	1049.3(2)	1049.6(2)			
8	1049.2(2)	1049.5(2)			
9	1049.1(1)				

Four additional diffuse bands have also been seen but not positively identified:

       1047.5(1)   1047.3(1)   1047.2(1)   1047.1(1).

IV.   System

Band heads, λ (Intensity) (70.10). These are diffuse and have not been positively identified.

v' | v''	0	1	2	3
v-1	943.1(8)	943.3(5)	943.5(3)	943.6(1)
v	941.5(10)	941.7(3)		

V.   System

Band heads, λ (Intensity) (70.10). These bands are diffuse and have not been positively identified.

v' | v''	0	1	2
v-3	931.7(7)		
v-2	930.2(3)		
v-1	929.2(6)	929.4(4)	929.6(2)
v	928.7(10)	929.0(6)	

Additional diffuse bands have been observed but not identified:

937.7(1)   936.8(10)   935.5(3)   924.2(9)   924.4(10)   924.6(4)

VI.   ## System

Band heads, λ (Intensity) (70.10). These bands have not been positively identified.

v' | v''	0	1	2	3	4
v-5	921. 7(1)	921. 9(1)			
v-4	920. 5(3)	920. 7(3)	920. 9(3)	921. 1(2)	921. 1(1)
v-3	919. 6(3)	919. 8(4)	919. 9(4)		
v-2	918. 9(8)	919. 1(5)	919. 3(3)		
v-1	918. 4(10)	918. 6(3)	918. 7(1)		
v	918. 0(7)	918. 2(5)			

VII.   ## System

Band heads, λ (Intensity) (70.10). These bands have not been positively identified.

v' | v''	0	1	2	3	4
v-5	901. 4(1)	901. 6(2)	901. 8(2)	901. 9(2)	902. 0(1)
v-4	900. 0(2)	900. 2(3)	900. 4(2)	900. 5(2)	900. 6(1)
v-3	898. 7(4)	898. 9(5)	899. 1(5)	899. 2(3)	899. 3(2)
v-2	897. 4(7)	897. 6(5)	897. 8(4)	897. 9(2)	898. 0(1)
v-1	896. 2(8)	896. 4(6)	896. 5(4)	896. 7(1)	
v	895. 0(10)	895. 2(7)	895. 4(4)	895. 5(1)	

VIII.   ## System

Band heads, λ (Intensity) (70.10). These bands have not been positively identified.

v' | v''	0	1	2
v-12	892. 5(6)		
v-11	891. 5(7)		
v-10	890. 5(8)		
v-9	889. 6(8)		
v-8	888. 7(8)	888. 9(1)	889. 1(1)
v-7	887. 9(8)	888. 1(2)	888. 2(2)
v-6	887. 1(8)	887. 3(4)	887. 4(2)
v-5	886. 3(7)	886. 5(5)	886. 7(2)
v-4	885. 6(6)	885. 8(6)	886. 0(2)
v-3	884. 9(2)	885. 1(7)	
v-2	884. 3(2)	884. 5(10)	884. 7(3)
v-1		884. 0(5)	884. 1(4)
v			883. 7(4)

IX.   <u>System</u>

Band heads, λ (Intensity) (70.10).   These bands are very diffuse and
have not been positively identified.

v' | v''	0	1
v-5	849. 8(2)	850. 0(1)
v-4	849. 0(2)	
v-3	848. 1(4)	
v-2	847. 4(6)	847. 5(4)
v-1	846. 7(8)	846. 9(1)
v	846. 1(10)	846. 3(5)

Ar_2

SPECTROSCOPIC CONSTANTS

State	T_e	ω_e	$x_e\omega_e$	B_e	$\alpha_e \times 10^3$	D_e	r_e	Remarks	Bibliography
$X^1\Sigma_g^+$	0	30.68	2.56	0.060	4.0	91.6	3.8		(73.19, 70.10)

Dissociation energy = 9.53×10^{-3} eV, 220 cal/mole, 76.9 cm^{-1} (70.10).

Perturbations and General Information

Laser action has been observed at 1261Å from the $^{3,1}\Sigma_u \rightarrow X^1\Sigma_g^+$ transition (74.23).

BIBLIOGRAPHY

(55. 1)   Y. Tanaka,
            J. Opt. Soc. 45, 710-3

(65. 2)   P. G. Wilkinson and E. T. Byram,
            Applied Optics 4, 581-8

(65. 3)   Emission in Condensed Discharge,
            R. E. Huffman, J. C. Larrabee, and Y. Tanaka,
            Applied Optics 4, 1581-8

(66. 4)   3100Å Continuum,
            J. F. Prince and W. W. Robertson,
            "Continuum Radiation in an Argon Positive Column,"
            J. Chem. Phys. 45, 2577-84

(67. 5)   P. G. Wilkinson,
            Can. J. Phys. 45, 1715-28

(67. 6)   T. L. Gilbert and A. C. Wahl,
            J. Chem. Phys. 47, 3425-38

(67. 8)   P. G. Wilkinson,
            "Absorption Spectrum of Argon in the 1070-1135Å Region,"
            Can. J. Phys. 46, 315-8

(70. 9)   L. W. Bruch and I. J. McGee,
            "Spectroscopic Data as a Test of Argon Intermolecular Potentials,"
            J. Chem. Phys. 53, 4711-3

(70. 10)   Band Systems I-IX Analysis,
            Y. Tanaka and K. Yoshino,
            "Absorption Spectrum of the Argon Molecule in the Vacuum-uv
            Region,"
            J. Chem. Phys. 53, 2012-30

(71. 11)   C. T. Chen and R. D. Present,
            "Vibrational Levels of Ar$_2$ and the Ar-Ar Pair Potential,"
            J. Chem. Phys. 54, 3645-6

(71. 12)   G. C. Maitland and E. B. Smith,
            "The Intermolecular Pair Potential of Argon,"
            Molec. Phys. 22, 861-8

(71. 13)   J. M. Parson, P. E. Siska, and Y. T. Lee,
           "Intermolecular Potentials from Crossed-Beam Differential Elastic
           Scattering Measurements.   IV.   Ar + Ar,"
           J. Chem. Phys. 56, 1511-1516

(72. 14)   O. Cheshnovsky, B. Raz, and J. Jortner,
           "Temperature Dependence of Rare Gas Molecular Emission in the
           Vacuum Ultraviolet,"
           Chem. Phys. Letters 15, 475-9

(72. 15)   R. J. LeRoy,
           "Improved Spectroscopic Dissociation Energy for Ground-State Ar$_2$,"
           J. Chem. Phys. 57, 573-4

(72. 16)   H. J. M. Hanley, J. A. Barker, J. M. Parson, Y. T. Lee, and M. Klein,
           "Comments on the Interatomic Potential for Argon,"
           Molec. Phys. 24, 11-15

(73. 17)   S. E. Harris, A. H. King, E. A. Stappaerts, and J. F. Young,
           "Stimulated Emission in Multi-Photon-Pumped Xenon and Argon
           Excimers,"
           Appl. Phys. Letters 23, 232-4

(73. 18)   A. Gedanken, B. Raz, and J. Jortner,
           "Emission Spectra of Homonuclear Diatomic Rate Gas Molecules in
           Solid Neon,"
           J. Chem. Phys. 59, 1630-3

(73. 19)   K. K. Docken and T. P. Schafer,
           "Spectroscopic Information on Ground-State Ar$_2$, Kr$_2$, and Xe$_2$ from
           Interatomic Potentials,"
           J. Molec. Spectrosc. 46, 454-9

(73. 20)   R. D. Present,
           "Collision Diameter and Well Depth of the Ar-Ar Interaction,"
           J. Chem. Phys. 58, 2659-60

(74. 21)   P. Lallemand, D. J. David, and B. Bigot,
           "Calculation of Polarizabilities for Atoms and Molecules.   I. Method
           and Application to Ar-Ar,"
           Molec. Phys. 27, 1029-43

(74. 22)   W. M. Hughes, J. Shannon, and R. Hunter,
           "126. 1-nm Molecular Argon Laser,"
           Appl. Phys. Letters 24, 488-90

As$_2$

Methods of Production and Experimental Technique

Absorption.

Fluorescence.

Emission in a discharge tube containing arsenic
  a. in the presence of hydrogen, neon, or helium
  b. continuous flow.

Excitation by rf discharge in arsenic and neon.

BAND SYSTEMS

System	Transition	Sources	Wavelength Limits	Degrading	Characteristic Bands, λ	Remarks	Bibliography
I	$c\left(^3\Sigma_u^+\right) \to X^1\Sigma_g^+$	Discharge	8400-5380	R			(74.19, 72.17, 67.8)
II	$e \to X^1\Sigma_g^+$	Discharge	6250-4440	R			(74.19, 72.17, 67.8)
III	$a\,^3\Sigma_u^+ \to X^1\Sigma_g^+$	Discharge	4500-3700	R	4317.9(2,5) 3870.5(5,1)		(74.19, 72.17, 70.14, 66.7, 37.4)
IV	$A^1\Sigma_u^+ \rightleftharpoons X^1\Sigma_g^+$	Discharge	5555-2240	R	2506.9(5,4) 2449.9(7,3)		(74.19, 72.17, 70.15, 65.6, 37.4, 35.3, 34.2)
V	$B^1\Sigma_u^+ \to X^1\Sigma_g^+$	Discharge	5530-2350	R	3150.6(2,24) 2998.3(1,19) 2554.4(0,4)		(74.19, 72.17, 70.15, 37.4, 35.3)
VI	$b\left(^3\Pi_u\right) \to X^1\Sigma_g^+$	Discharge	3840-2984	R	3235.9(0,18) 3198.2(0,2)	Single progression (v' = 0)	(74.19, 72.17, 37.4, 35.3)
VII	$d\left(^3\Pi_g\right) \to X^1\Sigma_g^+$	Discharge	3390-2980	R	3180 (2,0)		(74.19, 72.17, 67.9)
VIII	$F \leftarrow X^1\Sigma_g^+$	Absorption	2050-1800	V	1915.5(0,0)		(74.19, 72.17)
IX	$G \leftarrow X^1\Sigma_g^+$	Absorption	1915-1775	V	1832.5(0,0)		(74.19, 72.17)

BAND SYSTEMS

	System	Transition	Sources	Wavelength Limits	Degrading	Characteristic Bands, λ	Remarks	Bibliography
	X	$a\,^3\Sigma_u^+ \rightarrow c\left(^3\Sigma_u^+\right)$	Discharge	10000-7400	V	8506		(74.19, 67.9)
	XI	$d\left(^3\Pi_g\right) \rightarrow c\left(^3\Sigma_u^+\right)$	Discharge	7000-5600	V	6051.3(1,0) 5992.6(1,0)		(74.19, 72.17, 67.9)

I. $c\left(^3\Sigma_u^+\right) \to X^1\Sigma_g^+$ System

Distinct bands, λ (70.14):

v', v''	0	1	2	3	4	5	6	7
0						8126	8410	8713
1				7428	7667			
2			7047	7263	7491	7733		
3		6702		7106	7324	7556	7803	
4		6567	6755				7620	
5		6439			7012	7226		
6	6152	6317					7283	
7	6040	6200						

II. $e \to X^1\Sigma_g^+$ System

Distinct bands, λ (70.14):

v', v''	0	1	2	3	4	5	6	7
0						5633	5768	5910
1			5170	5286	5406			
2		4977	5085	5196				
3		4898	5002					
4	4725	4822						
5	4654							
6	4586							
7	4520							

III. $a^3\Sigma_u^+ \to X^1\Sigma_g^+$ System

Partial vibrational scheme, λ (70.14, 66.7, 37.4):

v', v''	0	1	2	3	4	5
0				4286.7	4365.1	4446.3
1		4079.8	4151.5	4225.1	4302.1	4381.3
2	3957.1	4024.5	4094.5			4317.9
3	3905.5	3972.2	4040.0			4256.8
4					4126.7	

IV. $A\,^1\Sigma_u^+ \rightleftarrows X\,^1\Sigma_g^+$ System

Partial vibrational scheme, λ (72.17, 65.6, 37.4):

v', v''	0	1	2	3	4	5	6	7
0								
...								
4						2551.6	2578.9	
5		2430.3	2455.0	2480.7	2506.9	2533.4		2587.9
6		2415.1	2439.4	2464.7	2490.6			2570.4
7	2376.5	2400.7	2424.9	2449.9				

V. $B\,^1\Sigma_u^+ \rightarrow X\,^1\Sigma_g^+$ System

Bands of greatest intensity, λ (37.4):

v', v''	2,24	1,23	2,23	1,22	1,19	0,7	6,10	6,9	0,4
λ	3150.6	3141.8	3113.8	3105.1	2998.3	2638.6	2615.6	2588.0	2554.4

VII. $d\left(^3\Pi_g\right) \rightarrow X\,^1\Sigma_g^+$ System

Partial vibrational scheme, λ (37.4):

v', v''	0	1	2	3	4	5
0	3248.9	3294.7	3341.4	3389.3	3438.2	3488.4
1	3214.1	3258.7		3351.2	3399.3	3448.3
2	3180.1		3268.7	3314.6		3409.2
3	3147.0		3233.8	3278.8	3324.6	3311.4
4	3114.8	3157.1			3288.5	

VIII. $F \leftarrow X\,^1\Sigma_g^+$ System

Partial vibrational scheme, λ (72.17):

v', v''	0	1	2	3	4	5	6	7
0	1915	1931	1947	1963	1980			
1	1902				1965			
2	1887					1966	1982	1998
3	1874				1935			1983
4	1862							

As$_2$

IX. $\quad$ G $\leftarrow$ X$^1\Sigma_g^+$ System

Partial vibrational scheme, λ (72.17):

v', v''	0	1	2	3	4	5	6	7
0	1833	1847	1862	1876	1892	1907		
1	1820	1834		1863	1878	1983	1908	
2	1808		1836			1880	1894	1909
3	1796							
4	1784							1883
5	1775						1858	1872

X. $\quad$ a$^3\Sigma_u^+ \rightarrow$ c$\left(^3\Sigma_u^+\right)$ System

Many bands have been seen for this system but as yet remain unclassified. Some of the more intense bands are (67.9):

8794, 8759, 8738, 8507, 8318, 8256

XI. $\quad$ d$\left(^3\Pi_g\right) \rightarrow$ c$\left(^3\Sigma_u^+\right)$ System

Partial vibrational scheme, λ (Intensity) (67.9):

v', v''	0	1	2	3	4	5	6
0	6118(3)	6236(3)	6359(2)	6486(1)	6616(0)	6748(0)	
	6180(3)	6299(4)	6425(3)	6554(2)		6824(1)	
1	5996(6)	6111(2)	6227(1)	6448(2)	6474(1)	6602(0)	6736(0)
	6054(6)	6171(2)	6289(2)	6313(3)	6541(2)	6673(1)	6809(1)
2	5879(7)	5988(0)	6101(3)	6217(2)		6462(1)	6588(0)
	5935(6)	6046(0)	6162(3)	6279(1)	6399(0)	6528(2)	6658(1)
3	5767(3)	5872(8)		6092(2)	6208(1)	6325(0)	
	5822(3)	5929(7)		6154(2)	6269(1)	6390(0)	
4		5762(3)		5974(0)		6199(0)	
		5816(3)		6031(0)		6260(0)	6380(0)
5			5756(2)				
			5810(0)				

SPECTROSCOPIC CONSTANTS

State	T_e	ω_e	$x_e\omega_e$	B_e	$\alpha_e \times 10^3$	$D_e \times 10^8$	r_e	Remarks	Bibliography
G	54590	385.0	3.2						(72.17)
F	52220								(72.17)
b($^3\Pi_u$)	42000								(70.14)
B$^1\Sigma_u^+$	40898	302	5.8	0.07112 [a]					(74.20, 70.15, 70.14, 69.12)
A$^1\Sigma_u^+$	40336	280	1.0	0.0797	0.31	2.6	2.374		(74.20, 74.19, 73.18, 72.17, 70.15, 70.14)
d($^3\Pi_g$)	30818	336.7	1.36	0.09222	0.33	2.8	2.209		(70.14)
a $^3\Sigma_u^+$	24641	337.0	0.83	0.08666	0.30	2.3	2.279	d	(70.14)
e	19914	330.0	0.90						(70.14)
c$^3\Sigma_u^+(0_u^-)$	14643	314.3	1.09	0.08492	0.35		2.302	c	(70.14)
c$^3\Sigma_u^+(1_u)$	14479	314.3	1.09	0.08472	0.35	1.6	2.305	c	(70.14)
X$^1\Sigma_g^+$	0	429.55	1.117	0.10179	0.333	1.4	2.104	b	(70.14)

[a] B_0; [b] $\gamma_e = -2.8 \times 10^{-7}$, $y_e\omega_e = 1.39 \times 10^{-4}$, $z_e\omega_e = 1.958 \times 10^{-5}$; [c] $\lambda = -1.38$; [d] $\lambda = -0.55$

Dissociation energy = 3.96 eV, 91.3 kcal/mole, 31950 cm^{-1}.

As$_2$

Perturbations and General Information

Potential energy curves (67.10):

State	v	$T(v)(cm^{-1})$	$r_{min}(\overset{\circ}{A})$	$r_{max}(\overset{\circ}{A})$
$X^1\Sigma_g^+$	0	214.3	2.243	2.332
	1	641.7	2.211	2.371
	2	1067.0	2.192	2.397
	3	1489.7	2.175	2.417
	4	1910.2	2.163	2.440
	5	2328.4	2.152	2.458
	6	2744.6	2.141	2.475
	7	3158.4	2.131	2.489
	8	3570.4	2.122	2.507
	9	3979.4	2.115	2.525
	10	4386.5	2.107	2.540
	11	4791.5	2.100	2.554
	12	5194.0	2.094	2.568
	13	5594.4	2.087	2.581
	14	5992.5	2.081	2.594
	15	6388.4	2.075	2.608
	16	6782.0	2.070	2.622
	17	7170.3	2.064	2.635
	18	7562.5	2.060	2.646
	19	7949.1	2.055	2.659
	20	8333.7	2.051	2.671
$a\ ^3\Sigma_u^+$	0	24810.9	2.464	2.568
	1	25146.8	2.430	2.610
	2	25480.7	2.408	2.640
	3	25812.9	2.391	2.668
	4	26143.1	2.376	2.691
	5	26471.5	2.362	2.712
	6	26798.1	2.351	2.731
	7	27122.7	2.340	2.751
$A^1\Sigma_u^+$	0	40474.3	2.745	2.858
	1	40736.7	2.710	2.918
	2	40989.4	2.687	2.958
	3	41205.0	2.672	3.010

Franck-Condon factors for a $^3\Sigma_u^+ \rightarrow X^1\Sigma_g^+$ (68.11):

v',v"	0	1	2	3	4	5	6	7	8
0	0.0029	0.0149	0.0459	0.1015	0.1394	0.1941	0.1653	0.1674	0.0853
1	0.0160	0.0582	0.1181	0.1460	0.0916	0.0318	0.0086	0.0901	0.1508
2	0.0437	0.1151	0.1321	0.0558	0.0011	0.0317	0.1012	0.0438	0.0017
3	0.0685	0.1206	0.0590	0.0000	0.0447	0.0818	0.0139	0.0147	0.0836
4	0.1086	0.1093	0.0089	0.0332	0.0638	0.0184	0.0248	0.0604	0.0145
5	0.1464	0.0471	0.0088	0.0626	0.0140	0.0160	0.0576	0.0053	0.0383
6	0.1498	0.0157	0.0362	0.0446	0.0004	0.0471	0.0102	0.0144	0.0425
7	0.1326	0.0000	0.0635	0.0071	0.0291	0.0337	0.0103	0.0468	0.0010

The perturbations that appear in (2,14) and (2,15) bands of $A^1\Sigma_u^+ \rightleftarrows X^1\Sigma_g^+$ system as well as in the (0,5) and (0,6) bands of $B^1\Sigma_u^+ \rightarrow X^1\Sigma_g^+$ system have been shown to be due to a mutual interaction between A(v = 2) and B(v = 0) levels, although this has been shown to be relatively weak (74.20, 70.15).

As$_2$

BIBLIOGRAPHY

(27. 1)   B. Rosen,
           Z. Physik 43, 69-130

(34. 2)   G. E. Gibson and A. MacFarlane,
           Phys. Rev. 46, 1059-68

(35. 3)   G. M. Almy and G. D. Kinzer,
           Phys. Rev. 47, 721-30

(37. 4)   G. D. Kinzer and G. M. Almy,
           Phys. Rev. 52, 814-21

(64. 5)   G. Pannetier, J. Guillaume, and P. Deschamps,
           J. Chim. Phys. 61, 1463-6

(65. 6)   J. D'Incan, P. Perdigon, and J. Janin,
           C. R. Acad. Sci. 261, 1639-41

(66. 7)   J. D'Incan, P. Perdigon, and J. Janin,
           C. R. Acad. Sci. 262, 951-3

(67. 8)   J. D' Incan, P. Perdigon, and J. Janin,
           C. R. Acad. Sci. 265, 141-3

(67. 9)   S. Mrozowski and C. Santaram,
           J. Opt. Soc. 57, 522-30

(67. 10)   P. Perdigon and R. Grandmontagne,
           J. Physique 28, 433-4

(68. 11)   P. Perdigon, J. D'Incan, and J. Janin,
           "New Analysis at High Resolution of the Many Bands of the D→X
           System of the As$_2$ Molecule,"
           C. R. Acad. Sci. B 267, 202-4

(69. 12)   P. Perdigon, J. D'Incan, and J. Sfeila,
           "Analysis of the Rotational Structure of the Many Bands of the d→X
           and B→X Systems of the As$_2$ Molecule,"
           C. R. Acad. Sci. B 268, 1432-5

(69. 13)   G. A. Ozin,
           "Gas Phase Raman Spectroscopy of Phosphorus, Arsenic, and
           Saturated Sulfur Vapours,"
           J. Chem. Soc. D 1969, 1325-7

(70.14) P. Perdigon and J. D'Incan,
"The Electronic Spectra of the As_2 Molecule,"
Can. J. Phys. **48**, 1140-50

(70.15) P. Perdigon, F. Martin, and J. D'Incan,
"The Mutual Perturbations of A $^1\Sigma_u^+$ (v = 2) and B $^1\Sigma_u^+$ (v = 0) Levels
of the As_2 Molecule,"
J. Molec. Spectrosc. **36**, 341-53

(70.16) I. R. Beattie, G. A. Ozin, and R. O. Berry,
"The Gas-Phase Raman Spectra of P_4, P_2, As_4, and As_2. The
Resonance Fluorescence Spectrum of $^{80}Se_2$. Resonance Fluorescence-
Raman Effects in the Gas-Phase Spectra of Sulphur and I_2. The Effect
of Pressure on the Depolarization Ratios for I_2."
J. Chem. Soc. A **12**, 2071-4

(72.17) A. Topouzkhanian and A. M. Sibai,
"Vacuum Ultraviolet Absorption Spectra of Diatomic Arsenic Vapors,"
Spectrochimica Acta **28A**, 2197-2207

(73.18) A. M. Sibai, A. Topouzkhanian, P. Perdigon, and J. D'Incan,
"New Spectroscopic Information Pertaining to the A($^1\Sigma_g^+$) State of the
As_2 Molecule,"
C. R. Acad. Sci. **276B**, 35-8

(74.19) A. M. Sibai, P. Perdigon, and A. Topouzkhanian,
"Absorption Spectrum of As_2. Further Analysis of the A $\rightarrow$ X System
and Observation of New Electronic States,"
Z. Naturforsch. **29a**, 429-35

(74.20) F. Martin, P. Perdigon, and J. D'Incan,
"An Extensive Rotational Analysis of A $\rightarrow$ X and B $\rightarrow$ X Systems in the
Emission Spectrum of As_2 Molecule,"
J. Molec. Spectrosc. **50**, 45-47

Au$_2$

Methods of Production and Experimental Technique

Emission at ~2100°C.

Absorption in a King furnace at ~2000°C.

BAND SYSTEMS

	System	Transition	Sources	Wavelength Limits	Degrading	Band Head, $\nu_{0,0}$	Remarks	Bibliography
	I	$AO_u^+ \rightleftarrows X^1\Sigma_g^+$	Absorption	6500-4800	R	19643.8		(67.8, 54.1)
	II	$BO_u^+ \rightleftarrows X^1\Sigma_g^+$	Absorption	4100-3800	R	25679.9		(67.8, 59.5, 54.2)

I. $\underline{AO^+ \rightleftharpoons X^1\Sigma_g^+ \text{ System}}$

Intense bands, λ (72.9, 54.1):

v', v''	0	1	2	3	4	5
0	5089.2	5138.9	5189.5	5240.7	5292.9	5345.8
1	5052.8	5112.2		5202.2	5253.4	5305.4
2	5017.2		5114.5	5164.5		
3	4982.3		5078.2		5177.2	5227.6
4	4947.0	4995.1				

II. $\underline{BO_u^+ \rightleftharpoons X^1\Sigma_g^+ \text{ System}}$

Intense bands, λ (59.5):

v', v''	0	1	2	3	4
0	3892.9	3921.9			
1	3866.0		3923.6	3952.8	
2	3839.7	3868.0			3954.5
3		3841.8			

Au_2

SPECTROSCOPIC CONSTANTS

State	T_e	ω_e	$x_e\omega_e$	$B_e \times 10^2$	$\alpha_e \times 10^5$	$D_e \times 10^4$	r_e	Remarks	Bibliography
BO_u^+	25679.87	179.85	0.680	2.7009	9.6		2.5174	$y_e\omega_e = +0.003$	(67.8, 59.5)
AO_u^+	19668.1	142.3	0.445	2.5958	9.0		2.5678	$y_e\omega_e = -0.0015$	(72.9, 67.8, 54.1)
$X^1\Sigma_g^+$	0	190.9	0.420	2.8013	7.2	1.68	2.4719	$y_e\omega_e = -0.0001$	(67.8, 65.7, 54.1)

Dissociation energy = 2.34 ± 0.1 eV, 53.9 kcal/mole, 18850 cm^{-1} (60.6).

Perturbations and General Information

Potential energy curves (72.9):

State	v	U+Te(cm^{-1})	r_{max}(Å)	r_{min}(Å)
XO$_g^+$	0	95.4	2.515	2.430
	2	475.3	2.573	2.384
	4	852.0	2.612	2.356
	6	1225.5	2.645	2.336
	8	1594.0	2.673	2.320
	10	1959.4	2.700	2.304
	12	2321.9	2.724	2.291
	14	2680.6	2.747	2.280
	16	3037.0	2.770	2.268
	18	3389.5	2.793	2.259
	20	3738.3	2.814	2.249
AO$_u^+$	0	19739.1	2.618	2.520
	2	20021.2	2.687	2.466
	4	20299.3	2.734	2.436
	6	20574.0	2.774	2.414
	8	20844.6	2.807	2.393
	10	21111.3	2.842	2.380
	12	21374.2	2.872	2.364
	14	21633.3	2.902	2.352
	16	21870.3	2.928	2.340
	18	22138.8	2.958	2.328
	20	22385.5	2.987	2.318

Au_2

Franck-Condon Factors and r-Centroids of the $AO_u^+ - XO_g^+$ System (72.9):

v', v''	0	1	2	3	4	5	6
0	0.107	0.237	0.266	0.300	0.117	0.057	0.016
	2.518	2.542	2.564	2.587	2.610	2.632	
	2	3	4	5	2	1	
1	0.234	0.161	0.008	0.049			
	2.500	2.523	2.547	2.570	2.592	2.615	
	3	2		1			
2	0.261	0.012	0.120	0.096			
	2.482	2.506	2.529	2.552	2.576	2.599	
	3		2	2			
3	0.198	0.042	0.236				
	2.464	2.488	2.512	2.535	2.558	2.580	
	2		3				
4	0.114	0.147					
	2.446	2.571	2.494	2.518	2.542	2.564	
	2	2					
5	0.065						
	1						

Top line = Franck-Condon factor
Middle line = r-Centroids
Bottom line = Intensity

BIBLIOGRAPHY

(54. 1)   B. Kleman, S. Lindqvist, and L. E. Selin,
Arkiv Fysik 8, 505-10

(54. 2)   J. Ruamps,
C. R. Acad. Sci. 238, 1489-91

(56. 3)   J. Drowart and R. E. Honig,
J. Chem. Phys. 25, 581-2

(57. 4)   P. Schissel,
J. Chem. Phys. 26, 1276-80

(59. 5)   J. Ruamps,
Ann. Phys. 4, 1111-57

(60. 6)   M. Ackerman, F. E. Stafford, and J. Drowart,
J. Chem. Phys. 33, 1789-9

(65. 7)   M. D. Dolgushin,
"Construction of Potential Curves for Diatomic Molecules of Copper,
Silver, and Gold,"
Optics and Spectrosc. 19, 519-23

(67. 8)   L. L. Ames and R. F. Barrow,
Trans. Faraday Soc. 63, 39-44

(72. 9)   T. V. R. Rao and S. V. J. Lakshman,
"True Potential Energy Curves, r-Centroids and Franck-Condon
Factors for the Bands of the Au$_2$ Molecule,"
Indian J. Pure Appl. Phys. 10, 69-70

(73. 10) Reference to Ground-State Au$_2$,
C. A. Stearns and F. J. Kohl,
"Mass Spectrometric Determination of the Dissociation Energies of
Gaseous Al$_2$, AlSi, and AlSiO,"
High Temp. Sci. 5, 113-127

Methods of Production and Experimental Technique

Emission from a high voltage ac discharge in He + BCl$_3$.

BAND SYSTEMS

	System	Transition	Sources	Wavelength Limits	Degrading	Characteristic Bands, λ	Remarks	Bibliography
	I	$^3\Sigma_u^- - X^3\Sigma_g^-$	Discharge	3300-3170	R	3272.8(0,0)		(74.7, 40.2, 40.1)

B_2

I. $\underline{{}^3\Sigma_u^- - {}^3\Sigma_g^- \text{ System}}$

Band heads, λ (40.1):

(v', v'')	(3, 3)	(2, 2)	(1, 1)	(0, 0)	(4, 3)	(3, 2)	(2, 1)	(1, 0)
λ	3300. 5	3292. 7	3283. 4	3272. 8	3204. 2	3196. 3	3178. 1	3176. 4
Intensity	2	8	9	10	0	1	2	3

SPECTROSCOPIC CONSTANTS

B_2

State	T_e	ω_e	$x_e\omega_e$	B_e	$\alpha_e \times 10^3$	$D_e \times 10^4$	r_e	Remarks	Bibliography
$^3\Sigma_u^-$	30573.4	937.4	2.6	1.160	11		1.625		(67.5, 59.3, 40.1)
$X^3\Sigma_g^-$	0	1051.3	9.4	1.212	14	5.13	1.589		(74.7, 67.5, 59.3, 40.1)

Dissociation energy = 2.9 ± 0.4 eV, 67 kcal/mole, 23400 cm^{-1}.

B_2

BIBLIOGRAPHY

(40. 1)   A. E. Douglas and G. Herzberg,
          Can. J. Phys. 18, 165-74

(40. 2)   A. E. Douglas and G. Herzberg,
          Phys. Rev. 57, 752

(59. 3)   A. A. Padgett and V. Griffing,
          J. Chem. Phys. 30, 1286-91

(64. 4)   G. Verhaegen, F. E. Stafford, and J. Drowart,
          J. Chem. Phys. 40, 1622-8

(67. 5)   C. F. Bender and E. R. Davidson,
          J. Chem. Phys. 46, 3313-9

(70. 6)   B. Kockel,
          "A Calculation of the $^5\Sigma_u^-$ State of the B_2 Molecule,"
          Z. Naturforsch. 25a, 595-8

(74. 7)   G. C. Lie and E. Clementi,
          "Study of the Electronic Structure of Molecules. XXII. Correlation
          Energy Corrections as a Function of the Hartree-Fock Type Density
          and its Application to the Homonuclear Diatomic Molecules of the
          Second Row Atoms,"
          J. Chem. Phys. 60, 1288-96

Be$_2$

It has been calculated that the ground state ($^1\Sigma_g^+$) is repulsive (74.9).

Dissociation energy = <0.7 eV, <16 kcal/mole, <5600 cm^{-1}.

BIBLIOGRAPHY

(57. 1)   Dissociation Energy,
            J. Drowart and R. E. Honig,
            J. Phys. Chem. 61, 980-5

(60. 2)   Theory,
            B. J. Ransil,
            Rev. Mod. Phys. 32, 245-54

(62. 3)   Theory,
            S. Fraga and B. J. Ransil,
            J. Chem. Phys. 36, 1127-42

(67. 4)   Theory,
            C. F. Bender and E. R. Davidson,
            J. Chem. Phys. 47, 4972-8

(67. 5)   J. R. De La Vega and H. F. Hameka,
            "Calculation of Magnetic Susceptibilities of Diatomic Molecules,"
            Physica 35, 313-22

(68. 6)   M. A. Marchetti and S. R. LaPaglia,
            "Theoretical $^1\Sigma_g^+$ - $^1\Sigma_u^+$ Dipole Strengths of Some Homonuclear
            Diatomic Molecules: Configuration Interaction,"
            J. Chem. Phys. 48, 434-9

(70. 7)   R. H. Ewing and A. M. Mellor,
            "Further Calculations of Equilibrium Dimer Properties,"
            J. Chem. Phys. 53, 2983-4

(72. 8)   W. Von Niessen,
            "Density Localization of Atomic and Molecular Orbitals,"
            Theoretica Chim. Acta 27, 9-23

(74. 9)   G. C. Lie and E. Clementi,
            "Study of the Electronic Structure of Molecules. XXII. Correlation
            Energy Corrections as a Function of the Hartree-Fock Type Density
            and its Application to the Homonuclear Diatomic Molecules of the
            Second Row Atoms,"
            J. Chem. Phys. 60, 1288-96

Bi_2

Methods of Production and Experimental Technique

Absorption.

Thermal Emission.

Fluorescence.

Microwave Discharge.

Mass Spectrometric Methods.

BAND SYSTEMS

	System	Transition	Sources	Wavelength Limits	Degrading	Characteristic Bands, λ	Remarks	Bibliography
	I	$AO_u^+ \rightleftharpoons X^1\Sigma_g^+$	Absorption T > 850°C	7900-4500	R	5587(4, 2) 5725(2, 3) 5883(1, 5)		(70.23, 65.19, 35.13, 33.11)
	II	$C \leftarrow X^1\Sigma_g^+$	Absorption T > 900°C	4000-2600		Continuous	Superimposed on System III	(70.23, 33.12)
	III	$D \leftarrow X^1\Sigma_g^+$	Absorption T > 1000°C	2900-2600	R	2731.6(1, 0)		(70.23, 35.13, 33.12)
	IV	$F \leftarrow X^1\Sigma_g^+$	Absorption T > 825°C	2270-2060	R	2188.8(0, 0)		(70.23, 33.12)
	V	$G \leftarrow X^1\Sigma_g^+$	Absorption	1930-1840	R	1857.2(0, 0)		(74.26)
	VI	$E \leftarrow A$	Absorption T > 1000°C	4200-4000	R	4128.0(0, 2) 4105.5(0, 1)		(70.23, 33.12)
	VII	$G \rightarrow A$	Discharge	8820-8030	R	8528(0, 1) 8451(1, 1) 8359(1, 0)		(70.23)
	VIII	$H \rightarrow A$	Discharge	7050-6730	V	6733(0, 0) 6794(0, 1)		(70.23)
	IX	$I \rightarrow A$	Discharge	6570-6290	V	6457(0, 0) 6452(1, 1)		(70.23)
	X	?	Absorption	2500-2470		2492 2482 2470	3 continuous bands	(33.12)

Bi$_2$

I. $AO_u^+ \rightleftarrows X^1\Sigma_g^+$ System

Band heads, λ (Intensity) (70.23):

v',v''	0	1	2	3	4	5	6	7	8
0					5871(4)	5929(4)	5989(6)	6049(6)	6111(8)
1			5712(4)	5768(4)	5823(6)	5883(8)	5942(6)	6002(6)	
2			5669(4)	5725(6)	5780(6)	5837(6)		5955(4)	6015(6)
3	5521(4)		5628(4)	5682(6)	5738(4)			5909(4)	
4	5482(4)	5533(4)	5587(6)				5807(4)	5864(4)	
5	5443(2)	5494(4)	5547(6)	5599(2)		5708(4)	5763(4)		
6	5405(2)	5456(4)	5508(4)						5833(4)
7	5368(2)	5418(4)						5734(4)	
8	5332(4)	5381(4)					5639(4)		

II. $C \leftarrow X^1\Sigma_g^+$ System

This system appears continuous with the upper state being a repulsive state in the region 4000-2600Å (70.23, 33.12).

III. $D \leftarrow X^1\Sigma_g^+$ System

Band heads, λ (Intensity):

v',v''	0	1	2	3	4	5	6
0		2755.9(4)	2768.9(7)	2782.2(4)			
1	2731.6(9)	2744.5(6)			2783.7(5)	2796.7(7)	2810.2(5)
2	2720.7(7)						
3	2710.3(5)						

IV. $F \leftarrow X^1\Sigma_g^+$ System

Bands unclassified, λ (Intensity):

2205.4(1) 2197.2(1) 2188.8(2) 2180.5(1) 2172.7(1) 2148.8(1) 2142.9(1) 2135.8(1)

V. $\quad$ $G \leftarrow X^1\Sigma_g^+$ System

Band heads, λ (Intensity) (74.26):

v', v''	0	1	2	3	4	5
0	1857.1	1866.5	1876.0	1885.5	1895.0	1904.6
1	1850.8	1860.1		1879.0	1888.4	1897.9
2	1844.1					

VI. $\quad$ E ← A System

Band heads, λ (Intensity):

v', v''	0	1	2	3
0	4083.0(1)	4105.5(4)	4128.0(5)	4150.7(4)
1	4065.5(2)	4086.8(1)		
2	4049.3(1)			

VII. $\quad$ G → A System

Band heads, λ (Intensity) (70.23):

v', v''	0	1	2	3	4
0		8528(4)	8623(6)	8720(8)	8818(4)
1	8359(6)	8451(4)	8545(4)		
2	8286(6)				
3	8214(2)	8303(6)			
4					
5		8161(2)			
6		8091(2)			
7			8110(6)	8194(4)	
8			8041(6)		8209(4)

VIII. $\quad$ H → A System

Band heads, λ (Intensity) (70.23):

v', v''	0	1	2	3	4	5
0	6733(6)	6794(8)	6856(8)	6919(8)	6982(8)	7046(6)

Bi$_2$

IX.   <u>I → A System</u>

Band heads, λ (Intensity) (70.23):

v', v''	0	1	2
0	6457(6)	6512(6)	6568(6)
1	6397(10)	6452(10)	6507(6)
2	6343(10)		
3	6295(6)		

In addition, three more bands have been seen but not identified:

6341(6)   6402(4)   6302(4)

X.   <u>System</u>

Three continuous bands have been observed (33.12) in the region 2500-2470Å and not identified.   They are:

2492   2482   2470

SPECTROSCOPIC CONSTANTS

State	T_e	ω_e	$x_e\omega_e$	B_e	$\alpha_e \times 10^4$	$D_e \times 10^4$	r_e	Remarks	Bibliography
F	46000								(70.23, 33.12)
E	42252	129	9.7						(70.23, 33.12)
D	36457	157	4.6						(70.23, 33.12)
I	33216.9	156.4	6.1						(70.23)
H	32591								(70.23)
C	32000							Repulsive state	(70.23, 33.12)
G	29609.0	107.0	0.2						(70.23)
AO_u^+	17742.3	132.49	0.302	0.01968	0.53	1.452267	2.8629		(70.24, 70.23, 65.9, 35.13, 33.12)
X	0	172.71	0.341	0.022806	0.50	2.310872	2.6594		(70.24, 70.23, 65.19, 35.13, 33.12)

Dissociation energy = 1.85 ± 0.1 eV, 43 kcal/mole, 15040 cm^{-1}.

Bi$_2$

Perturbations and General Information

Potential energy curves (70.24):

State	v	Te+U(cm^{-1})	r$_{min}$(Å)	r$_{max}$(Å)
X$^1\Sigma_g^+$	0	86.3	2.617	2.704
	1	258.3	2.587	2.737
	2	430.5	2.568	2.762
	3	601.0	2.553	2.783
	4	771.5	2.540	2.801
	5	941.2	2.529	2.818
	6	1110.1	2.518	2.833
	7	1277.9	2.509	2.848
	8	1443.4	2.500	2.862
	9	1610.5	2.492	2.876
	10	1775.4	2.484	2.889
	11	1939.2	2.478	2.901
	12	2103.0	2.470	2.911
	13	2266.1	2.464	2.923
	14	2428.4	2.458	2.935
	15	2590.2	2.452	2.946
	16	2750.6	2.447	2.959
	17	2910.2	2.442	2.971
	18	3069.0	2.437	2.982
	19	3227.6	2.432	2.992
	20	3385.1	2.427	3.003
AO$_u^+$	0	17808.3	2.813	2.912
	1	17939.3	2.781	2.953
	2	18070.3	2.760	2.982
	3	18200.5	2.742	3.006
	4	18329.5	2.728	3.027
	5	18458.5	2.715	3.047
	6	18587.1	2.704	3.065
	7	18715.2	2.694	3.082
	8	18842.6	2.685	3.099
	9	18969.8	2.676	3.114
	10	19095.8	2.668	3.129
	11	19221.1	2.661	3.144
	12	19345.3	2.653	3.159
	13	19470.6	2.646	3.173
	14	19594.1	2.639	3.186
	15	19716.6	2.633	3.200
	16	19838.4	2.627	3.214
	17	19960.0	2.621	3.227
	18	20080.8	2.615	3.240
	19	20201.1	2.610	3.253
	20	20321.7	2.606	3.264

BIBLIOGRAPHY

(23. 1)   Absorption,
W. Grotrian,
Z. Physik 18, 169-82

(25. 2)   Absorption,
A. L. Narayan and K. R. Rao,
Philos. Mag. 50, 645-9

(25. 3)   Absorption,
K. R. Rao,
Proc. Roy. Soc. A 107, 760-2

(26. 4)   Absorption,
J. G. Frayne and A. W. Smith,
Philos. Mag. 1, 732-7

(27. 5)   Absorption,
J. C. MacLennan, I. Walerstein, and H. G. Smith,
Philos. Mag. 3, 390-5

(30. 6)   Absorption,
S. Barratt and A. R. Bonar,
Philos. Mag. 9, 519-24

(30. 7)   Absorption,
F. Charola,
Z. Physik 31, 457-63

(31. 8)   Fluorescence,
J. Parys,
Z. Physik 71, 807-13

(32. 9)   Fluorescence,
J. Parysowna,
Acta Phys. Polon. 1, 93-101

(33. 10)   Absorption,
A. Trojecka,
Acta Phys. Polon. 2, 245-52

(33. 11)   G. M. Almy and F. M. Sparks,
Phys. Rev. 43, 1043

Bi_2

(33. 12)   Absorption and Thermal Excitation,
G. M. Almy and F. M. Sparks,
Phys. Rev. **44**, 365-75

(35. 13)   Absorption,
G. Nakamura and T. Shidei,
Japan J. Phys. **10**, 11-25

(37. 14)   G. M. Almy,
J. Phys. Chem. **41**, 47-56

(37. 15)   C. H. D. Clark,
Trans. Faraday Soc. **33**, 1398-401

(37. 16)   C. H. D. Clark and C. W. Scaife,
Trans. Faraday Soc. **33**, 1394-8

(49. 17)   Dissociation Energy,
L. Gerö and C. Fono,
J. Chem. Phys. **17**, 345-6

(60. 18)   Dissociation Energy,
R. F. Porter and C. W. Spencer,
J. Chem. Phys. **32**, 943-4

(65. 19)   A - X System, Rotational Analysis,
N. Aslund, R. F. Barrow, W. G. Richards, and D. N. Travis,
Arkiv Fysik **30**, 171-85

(66. 20)   Dissociation Energy,
J. H. Kim and A. Cosgarea, Jr.,
J. Chem. Phys. **44**, 806-9

(67. 21)   Dissociation Energy,
F. J. Kohl, O. M. Uy, and K. D. Carlson,
J. Chem. Phys. **47**, 2667-76

(67. 22)   Dissociation Energy,
L. Rovner, A. Drowart, and J. Drowart,
Trans. Faraday Soc. **63**, 2906-12

(70. 23)   S. P. Reddy and M. K. Ali,
"The Emission Spectrum of Diatomic Bismuth,"
J. Molec. Spectrosc. **35**, 285-97

(70. 24)   T. V. R. Rao and S. V. J. Lakshman,
"True Potential Energy Curves for the Bands of the Bi_2 Molecule,"
Indian J. Pure Appl. Phys. **8**, 785-7

(70.25) J. Singh and R. K. Pandey,
"Potential Energy Curves and Dissociation Energies of Bi$_2$ and AgH Molecules,"
Indian J. Phys. 46, 136-42

(74.26) A. Topouzkhanian, A. M. Sibai, and J. D'Incan,
"Observation in Absorptions and Vibrational Analysis of New Systems in the Ultraviolet of the Sb$_2$ and Bi$_2$ Molecules,"
Z. Naturforsch. 29a, 436-9

Br$_2$

Methods of Production and Experimental Technique

Absorption

Emission from a discharge, thermoluminescence, chemiluminescence, and flames

BAND SYSTEMS

	System	Transition	Sources	Wavelength Limits	Degrading	Band Head, $\nu_{0,0}$	Remarks	Bibliography
	I	$A^3\Pi_u \rightleftharpoons X^1\Sigma_g^+$ (1_u) (0_g^+)	Absorption, Emission, Chemiluminescence	8180-6450	R	13815		(67.58)
	II	$B^3\Pi_u \rightleftharpoons X^1\Sigma_g^+$ (0_u^+) (0_g^+)	Absorption, Emission	8672-5510	R	15814.3	Converges at λ ~ 5110Å	(67.60, 47.39, 47.42, 37.22, 26.1)
	III	?	Discharge	6700-5100	R	16105		(28.2)
	IV	?	Discharge	6700-5100	R	17325.80		(28.2)
	V	?	Absorption	<5100	Continuum		Maximum at λ: 4950/4150	(47.40, 47.39, 34.16)
	VI	$C^3\Sigma^+ \rightarrow$ dissociation (1_u)	Emission	-	Continuum	31104.4	Dissociates to $^2P + ^2P$ atoms	(67.63)
	VII	$D^3\Sigma_g^- \rightleftharpoons B^3\Pi_u$ $(1_g$ or $0_g^+)$	Emission	2950-2670	R	32620.70		(57.51)

BAND SYSTEMS

	System	Transition	Sources	Wavelength Limits	Degrading	Band Head, $\nu_{0,0}$	Remarks	Bibliography
	VIII	$E\,^1\Sigma_g^+ \to B\,^3\Pi_u$ (0_g^+)	Emission	3150-2970	R	35900.7		(57.52)
	IX	$H\,^3\Pi_g \to B\,^3\Pi_u$ $(1_g$ or $0_g^+)$	Emission	3150-2970		40854.7		(57.52, 57.51)
	X	$F\,^3\Sigma^- \to X\,^1\Sigma_g^+$ (1_u)	Emission	2100-1850	R	52090		(58.53)
	XI	$a\Pi_g \to$ dissociation	Emission	-	Continuum	39638.4	Dissociates to $^2P + {}^2P$ atoms	(47.40)
	XII	$G\,^3\Sigma^- \to X\,^1\Sigma_g^+$ (0_u^+)	Emission	1850-1700	R	56303		(58.53)
	XIII	$J\,^1\Pi_g \to$ dissociation (1_g)	Emission	-	Continuum	45548.4	Dissociates to $^2P + {}^2P$ atoms	(47.40)
	XIV	$d\Pi \gets X\,^1\Sigma_g^+$ (1_u)	Absorption	1510-1485	-	66227		(67.63)

	System	Transition	Sources	Wavelength Limits	Degrading	Band Head, $\nu_{0,0}$	Remarks	Bibliography
	XV	$e\Delta \gets X\,^1\Sigma_g^+$ (1_u)	Absorption	1505	-	66473		(67.63)
	XVI	$f\Pi \to$ dissociation $(2_g, 1_g)$	Emission	-	Continuum	50604.4	Dissociates to $^2P + {}^2P$ atoms	(67.63, 47.40)
	XVII	$K\left({}^1\Sigma_u^+\right) \gets X\,^1\Sigma_g^+$ (0_u^+)	Absorption	1700-1500	R	(62500)		(67.63)
	XVIII	$g\Sigma^- \gets X\,^1\Sigma_g^+$ (0_u^-)	Absorption	1458	-	68608		(67.63)
	XIX	$h\Sigma^+ \gets X\,^1\Sigma_g^+$ (0_u^+)	Absorption	1465-1410	-	68651		(67.63)
	XX	$i\Pi \gets X\,^1\Sigma_g^+$ (1_u)	Absorption	1460-1445	-	68814		(67.63)
	XXI	$j\Pi \gets X\,^1\Sigma_g^+$ (0_u^+)	Absorption	1442-1400	-	69396		(67.63)

BAND SYSTEMS

	System	Transition	Sources	Wavelength Limits	Degrading	Band Head, $\nu_{0,0}$	Remarks	Bibliography
	XXII	$k\Pi \leftarrow X^1\Sigma_g^+$ (1_u)	Absorption	1425-1410		70913		(67.63)
	XXIII	$1\Sigma^+ \leftarrow X^1\Sigma_g^+$ (1_u)	Absorption	(1399)		(71481)		(67.63)
	XXIV	$m\Pi \leftarrow X^1\Sigma_g^+$ (1_u)	Absorption	1408-1380		71705		(67.63)
	XXV	$n\Sigma^- \leftarrow X^1\Sigma_g^+$ $\left(0_u^+ \text{ or } 1_u\right)$	Absorption	1398-1367		71876		(67.63)
	XXVI	$L^3\Pi_u \leftarrow X^1\Sigma_g^+$ (1_u)	Absorption	1383-1368		72674		(67.63)
	XXVII	$o\Delta \leftarrow X^1\Sigma_g^+$ (1_u)	Absorption	(1365)		(73240)		(67.63)
	XXVIII	$p\Sigma^+ \leftarrow X^1\Sigma_g^+$ $\left(0_u^+\right)$	Absorption	1374-1360		73459		(67.63)

	System	Transition	Sources	Wavelength Limits	Degrading	Band Head, $\nu_{0,0}$	Remarks	Bibliography
	XXIX	$M^3\Pi_u \leftarrow X^1\Sigma_g^+$ $\left(0_u^+\right)$	Absorption	1369-1325		74018		(67.63)
	XXX	$q\Pi \leftarrow X^1\Sigma_g^+$ $\left(0_u^+, 1_u\right)$	Absorption	1355-1329		74161		(67.63)
	XXXI	$r\Pi \leftarrow X^1\Sigma_g^+$ $\left(0_u^+\right)$	Absorption	1349-1330		74455		(67.63)
	XXXII	$s\Pi \leftarrow X^1\Sigma_g^+$ (1_u)	Absorption	1352-1328		74651		(67.63)
	XXXIII	$t\Pi \leftarrow X^1\Sigma_g^+$ $\left(0_u^+\right)$	Absorption	1344-1320		74768		(67.63)
	XXXIV	$N^1\Pi_u \leftarrow X^1\Sigma_g^+$ (1_u)	Absorption	1325-1270		76491		(67.63)
	XXXV	$u\Sigma^+ \leftarrow X^1\Sigma_g^+$ (1_u)	Absorption	1296-1278		77491		(67.63)

BAND SYSTEMS

	System	Transition	Sources	Wavelength Limits	Degrading	Band Head, $\nu_{0,0}$	Remarks	Bibliography
	XXXVI	$\nu\Sigma^- \leftarrow X^1\Sigma^+_g$ $\left(0^+_u, \ 1_u\right)$	Absorption	1294-1283		77639		(67.63)

Br_2

I. $\underline{A^3\Pi(1_u) \rightleftarrows X^1\Sigma_g^+ \text{ System}}$

Band heads of $^{79}Br_2$, λ (72.73):

v', v''	3	4	5
7		7430	
8	7202	7370	
9	7149	7315	
10	7101	7264	7434
11	7056	7215	7385
12	7016	7175	
13	6979	7137	
14	6946	7103	

II. $\underline{B^3\Pi\left(0_u^+\right) \leftarrow X^1\Sigma_g^+ \text{ System}}$

Band heads of $^{79}Br_2$, λ (71.70):

v', v''	4	5	6	7	8	9	10
1						7599	
2					7339	7508	7682
3				7097	7255	7420	7591
4		6727	6871	7020	7175	7336	
5	6523	6659	6800	6947	7099		
6	6461	6595	6733	6877			
7		6533	6669	6810			
8		6465	6607				
9	6290		6549				

III. and IV. $\underline{6700\text{-}5100\text{Å Systems}}$ — the two systems are superimposed on one another (28.2).

a. $v = 17325.80 + 191.45\, u' - 1.05\, u'^2 - 360.65\, u'' + 0.65\, u''^2$

b. $v = 16105.00 + 152.40\, u' - 0.40\, u'^2 - 361.70\, u'' + 1.62\, u''^2$

u',u''	System	λ	u',u''	System	λ	u',u''	System	λ
		6664	1,4	a	}6217.0	7,3	a	5699.9
		6646	7,3	b	}6217.0	10,0	b	5684.7
3,4	b	6604.1	5,2	b	6194.4			5661.1
0,6	a	6579.0	2,4	a	6144.8	8,3	a	5644.7
4,4	b	6540.7	1,3	a	}6083.1	7,2	a	5586.1
2,3	b	6519.7	7,2	b	}6083.1	8,2	a	5532.4
0,2	b	6499.4	3,4	a	6074.6	6,1	a	5529.2
5,4	b	6475.6	8,2	b	6027.8	7,1	a	5476.5
3,3	b	6455.1	4,4	a	6004.6			5463.9
1,2	b	6435.9	9,2	b	5975.8	10,2	a	5428.6
2,6	a	6421.2	3,3	a	5945.7			5382.6
4,3	b	6392.8	10,2	b	5924.3	10,1	a	5324.9
2,2	b	6372.1	8,1	b	5905.0	11,1	a	5277.1
5,3	b	6332.8	6,4	a	5880.6	12,1	a	5230.7
		6315.4	9,1	b	5852.1		a	5205.6
2,5	a	6282.0	5,3	a	5819.7		a	5168.8
		6260.0	10,1	b	5804.5	12,0	a	5134.2
			6,3	a	5758.3	15,1	a	5100.7
			9,0	b	5731.6			

V., VI., XI., XIII., and XVI. Dissociative Systems

The transitions have not been positively identified, λ (Intensity) (67.63, 47.40):

3549.4(10)	2900.4(10)	2732.4(7)	2526.9(6)
3366.8(6)	2872.5(8)	2709.8(7)	2510.9(6)
3336.6(10)	2780.6(7)	2638.8(6)	2494.2(6)
2923.8(8)	2753.6(8)	2623.1(6)	2478.8(7)

VII. $D^3\Sigma_g^- \to B^3\Pi_u$ System

Band heads, λ (67.63, 57.51):

2813.3(0,2)│2800.3(0,1)│2792.1(0,0)│2788.9(1,1)│2777.6(2,1)

VIII. $E^1\Sigma_g^+ \to B^3\Pi_u$ System

Band heads, λ (72.71, 57.52):

3098.4(0,2)│3083.6(0,1)│3067.9(1,1)│3067.9(0,0)│3053.0(2,1)

XIV. $d\Pi(1_u) \leftarrow X^1\Sigma_g^+$ System

First member of a Rydberg Series represented by:

$$\nu = 85165 - R/(n - 2.593)^2, \quad n = 5, 6, 7, \cdots 20 \quad (67.63)$$

(v', v''), λ (Intensity) for n = 5 level

(0, 0)	1505. 96(10)
(1, 0)	1502. 43(10)
(2, 0)	1495. 10(8)

(3, 0) $\left\{\begin{array}{l} 1488.\ 12 \\ 1488.\ 03 \\ 1487.\ 96 \\ 1487.\ 90 \end{array}\right\}$ (4)

XV. $e\Delta(1_u) \leftarrow X^1\Sigma_g^+$ System

Weak system, probably a forbidden transition.

v', v''	0	1
0	1504. 37(4)	1511. 53
1	1495. 82(4)	1503. 08(4)
2	1487. 01(4)	

XVIII. $g\Sigma^-\left(0_u^-\right) \leftarrow X^1\Sigma_g^+$ System

Weak system, probably a forbidden transition (67.63):

(v', v'')	λ (Intensity)
0, 1	1464. 43(2)
0, 0	1457. 56(2)

1, 0 $\left\{\begin{array}{l} 1450.\ 43 \\ 1450.\ 49 \\ 1450.\ 39 \end{array}\right\}$ (2)

XIX. $h\Sigma^+\left(0_u^+\right) \leftarrow X^1\Sigma_g^+$ System

First member of a Rydberg Series represented by:

$$\upsilon = 85165 - R/(n - 2.422)^2, \quad n = 5, 6, 7, \cdots 20 \quad (67.63)$$

(v', v''), λ (Intensity) for n = 5 level

v', v''	0	1	2
0	1456.64(10)	1463.49(4)	
1	1445.63(8)	1456.41(6)	1463.25(4)
2	1441.82(8)		
3	1438.25(8)		
4	1430.98(8)		
5		1430.45(6)	
		1430.35(6)	
		1430.27(6)	
6	1417.25(6)	1423.95(6)*	
	1417.13(6)	1423.81(6)*	
	1416.99(6)	1423.69(6)*	
7	1410.89(6)	1417.07(4)	1423.55(4)
	1410.76(6)		1423.45(4)
	1410.66(6)		
	1410.48(6)		

*Has also tentatively been classified as part of the (5, 0) system.

XX. $i\Pi(1_u) \leftarrow X^1\Sigma_g^+$ System

First member of a Rydberg Series represented by:

$$\upsilon = 88306 - R/(n - 2.629)^2, \quad n = 5, 6, 7 \quad (67.63)$$

(v', v''), λ (Intensity) for n = 5 level

v', v''	0	1
0	1453.28(8) ⎫	1460.05(2)
	1453.21(8) ⎬ (a)	
	1433.17(8) ⎭	
1	1446.55(6) ⎫	
	1446.51(6) ⎬ (b)	
	1446.47(6) ⎭	
2	1435.76(2)	

(a) Has also tentatively been identified as the (1,1) level.

(b) Has also tentatively been identified as the (2,1) level.

XXI. $j\Pi\left(0_u^+\right) \leftarrow X\,^1\Sigma_g^+$ System

First member of a Rydberg Series represented by:

$$\upsilon = 88306 - R/(n - 2.591)^2, \quad n = 5, 6, 7 \ (67.63)$$

(v', v''), λ (Intensity) for $n = 5$ level

v', v''	0	1	2
0	1441.05(6)		
	1441.01(6)		
	1440.97(6)		
1	1434.37(6)		
	1434.33(6)		
	1434.29(6)		
	1434.25(6)		
2	1427.51(8)	1434.04(6)	1440.55(2)
	1427.39(8)	1433.96(6)	
	1427.33(8)	1433.88(6)	
	1427.27(8)		
3	1420.64(4)	1427.21(6)	1433.75(4)
	1420.54(4)	1427.13(6)	1433.65(4)
	1420.47(4)	1427.06(6)	1433.57(4)
	1420.41(4)		
4	1414.23(4)	1420.35(4)	1426.90(4)
	1414.09(6)	1420.31(4)	1426.84(4)
	1413.59(4)	1420.29(4)	1426.80(4)
	1413.83(6)		
5	1407.36(2)*	1413.67(4)	1420.15(2)
	1407.28(2)*	1413.61(4)	
	1407.20(2)*	1413.55(4)	
6	1400.62(2)		

*Has also been tentatively identified as part of the m_5 (0,2) level.

XXII. $k\Pi(1_u) \leftarrow X^1\Sigma_g^+$ System

Bands with $v' \geq 1$ have not been observed.   First member of a Rydberg Series represented by:

$$\nu = 85165 - R/(n - 2.225)^2, \quad n = 5, \ 6, \ 7, \ \cdots \ 20 \ (67.63)$$

(v', v''), λ (Intensity) for $n = 5$ level

v', v''	0	1	2
0	1410.18(10)	1416.71(6)	1423.04(4)
	1410.04(10)	1416.61(6)	1422.96(4)
		1416.51(6)	1422.88(4)
		1416.45(6)	1422.80(4)
		1416.35(6)	

XXIII. $1\Sigma^+(1_u) \leftarrow X^1\Sigma_g^+$ System

First member of a Rydberg Series represented by:

$$\nu = 88306 - R/(n - 2.446)^2, \quad n = 5, 6, 7 \ (67.63)$$

(v', v''), λ (Intensity) for $n = 5$ level

v', v''	0	1
0	1399.15(6)	1405.46(4)
	1399.09(6)	1405.32(4)
	1398.97(6)	1405.24(4)
	1398.88(6)	1405.15(4)
	1398.62(6)	1405.05(4)
	1398.58(6)	1404.93(4)
	1398.52(6)	

Br_2

XXIV. $m\Pi(1_u) \leftarrow X^1\Sigma_g^+$ System

Member of a Rydberg Series (67.63)

(v', v''), λ (Intensity) for $n = 5$ level

v', v''	0	1	2
0	1394. 64(8)	1400. 89(4)	1407. 36(2)*
	1394. 60(8)		1407. 28(2)*
	1394. 56(8)		1407. 20(2)*
	1394. 51(8)		
	1394. 41(8)		
1	1388. 45(6)		
	1388. 35(6)		
	1388. 19(6)		
	1388. 12(6)		

*Has also been tentatively described as the (5,0) level of the j_5 state.

XXV. $n\Sigma^-\left(0_g^+ \text{ or } 1_u\right) \leftarrow X^1\Sigma_g^+$ System

First member of a Rydberg Series represented by:

$$\upsilon = 88306 - R/(n - 2.416)^2, \quad n = 5, 6, 7 \quad (67.63)$$

(v', v''), λ (Intensity) for $n = 5$ level

v', v''	0	1
0	1391. 28(4)	1397. 57(4)
1	1385. 33(4)*	1391. 36(4)
	1385. 20(4)*	
	1385. 06(4)*	
2	1379. 12(4)	
	1379. 06(4)	
	1378. 99(4)	
3	1373. 19(4)	
	1373. 14(4)	
	1373. 08(4)	
4	1367. 41(4)	1373. 44(2)

*May also be tentatively assigned to the (2,1) level.

XXVI. $\underline{L\,^3\Pi_u(1_u) \leftarrow X\,^1\Sigma_g^+ \text{ System}}$

Band heads, λ (Intensity) (67.63):

v', v''	0	1
0	1376.01(6)	1382.15(4)
1	1372.01(6)	1378.11(4)
2	1368.14(4)	1374.04(2)

XXVII. $\underline{o\,\Delta(1_u) \leftarrow X\,^1\Sigma_g^+ \text{ System}}$

Represents a forbidden transition. Only the (0,0) and (0,1) bands have been identified (67.63).

v', v''	λ (Intensity)
(0, 0)	1365.37(2)
(0, 1)	1371.38(3)

XXVIII. $\underline{p\Sigma^+\!\left(0_u^+\right) \leftarrow X\,^1\Sigma_g^+ \text{ System}}$

First member of a Rydberg Series represented by:

$$\nu = 85165 - R/(n - 1.938)^2, \quad n = 5, 6, 7 \cdots 18 \ (67.63)$$

v', v''	λ (Intensity) for n = 5 level
(0, 0)	1361.30(6)
	1361.10(6)
(0, 1)	1367.28(4)

XXIX. $\underline{M^3\Pi_u\left(0_u^+\right) \leftarrow X^1\Sigma_g^+ \text{ System}}$

Band heads, λ (Intensity) (67.63):

v', v''	0	1	2	3
0	1351.00(8)	1356.96(6)	1363.97(6)	1368.85(2)
			1362.90(6)	
			1362.82(6)	
1	1346.64(10)	1352.52(6)	1358.44(4)	
2	1342.35(8)	1348.29(6)		
	1342.30(8)	1348.20(6)		
	1342.26(8)	1348.11(6)		
3	1338.01(8)			
4	1333.87(4)			
	1333.80(4)			
	1333.74(4)			
5	1329.58(6)			
	1329.50(6)			
	1329.43(6)			
6	1325.26(2)			

XXX. $\underline{q\Pi\left(0_u^+ \text{ or } 1_u\right) \leftarrow X^1\Sigma_g^+ \text{ System}}$

First member of a Rydberg Series represented by:

$$v = 85165 - R/(n - 1.843)^2, \quad n = 5, 6, 7 \cdots 12 \ (67.63):$$

(v', v''), λ (Intensity) for $n = 5$ level

v', v''	0	1
0	1348.47(4)	1354.30(2)
	1348.42(4)	
	1348.36(4)	
1	1342.16(2)	
2	1336.01(4)	
	1335.95(4)	
	1335.86(4)	
3	1329.80(4)	1335.61(6)
		1335.54(6)
		1335.43(6)
		1334.94(6)

XXXI. $r\Pi\left(0_u^+\right) \leftarrow X^1\Sigma_g^+$ System

Probably a Rydberg transition. Assignment is tentative (67.63).

(v', v''), λ (Intensity) for $n = 5$ level

v', v''	0	1
0	1343.09(4)	
1	1337.15(6)	1342.82(4)
	1337.08(6)	
	1336.97(6)	
2	1331.01(2)	1336.72(2)

XXXII. $s\Pi(1_u) \leftarrow X^1\Sigma_g^+$ System

Represents a Rydberg transition (67.63).

(v', v''), λ (Intensity) for $n = 5$ level

v', v''	0	1	2
0	1339.33(6)	1345.35(2)	1351.22(4)
	1339.62(6)		
	1339.75(6)		
1	1334.19(6)	1340.14(6)	
		1339.98(6)	
		1339.94(6)	
2	1328.69(4)	1334.43(6)	
		1334.40(6)	
3		1328.96(6)	
		1328.78(6)	

XXXIII. $t\Pi\left(0_u^+\right) \leftarrow X^1\Sigma_g^+$ System

Represents a Rydberg transition (67.63).

(v', v''), λ (Intensity) for $n = 5$ level

v', v''	0	1
0	1337.52(6)	1343.27(2)
	1337.47(6)	
	1337.43(6)	
	1337.33(6)	
	1337.29(6)	
1	1332.16(8)	1337.94(8)
		1337.86(8)
2	1326.96(4)	1332.69(8)
3	1321.56(4)	

XXXIV. $\underline{N\,^1\Pi_u(1_u) \leftarrow X\,^1\Sigma_g^+ \text{ System}}$

Band heads, λ (Intensity) (67.63):

v', v''	0	1	2
0	1307. 34(6)	1312. 85(4)	1318. 37(4)
1	1303. 51(10)	1308. 97(8)	1314. 47(6)
			1314. 44(6)
2	1299. 60(8)	1304. 94(8)	1310. 39(6)
	1299. 49(8)		
	1299. 41(8)		
3	1295. 44(8)	1300. 83(6)	1306. 25(4)
4	1291. 61(8)	1296. 97(4)	
	1291. 46(8)	1296. 83(4)	
	1291. 32(8)		
5	1287. 55(8)	1292. 96(4)	
6	1283. 61(6)	1288. 93(2)	
7	1279. 84(6)	1285. 08(4)	1290. 42(6)
	1279. 59(6)		
	1279. 30(6)		
8	1276. 54(6)	1278. 42(4)	1287. 13(4)
9	1273. 25(6)		1283. 75(4)
10	1270. 00(4)	1275. 17(4)	1280. 39(4)

XXXV. $\underline{u\Sigma^+(1_u) \leftarrow X\,^1\Sigma_g^+ \text{ System}}$

Represents a Rydberg transition. Assignment is tentative (67.63).

(v', v''), λ (Intensity) for n = 5 level

v', v''	0	1
0	1295. 87(6)	
	1295. 82(6)	
	1295. 76(6)	
	1290. 47(6)	
1	1284. 27(6)	1289. 64(2)
2	1278. 43(6)	1283. 43(4)
	1278. 20(6)	
	1278. 13(6)	

XXXVI. $v\Sigma^-\left(0_u^+, 1_u\right) \leftarrow X^1\Sigma_g^+$ System

Represents a Rydberg transition.   Assignment is tentative (67.63).

(v', v''), λ (Intensity) for $n = 5$ level

v', v''	0	1
0	1288.08(8)	1293.16(2)
	1288.01(8)	
	1287.95(8)	
1	1283.01(2)	1288.36(6)

SPECTROSCOPIC CONSTANTS

State	T_o	ω_o	$x_o\omega_o$	B_e	$\alpha_e \times 10^3$	$D_e \times 10^8$	r_e	Remarks	Bibliography
$v\Sigma^-(0_u^+, 1_u)$	77639	303						Rydberg	(67.63)
$u\Sigma^+(1_u)$	77491	374						Rydberg	(67.63)
$N^1\Pi_u(1_u)$	76491	230						Rydberg	(67.63)
$t\Pi(0_u^+)$	74768	299	~ 1					Rydberg	(67.63)
$s\Pi(1_u)$	74651	303						Rydberg	(67.63)
$r\Pi(0_u^+)$	74455	341	(1.0)					Rydberg	(67.63)
$q\Pi(0_u^+, 1_u)$	74161	346						Rydberg	(67.63)
$m^3\Pi_u(0_u^+)$	74018	241	(0.3)					Rydberg	(67.63)
$p\Sigma^+(0_u^+)$	73459							Rydberg	(67.63)
$o\Delta(1_u)$	(73240)								(67.63)
$L^3\Pi_u(1_u)$	72674	215	~ 3						(67.63)

SPECTROSCOPIC CONSTANTS

State	T_o	ω_o	$x_o\omega_o$	B_e	$\alpha_e \times 10^3$	$D_e \times 10^8$	r_e	Remarks	Bibliography
$n\Sigma^-(0_u^+, 1_u)$	71876	313						Rydberg	(67.63)
$m\Pi(1_u)$	71705	323	(1.0)					Rydberg	(67.63)
$l\Sigma^+(1_u)$	(71481)							Rydberg	(67.63)
$k\Pi(1_u)$	70913	338						Rydberg	(67.63)
$j\Pi(0_u^+)$	69396	324	(1.0)					Rydberg	(67.63)
$i\Pi(1_u)$	68814	330						Rydberg	(67.63)
$h\Sigma^+(0_u^+)$	68651							Rydberg Perturbed, possibly predissociated	(67.63)
$g\Sigma^-(0_u^-)$	68608	339							(67.63)
$K(^1\Sigma_u^+)(0_u^+)$	(62500)	(293)							(67.63, 58.53)
$f\Pi(2, 1_g)$	66500	480							(67.63, 47.40)
$e\Delta(1_u)$	66473	381							(67.63)

SPECTROSCOPIC CONSTANTS

State	T_o	ω_o	$x_o\omega_o$	B_e	$\alpha_e \times 10^3$	$D_e \times 10^8$	r_e	Remarks	Bibliography
$d\Pi(1_u)$	66227	335	3.0					Rydberg	(67.63)
$J^1\Pi(1_g)$	61444	220							(67.63, 47.40)
$H^3\Pi_g$ $(1_g, 0_g^+)$	56669	106.5	1.5						(67.63, 57.51)
$G^3\Sigma^-(0_u^+)$	56303	255							(67.63, 58.53)
$a\Pi_g(1_g)$	55534	330							(67.63, 47.40)
$F^3\Sigma^-$	52090	120							(67.63, 58.53)
$E^1\Sigma_g^+(0_g^+)$	51715	149.2	1.15						(67.63, 57.51)
$D^3\Sigma_g^-$ $(1_g, 0_g^+)$	48435	161.7	0.29						(67.63, 63.55, 57.52, 57.51)
$C^3\Sigma^+(1_u)$	(47000)								(67.63, 47.40)

SPECTROSCOPIC CONSTANTS

State	T_o	ω_o	$x_o\omega_o$	B_e	$\alpha_e \times 10^3$	$D_e \times 10^8$	r_e	Remarks	Bibliography
$B^3\Pi_u(0_u^+)$	15840	166.1	1.84	0.0585	0.41	2.8	2.686		(67.63, 67.60, 37.22, 32.11)
$A^3\Pi_u(1_u)$	13815	150	2.7						(67.63, 67.59, 67.58, 37.22, 32.11)
$X^1\Sigma_g^+(0_g^+)$	0	323.07	1.167	0.081101	0.321	2.05	2.2809		(67.63, 67.60)

Dissociation energy = 1.970 eV, 45.44 kcal/mole, 15893 ± 2 cm^{-1}.

Br_2

Perturbations and General Information

The $B^3\Pi\left(0_u^+\right)$ state appears to be predissociated (73.75).

The $h\Sigma^+\left(0_u^+\right)$ state appears to be perturbed and possibly predissociated in the region $v'' = 1$ to 4 (67.63).

Lifetimes — $B^3\Pi\left(0_u^+\right)$ state $v = 1, 3, 5$ $\tau = 0.28 \pm 0.1$ μsec (71.69)

$\qquad\qquad$ $A^3\Pi(1_u)$ state $\qquad\qquad\qquad$ $\tau = 1$ msec (72.74)

Potential energy curve for the $X^1\Sigma_g^+$ state using RKR potential (71.70):

$T_e = 0$

v	$E(v)(cm^{-1})$	r_{min}	r_{max}
0	162.38	2.2316	2.3342
1	485.52	2.1979	2.3762
2	806.49	2.1759	2.4067
3	1125.28	2.1586	2.4325
4	1441.87	2.1441	2.4555
5	1756.25	2.1315	2.4768
6	2068.39	2.1202	2.4967
7	2378.28	2.1101	2.5157
8	2685.93	2.1007	2.5339
9	2991.31	2.0921	2.5515
10	3294.38	2.0841	2.5685

Potential energy curve for the $A^3\Pi(1_u)$ state ($^{79}Br_2$) using RKRV potential (72.71):

$T_e = 13901 \text{ cm}^{-1}$

v	E(v)	r_{min}	r_{max}
0	76.2	2.627	2.777
1	224.3	2.583	2.848
2	367.0	2.556	2.903
3	504.0	2.535	2.952
4	635.5	2.518	3.000
5	761.4	2.504	3.046
6	881.7	2.492	3.092
7	996.4	2.482	3.139
8	1105.23	2.473	3.188
9	1207.65	2.465	3.239
10	1303.43	2.457	3.293
11	1392.31	2.451	3.351
12	1474.07	2.445	3.414
13	1548.61	2.440	3.483
14	1615.98	2.436	3.558
15	1676.44	2.432	3.641
16	1730.50	2.428	3.730
17	1778.84	2.425	3.825
18	1822.21	2.423	3.925
19	1861.31	2.420	4.029
20	1896.72	2.418	4.138

Br$_2$

Potential energy curve for the $B^3\Pi\left(0_u^+\right)$ state of (^{79}Br$_2$) using RKR potentials (71.70):

T_e = 15902. 51

v	E(v')(cm^{-1})	r$_{min}$	r$_{max}$
0	83. 37	2. 6107	2. 7541
1	247. 63	2. 5667	2. 8172
2	408. 58	2. 5386	2. 8648
3	566. 11	2. 5171	2. 9065
4	720. 17	2. 4994	2. 9450
5	870. 71	2. 4842	2. 9816
6	1017. 67	2. 4710	3. 0171
7	1160. 99	2. 4592	3. 0519
8	1300. 62	2. 4485	3. 0863
9	1436. 47	2. 4388	3. 1206
10	1568. 51	2. 4299	3. 1550
11	1696. 70	2. 4218	3. 1898
12	1820. 99	2. 4143	3. 2250
13	1941. 34	2. 4074	3. 2608
14	2057. 69	2. 4012	3. 2976
15	2169. 99	2. 3954	3. 3351
16	2278. 21	2. 3897	3. 3734
17	2382. 33	2. 3846	3. 4131
18	2482. 33	2. 3796	3. 4539
19	2578. 20	2. 3751	3. 4962

Br_2

Franck-Condon factors and r-Centroids for the $A^3\Pi - X^1\Sigma_g^+$ system of $^{79}Br^{81}Br$ using RKRV potential (72.74):

v', v''	0	1	2	3	4	5	6	7
0	-	-	-	-	-	-	0.00005	0.00017
	2.456	2.467	2.479	2.491	2.503	2.515	2.527	2.540
1	-	-	-	-	0.00002	0.00009	0.00035	0.0011
	2.451	2.462	2.474	2.485	2.497	2.509	2.521	2.533
2	-	-	-	-	0.00008	0.00038	0.0013	0.0040
	2.445	2.457	2.468	2.480	2.492	2.504	2.516	2.528
3	-	-	-	0.00005	0.00028	0.0011	0.0036	0.0093
	2.441	2.452	2.463	2.475	2.486	2.498	2.510	2.522
4	-	-	0.00002	0.00015	0.00072	0.0026	0.0075	0.0171
	2.436	2.447	2.459	2.470	2.481	2.493	2.505	2.517
5	-	-	0.00005	0.00034	0.0015	0.0051	0.0129	0.0254
	2.432	2.443	2.454	2.465	2.477	2.488	2.500	2.512
6	-	-	0.00012	0.00069	0.0028	0.0084	0.0189	0.0320
	2.428	2.439	2.450	2.461	2.472	2.484	2.495	2.507
7	-	0.00003	0.00022	0.0012	0.0045	0.0122	0.0242	0.0350
	2.424	2.434	2.446	2.457	2.468	2.479	2.491	2.502
8	-	0.00005	0.00039	0.0019	0.0066	0.0160	0.0279	0.0338
	2.420	2.431	2.442	2.453	2.464	2.475	2.487	2.498
9	-	0.00008	0.00061	0.0028	0.0088	0.0193	0.0292	0.0290
	2.416	2.427	2.438	2.449	2.460	2.471	2.483	2.494
10	-	0.00012	0.00088	0.0038	0.0110	0.0215	0.0282	0.0224
	2.413	2.424	2.435	2.446	2.457	2.468	2.479	2.490

Top line = Franck-Condon factor
Bottom line = r-Centroid

Br_2

Franck-Condon factors and r-Centroids for the $B^3\Pi - X^1\Sigma_g^+$ system of $^{79}Br^{81}Br$ (72.71) using exact Morse potentials.

v', v''	0	1	2	3	4	5	6	7
0	3. -10 2.4511	8. -9 2.4629	1. -7 2.4747	1. -6 2.4867	0.00001 2.4989	0.00003 2.5112	0.00013 2.5236	0.00046 2.5362
1	4. -9 2.4451	1. -7 2.4568	1. -6 2.4685	0.00001 2.4804	0.00006 2.4925	0.00028 2.5046	0.00102 2.5169	0.00307 2.5294
2	3. -8 2.4394	7. -7 2.4509	8. -6 2.4625	0.00006 2.4743	0.00030 2.4862	0.00122 2.4982	0.00391 2.5104	0.01017 2.5227
3	1. -7 2.4337	3. -6 2.4451	0.00003 2.4567	0.00022 2.4683	0.00102 2.4801	0.00364 2.4920	0.01015 2.5041	0.02243 2.5163
4	6. -7 2.4283	0.00001 2.4396	0.00011 2.4510	0.00063 2.4626	0.00265 2.4742	0.00832 2.4860	0.01992 2.4979	0.03672 2.5100
5	2. -6 2.4320	0.00003 2.4342	0.00028 2.4455	0.00151 2.4570	0.00564 2.4685	0.01542 2.4802	0.03131 2.4920	0.04690 2.5038
6	5. -6 2.4178	0.00008 2.4290	0.00064 2.4402	0.00309 2.4515	0.01022 2.4630	0.02411 2.4745	0.04075 2.4862	0.04775 2.4979
7	0.00001 2.4129	0.00018 2.4239	0.00128 2.4351	0.00557 2.4463	0.01617 2.4576	0.03257 2.4690	0.04470 2.4805	0.03852 2.4920
8	0.00003 2.4081	0.00036 2.4190	0.00232 2.4301	0.00899 2.4412	0.02281 2.4525	0.03863 2.4637	0.04153 2.4751	0.02350 2.4862
9	0.00005 2.4034	0.00065 2.4143	0.00383 2.4253	0.01322 2.4363	0.02906 2.4474	0.04059 2.4586	0.03243 2.4697	0.00930 2.4802

Top line = Franck-Condon factor followed by factor of ten
Bottom line = r-Centroid

BIBLIOGRAPHY

(26. 1)   Visible Absorption,
          H. Kuhn,
          Z. Physik 39, 77-91

(28. 2)   Emission,
          Y. Uchida and Y. Ota,
          Japan J. Phys. 5, 59-66

(29. 3)   M. B. Hays,
          J. Franklin Inst. 208, 363-9

(29. 4)   A. E. Gillam and R. A. Morton,
          Proc. Roy. Soc. A 124, 604-6

(29. 5)   R. W. Armour and E. B. Ludlam,
          Proc. Roy. Soc. Edinburgh 49, 91-101

(30. 6)   Far Ultraviolet Absorption,
          H. Cordes and H. Sponer,
          Z. Physik 63, 334-44

(31. 7)   N. Demassieux and V. Henri,
          C. R. Acad. Sci. 193, 591-2

(31. 8)   W. G. Brown,
          Phys. Rev. 37, 1007-8

(31. 9)   W. G. Brown,
          Phys. Rev. 37, 1179-86

(32. 10) W. G. Brown,
          Phys. Rev. 39, 190

(32. 11) Visible Absorption, Vibrational Analysis,
          W. G. Brown,
          Phys. Rev. 39, 777-87

(33. 12) Resonance Lines,
          H. J. Plumley,
          Phys. Rev. 43, 495

(34. 13) L. Dabrowski,
          Acta Phys. Polon. 3, 301-5

Br$_2$

(34. 14) J. Patkowski,
          Acta Phys. Polon. 3, 385-91

(34. 15) J. F. H. Custers and J. H. deBoer,
          Physica 1, 265-70

(34. 16) Continuous Absorption,
          A. P. Acton, R. G. Aickin, and N. S. Bayliss,
          J. Chem. Phys. 4, 474-9

(34. 17) Induced Predissociation,
          L. Avramenko and V. Kondratjew,
          Phys. Z. Soviet Union 10, 741-50

(36. 18) Thermoluminescence,
          Y. Uchida,
          Sci. Papers Inst. Phys. Chem. Res. 30, 71-82

(37. 19) Predissociation,
          L. Avramenko and V. Kondratjew,
          J. Exper. Theor. Phys. 7, 249-55

(37. 20) Theory,
          R. S. Mulliken,
          J. Phys. Chem. 41, 5-45

(37. 21) Theory,
          N. S. Bayliss,
          Proc. Roy. Soc. A 158, 551-61

(37. 22) Visible Absorption, Vibrational Analysis,
          O. Darbyshire,
          Proc. Roy. Soc. A 159, 93-109

(38. 23) Emission Bands in the Visible,
          W. M. Vaidya,
          Proc. Indian Acad, Sci. A 7, 321-6

(38. 24) Themoluminescence,
          T. Kitagawa,
          Rev. Phys. Chem. Japan 12, 135-47

(38. 25) Continuous Ultraviolet Absorption,
          R. G. Aickin and N. S. Bayliss,
          Trans. Faraday Soc. 34, 1371-4

(39.26) Potential Energy Curves, Perturbations,
N. S. Bayliss and A. L. G. Rees,
J. Chem. Phys. 7, 854-5

(39.27) Collisions,
N. S. Bayliss and A. L. G. Rees,
Nature 143, 560

(39.28) Collisions,
N. S. Bayliss and A. L. G. Rees,
Trans. Faraday Soc. 35, 792-800

(40.29) Absorption in Solution,
N. S. Bayliss and A. L. G. Rees,
J. Chem. Phys. 8, 377-81

(40.30) Theory,
R. S. Mulliken,
J. Chem. Phys. 8, 382-95

(40.31) Absorption in Solution,
A. L. G. Rees,
J. Chem. Phys. 8, 429-30

(40.32) Theory,
G. B. B. M. Sutherland,
J. Chem. Phys. 8, 161-4

(40.33) Theory,
R. S. Mulliken,
Phys. Rev. 57, 500-15

(41.34) Theory,
H. M. Hulbert and J. O. Hirschfelder,
J. Chem. Phys. 9, 61-9

(41.35) Recombination of Atoms,
O. K. Rice,
J. Chem. Phys. 9, 258-61

(42.36) Absorption in Solution,
A. F. Prikhotko,
Acta Physicochim. 16, 125-31

(46.37) Influence of Collisions on Absorption,
G. Kortüm and D. Müller,
Z. Naturforsch. A 1, 637-46

Br_2

(47.38)   Band Emission in the Ultraviolet,
          E. H. Coleman and A. G. Gaydon,
          Discuss. Faraday Soc. 2, 177-76

(47.39)   Absorption and Emission,
          J. Romand and B. Vodar,
          C. R. Acad. Sci. 226, 238-40

(47.40)   Band Emission Between 4250-2000Å,
          P. Venkateswarlu,
          Proc. Indian Acad. Sci. A 25, 138-50

(47.41)   Theory,
          A. L. G. Rees,
          Proc. Phys. Soc. 59, 998-1008

(47.42)   Visible Absorption,
          A. L. G. Rees,
          Proc. Phys. Soc. 59, 1008-10

(50.43)   Raman Effect,
          H. Stammreich,
          Phys. Rev. 78, 79-80

(51.44)   G. Kortüm and W. Luck,
          Z. Naturforsch. A 6, 305-12

(54.45)   Shows that the experiment of (54.47) is a phenomena due to fluorine.
          R. F. Barrow,
          J. Chem. Phys. 22, 1775

(54.46)   Vacuum Ultraviolet Absorption,
          N. S. Bayliss and J. V. Sullivan,
          J. Chem. Phys. 22, 1615-6

(54.47)   Raman Effect,
          H. Stammreich and R. Fornesis,
          J. Chem. Phys. 22, 1624

(55.48)   Absorption in Inert Solvents,
          D. F. Evans,
          J. Chem. Phys. 23, 1426-8

(57.49)   Amount of Dissociation in a Shock Wave,
          H. B. Palmer and D. F. Hornig,
          J. Chem. Phys. 26, 98-105

(57. 50) Ionization Potential and Critical Discussion,
K. Watanabe,
J. Chem. Phys. <u>26</u>, 542-7

(57. 51) Absorption in Argon,
P. Venkateswarlu and R. D. Verma,
Proc. Indian Acad. Sci. A <u>46</u>, 251-64

(57. 52) Absorption in Argon,
P. Venkateswarlu and R. D. Verma,
Proc. Indian Acad. Sci. A <u>46</u>, 416-21

(58. 53) Emission,
P. B. V. Haranath and P. T. Rao,
J. Molec. Spectrosc. <u>2</u>, 428-63

(58. 54) Emission,
R. D. Verma,
Proc. Indian Acad. Sci. A <u>47</u>, 196-200

(63. 55) A. G. Briggs and R. G. W. Norrish,
Proc. Roy. Soc. A <u>276</u>, 51

(64. 56) Spectral Resonance in the Vacuum Ultraviolet,
Y. V. Rao and P. Venkatesvarlu,
J. Molec. Spectrosc. <u>13</u>, 288-95

(64. 57) A. G. Briggs and R. G. W. Norrish,
Proc. Roy. Soc. A <u>278</u>, 27-34

(67. 58) Emission by Chemiluminescence,
M. A. A. Clyne and J. A. Coxon,
J. Molec. Spectrosc. <u>23</u>, 258-71

(67. 59) $A\,^3\Pi_u - X\,^1\Sigma_g^+$ System, Rotational Analysis,
J. A. Horsley,
J. Molec. Spectrosc. <u>22</u>, 469-71

(67. 60) $^3\Pi(0_u^+) - \,^1\Sigma_g^+$ in Absorption, Vibrational and Rotational Analysis,
J. A. Horsley and R. F. Barrow,
Trans. Faraday Soc. <u>63</u>, 32-8

(67. 61) R. K. R. Curve for $^3\Pi(0_u^+)$,
J. A. C. Todd, W. G. Richards, and M. A. Byrne,
Trans. Faraday Soc. <u>63</u>, 2081-4

(67. 62)   A. A. Passchier, J. D. Christian, and N. W. Gregory,
"The Ultraviolet-Visible Absorption Spectrum of Bromine,"
J. Phys. Chem. 4, 937-42

(67. 63)   P. Venkateswarlu,
"The Vacuum Ultraviolet Spectrum of the Bromine Molecule,"
Can. J. Phys. 47, 2525-38

(70. 64)   J. A. Coxon and M. A. A. Clyne,
"Vibrational Numbering of the A $^3\Pi(1_u)$ State of Bromine,"
J. Phys. B 3, 1164-5

(70. 65)   W. Kiefer and H. W. Schrötter,
"Simultaneous Observation of Resonance Fluorescence and
Resonance Raman Effect in Gaseous Bromine,"
J. Chem. Phys. 53, 1612-3

(70. 66)   W. Holzer, W. F. Murphy, and H. J. Bernstein,
"Resonance Fluorescence of Iodine, Bromine, and Chlorine Gases
Obtained With Argon-Ion Laser Excitation,"
J. Chem. Phys. 52, 469-70

(70. 67)   W. Holzer, W. F. Murphy, and H. J. Bernstein,
"Resonance Raman Effect and Resonance Fluorescence in Halogen
Gases,"
J. Chem. Phys. 52, 399-407

(70. 68)   R. J. LeRoy and R. B. Bernstein,
"Dissociation Energies of Diatomic Molecules From Vibrational
Spacings of Higher Levels Application to the Halogens,"
Chem. Phys. Letters 5, 42-5

(71. 69)   G. Capelle, K. Sakurai, and H. P. Broida,
"Lifetimes and Self-Quenching Cross Sections of Vibrational Levels
in the B State of Bromine Excited by a Tunable Dye Laser,"
J. Chem. Phys. 54, 1728-30

(71. 70)   J. A. Coxon,
"The B$^3\Pi(0_u^+)$ ← X$^1\Sigma_g^+$ System of ^{79}Br^{79}Br,"
J. Molec. Spectrosc. 37, 39-62

(72. 71)   K. Wieland, J. B. Tellinghuisen, and A. Nobs,
"The Band Systems E→B (4000-4360Å) and F→X (2530-2740Å) of
^{127}I$_2$ and ^{129}I$_2$, and the Corresponding System E⇄B of Br$_2$ and Cl$_2$,"
J. Molec. Spectrosc. 41, 69-83

(72. 72) J. A. Coxon,
"Franck-Condon Factors and r-Centroids for Halogen Molecules II.
The $B^3\Pi(0_u^+)$ - $X^1\Sigma_g^+$ System of $^{79}Br^{81}Br$,"
J. Quant. Spectrosc. Radiat. Transfer. 12, 639-50

(72. 73) J. A. Coxon,
"The Extreme Red Absorption Spectrum of Br_2, $A^3\Pi(1_u) \leftarrow X^1\Sigma_g^+$,"
J. Molec. Spectrosc. 41, 548-65

(72. 74) J. A. Coxon,
"Franck-Condon Factors and r-Centroids for Halogen Molecules III.
The $A^3\Pi(1_u)-X^1\Sigma^+$ System of $^{79}Br^{81}Br$,"
J. Molec. Spectrosc. 41, 566-76

(73. 75) M. S. Child and R. B. Bernstein,
"Diatomic Interhalogens: Systematics and Implications of Spectroscopic Interatomic Potentials and Curve Crossings,"
J. Chem. Phys. 59, 5916-25

(73. 76) R. F. Barrow, D. F. Broyd, L. B. Pederson, and K. K. Yee,
"The Dissociation Energies of Gaseous Br_2 and I_2,"
Chem. Phys. Letters 18, 357-8

(73. 77) K. K. Yee and T. J. Stone,
"Analysis of RKR Long-Range Potentials of the $B\,^3\Pi(0_u^+)$ States of Br_2 and Cl_2,"
Molec. Phys. 26, 1169-76

(73. 78) R. J. LeRoy,
"Long Range Potential Coefficients From RKR Turning Points: C_6 and C_8 for $B\,^3\Pi(0_u^+)$ State Cl_2, Br_2, and I_2,"
Can. J. Phys. 52, 246-56

(74. 79) R. C. Oldenborg, J. L. Gole, and R. N. Zare,
"Chemiluminescent Spectra of Alkali-Halogen Reactions,"
J. Chem. Phys. 60, 4032-42

(74. 80) R. S. Eng and J. T. LaTourrette,
"Hyperfine Spectra of Bromine Vapor Near 633 nm,"
J. Molec. Spectrosc. 52, 269-76

(74. 81) R. F. Barrow, T. C. Clark, J. A. Coxon, and K. K. Yee,
"The $^3\Pi(0_u^+)$ - $X^1\Sigma_g^+$ System of Br_2: Rotational Analysis, Franck-Condon Factors, and Long Range Potential in the $B\,^3\Pi(0_u^+)$ State,"
J. Molec. Spectrosc. 51, 528-49

Br$_2$

(74. 82)   M. Brith, O. Schnepp, and P. Stephens,
           "Magnetic Circular Dichroism Spectra of the Halogen Molecules
           I$_2$, Br$_2$, and Cl$_2$.   Resolution of Overlapping $0_u^+(^3\Pi)$ and $^1\Pi$ Bands,"
           Chem. Phys. Letters 26, 549-52

Methods of Production and Experimental Technique

Absorption in a King furnace.

Absorption from flash-photolysis or flash-discharge into mixtures of hydrocarbons and rare gases.

Emission from flames, arcs, and discharges with C, CO, CO_2, or hydrocarbon shock waves with and without rare gases.

Astrophysics: Absorption in stellar (type R and N) atmospheres.

Emission from comets.

BAND SYSTEMS

	System	Transition	Sources	Wavelength Limits	Degrading	Characteristic Bands, λ	Remarks	Bibliography
Ballik-Ramsay	I	$b^3\Sigma_g^- \rightarrow a^3\Pi_u$	King furnace	27000-11000	R	17675(0,0)	Perturbed	(63.41, 63.40)
Phillips	II	$A^1\Pi_u \rightleftharpoons X^1\Sigma_g^+$	Discharge	15490-6720	R	8750.8(2,0)	Perturbed	(70.126, 63.41)
Swan	III	$d^3\Pi_g \rightleftharpoons a^3\Pi_u$	Numerous sources	7850-3400	V	5635.5(0,1) 5165.2(0,0) 4737.1(1,0)	Perturbed	(73.135, 69.112, 63.43)
Deslandres-d'Azambuja	IV	$C^1\Pi_g \rightarrow A^1\Pi_u$	Discharge	4110-3390	V	4102.3(0,1) 3852.2(0,0) 3607.3(,0)	Perturbed	(67.80, 40.6, 30.4)
Messerle-Krauss	V	$C'^1\Pi_g \rightarrow A^1\Pi_u$	Discharge	3780-3390	V	(a)	Perturbed	(67.80)
Fox-Herzberg	VI	$e^3\Pi_g \rightarrow a^3\Pi_u$	Discharge	3285-2370	R	2987(0,4) 2855(0,3)		(68.100, 49.8)

C_2

BAND SYSTEMS

	System	Transition	Sources	Wavelength Limits	Degrading	Characteristic Bands, λ	Remarks	Bibliography
Mulliken	VII	$D^1\Sigma_u^+ \rightleftharpoons X^1\Sigma_g^+$	Discharge, Flames	2415-2310	R	2312.7(0,0)		(68.100, 39.5)
Freymark	VIII	$E^1\Sigma_g^+ \rightarrow A^1\Pi_u$	Discharge in C_2H_2	2220-2070	V	2218.2(0,1) 2142.9(0,0) 2072.4(1,0)	Perturbed	(51.10)
Ultraviolet Region	IX	$f^3\Sigma_g^- \leftarrow a^3\Pi_u$	Absorption from flash discharge into mixtures of hydrocarbons and rare gases	1425-1370	R	1424.34(0,0) 1397.8 (1,0) 1372.87(2,0)		(69.109)
	X	$g^3\Delta_g \leftarrow a^3\Pi_u$		1400-1365	R	1395.21(0,0) 1367.47(1,0)		(69.109)
	XI	$F^1\Pi_u \leftarrow X^1\Sigma_g^+$		1345-1310	R	1341.38(0,0) 1313.97(1,0)		(69.109)

(a) This band system has been observed by analysis of the perturbations of the $C^1\Pi_g \rightarrow A^1\Pi_u$ System.

Redesignation of the electronic states of C_2 based on work by Herzberg et al. (69.109). In assigning the new designation, the letter B has been left out since two predicted low lying singlet states have not yet been found.

Old	New	
$a\,^1\Sigma_g^+$, $x\,^1\Sigma_g^+$	$X\,^1\Sigma_g^+$	Lower state of Mulliken and Phillips bands
$X\,^3\Pi_u$, $X'\,^3\Pi_u$	$a\,^3\Pi_u$	Lower state of Swan bands
$A'\,^3\Sigma_g^-$	$b\,^3\Sigma_g^-$	Upper state of Ballik-Ramsay bands
$b\,^1\Pi_u$	$A\,^1\Pi_u$	Lower state of Deslandres-d'Azambuja bands
$A''\,^3\Sigma_g^+$	$c\,^3\Sigma_u^+$	From perturbations in upper state of Phillips bands
$A\,^3\Pi_g$	$d\,^3\Pi_g$	Upper state of Swan bands
$c\,^1\Pi_g$	$C\,^1\Pi_g$	Upper state of Deslandres-d'Azambuja bands
$c'\,^1\Pi_g$	$C'\,^1\Pi_g$	State perturbing $C\,^1\Pi_g$
$B\,^3\Pi_g$	$e\,^3\Pi_g$	Upper state of Fox-Herzberg bands
$d\,^1\Sigma_u^+$	$D\,^1\Sigma_u^+$	Upper state of Mulliken bands
$e\,^1\Sigma_u^+$	$E\,^1\Sigma_g^+$	Upper state of Freymark bands
	$f\,^3\Sigma_g^-$	Upper state of ultraviolet bands
	$g\,^3\Delta_g$	Upper state of ultraviolet bands
	$F\,^1\Pi_u$	Upper state of ultraviolet bands

C_2

I. $b^3\Sigma_g^- \rightarrow a^3\Pi_u$ System (Ballik-Ramsay)

Band heads, λ (Intensity) (63.41, 63.40, 62.36):

(v', v'')	(0, 1)	(0, 0)	(1, 0)	(2, 0)
λ (Intensity)	24745(2)	17675(4)	14075(3)	11726(1)

II. $A^1\Pi_u \rightleftharpoons X^1\Pi_g$ System (Phillips)

Band heads, λ (70.126, 67.76, 66.56, 64.45, 63.41):

(v', v'')	(0, 1)	(0, 0)	(2, 0)	(6, 3)	(5, 2)	(4, 1)
λ	15484.06	12070.21	8750.86	8315.86	8108.17	7907.69

(v', v'')	(3, 0)	(8, 4)	(7, 3)	(6, 2)	(5, 1)	(4, 1)	(8, 3)
λ	7714.58	7612.97	7428.12	7249.03	7076.79	6909.37	6722.59

III. $d^3\Pi_g \rightleftharpoons a^3\Pi_u$ System (Swan)

Band heads, λ (74.141, 74.140, 73.135, 71.130, 70.123, 69.112, 69.108, 67.83, 66.67, 66.58, 65.53):

v', v''	0	1	2	3	4	5	6
0	5165.2	5635.5	6191.3				
1	4737.1	5129.4	5585.5	6122.2	6762.4		
2	4382.2	4715.4	5097.7	5540.7	6059.7	6675.9	
3		4371.4	4697.6	5070.9	5501.9	6004.8	6599.1
4		4364.9	4684.9			5470.3	5959.0
5					4678.6		5447.7
6	3419	3619.5		4093	4368.8	4680.2	5030

IV. $\underline{C\,^1\Pi_g \rightarrow A\,^1\Pi_u}$ System (Deslandres-d'Azambuja)

Band heads, λ (Intensity) (67.80, 40.6, ·30.4):

v', v''	0	1	2	3	4	5
0	3852.2(10)	4102.3(9)				
1	3607.3(8)	3825.6(5)	4068.1(6)			
2	3399.8(5)	3592.9(7)		4041.9(3)		
3		3398.1(5)	3587.6(7)		4026.9(1)	
4			3405.1	3599.3		
5				3431.9	3617.9	
6						3689.0

Isotope shift studies (69.107).

V. $\underline{C'\,^1\Pi_g \rightarrow A\,^1\Pi_u}$ System (Messerle-Krauss)

Bands determined by perturbations in $C\,^1\Pi_g \rightarrow A\,^1\Pi_u$ System, λ (67.80):

v',v''	(4, 6)	(3, 5)	(5, 6)	(2, 4)	(0, 2)	(1, 2)	(0, 1)
λ	3779.64	3691.54	3672.73	3627.03	3586.02	3405.66	3396.09

VI. $\underline{e\,^3\Pi_g \rightarrow a\,^3\Pi_u}$ System (Fox-Herzberg)

Band heads, λ (49.8):

v', v''	0	1	2	3	4	5	6
0			2731.6	2855	2987	3129	3283
1			2656.3	2772.1	2896.4		
2		2486.3	2589.0	2698.8			
3		2429.9	2527.9				
4		2378.2					

VII. $\underline{D\,^1\Sigma_u^+ \rightleftarrows X\,^1\Sigma_g^+}$ System (Mulliken)

Band heads, λ (68.100, 39.5):

(v', v'')	(0, 1)	(3, 3)	(2, 2)	(1, 1)	(0, 0)
λ	2414.8	2316.8	2315.4	2314.0	2312.7

C_2

VIII. $\underline{E\,^1\Sigma_g^+ \rightarrow A\,^1\Pi_u}$ System (Freymark)

Band heads, λ (Intensity) (51.10):

v', v''	0	1	2	3	4
0	2142.6(10)	2218.2(7)			
1	2072.4(6)	2142.9(1)	2216.6(9)		
2		2075.6(6)			
3			2081.2(5)		
4				2087.1(5)	
5					2096.9(4)

IX. $\underline{f\,^3\Sigma_g^- \leftarrow a\,^3\Pi_u}$ System (Ultraviolet)

Band heads, λ (69.109):

(v', v'')	(0, 0)	(1, 0)	(2, 0)
λ	1424.34	1397.86	1372.87

X. $\underline{g\,^3\Delta_g \leftarrow a\,^3\Pi_u}$ System (Ultraviolet)

Band heads, λ (69.109):

(v', v'')	(0, 0)	(1, 0)
λ	1395.21	1367.47

XI. $\underline{F\,^1\Pi_u \leftarrow X\,^1\Sigma_g^+}$ System (Ultraviolet)

Band heads, λ (69.109):

(v', v'')	(0, 0)	(1, 0)
λ	1341.38	1313.97

SPECTROSCOPIC CONSTANTS

State	T_e	ω_e	$x_e\omega_e$	B_e	$\alpha_e \times 10^3$	$D_e \times 10^6$	r_e	Remarks	Bibliography
$F\ ^1\Pi_u$	74532.9[a]	1557.5[b]		1.645	19		1.307	Rydberg	(69.109)
$g\ ^3\Delta_g$	71649.6[a]	1458.06[b]		1.5238	17.0		1.358	Rydberg	(69.109)
$f\ ^3\Sigma_g^-$	70188.4[a]	1360.5	14.8	1.448	40		1.393	Rydberg	(69.109)
$E\ ^1\Sigma_g^+$	55034.6	1671.5	40.02	1.7930	42.1	8.3	1.2517	$y_e\omega_e = +0.248$ $\beta_e = 0.6 \times 10^{-6}$ Rydberg	(69.109, 51.10)
$D\ ^1\Sigma_u^+$	43240.23	1829.57	13.97	1.8334	20.4		1.2378		(39.5)
$e\ ^3\Pi_g$	40797.65	1106.56	39.26	1.1922	24.2	6.3	1.5350	$y_e\omega_e = +2.805$ $z_e\omega_e = -0.1271$ Perturbed	(49.8)
$C'\ ^1\Pi_g$	(37450)[c]	(1697)[c]		(1.7113)[c]		(11.25)[c]	(1.2813)[c]		(67.80)
$C\ ^1\Pi_g$	34261.9	1809.1	15.81	1.7834	18.0	7.1	1.2552	$y_e\omega_e = -4.02$	(63.41, 40.6)
$d\ ^3\Pi_g$	20022.50	1788.22	16.440	1.7527	16.08		1.2660	$y_e\omega_e = -0.5067$ $\gamma_e = -1.274 \times 10^{-3}$ Perturbed	(68.97, 63.43, 63.42, 63.41)

SPECTROSCOPIC CONSTANTS

State	T_e	ω_e	$x_e\omega_e$	B_e	$\alpha_e \times 10^3$	$D_e \times 10^6$	r_e	Remarks	Bibliography
$c\,^3\Sigma_u^+$	13312.1(d)	1961.6(d)	13.65(d)	1.87(d)			1.23(d)		(63.41)
$A\,^1\Pi_u$	8391.00	1608.35	12.078	1.6163	16.86	6.44	1.3184	$y_e\omega_e = -0.01$ $\beta_e = 3.6\times10^{-8}$ $\gamma_e = -5.4\times10^{-5}$ $q = -2.3\times10^{-4}$	(63.41)
$b\,^3\Sigma_g^-$	6434.27	1470.45	11.19	1.4985	16.34	6.22	1.3692	$y_e\omega_e = +0.02$	(63.41, 63.40)
$a\,^3\Pi_u$	714.24	1641.35	11.67	1.6324	16.61	6.44	1.3119		(63.40)
$X\,^1\Sigma_g^+$	0	1854.71	13.34	1.8198	17.65	6.92	1.2425	$y_e\omega_e = -1.17$ $\beta_e = 8.1\times10^{-8}$ $\gamma_e = -2.3\times10^{-4}$	(74.138, 63.41)

(a) Values of $\nu_o(0,0)$; (b) Values of $\Delta G_{1/2}$; (c) Approximate values determined by the study of the perturbations between the C and C' states; (d) Values determined starting with the perturbations between the C and A states.

Dissociation energy = 6.25 ± 0.2 eV, 144 kcal/mole, 50400 cm^{-1}.

Perturbations and General Information

The $e^3\Pi_g$ state is perturbed (64.44, 62.33).

The $C'^1\Pi_g$ state has been identified by study of the perturbation of the $C^1\Pi_g$ state (67.80).

The $d^3\Pi_g$ state is perturbed (68.97, 63.43, 63.42).

The $A^1\Pi_u$ state is perturbed by the $c^3\Sigma_u^+$ state (63.41).

The $X^1\Sigma_g^+$ state is perturbed by the $b^3\Sigma_g^-$ state (63.40).

Franck-Condon factors, r-Centroids for the $d^3\Pi_g \rightleftharpoons a^3\Pi_u$ System using Klein-Dunham potentials (74.140):

v', v''	0	1	2	3	4	5	6	7
0	1.294	1.225	1.169	1.120	1.076			
	7.22-1	2.21-1	4.69-2	8.55-3	1.48-3			
1	1.368	1.307	1.232	1.177	1.127	1.083		
	2.48-1	3.33-1	2.84-1	1.02-1	2.62-2	5.78-3		
2	1.462	1.378	1.328	1.239	1.184	1.134	1.089	
	2.84-2	3.75-1	1.30-1	2.61-1	1.40-1	4.78-2	1.31-2	
3		1.475	1.390	1.371	1.245	1.190	1.140	1.096
		6.89-2	4.23-1	3.86-2	2.08-1	1.60-1	6.90-2	2.28-2
4		1.488	1.402	1.520	1.250	1.196	1.145	
		1.12-1	4.31-1	5.72-3	1.51-1	1.62-1	8.61-2	
5			1.502	1.416		1.252	1.202	
			1.52-1	4.25-1		9.99-2	1.52-1	
6				1.518	1.432	0.839	1.246	
				1.86-1	4.22-1	2.92-3	6.02-2	
7					1.538	1.450	0.725	
					2.08-1	4.35-1	3.66-3	
8						1.564	1.470	
						2.08-1	4.77-1	
9							1.601	
							1.72-1	

Top line   = r-Centroid
Bottom line = Franck-Condon factor followed by factor of ten

C_2

Franck-Condon factors for the C_2 Phillips Band System $A^1\Pi_u \rightleftharpoons X^1\Sigma_g^+$ using Morse potentials (64.45):

v', v''	0	1	2	3	4	5	6	7	8	9
0	0.41016	39989	15563	03084	00328	00018	00000	00000	00000	00000
1	0.33151	00532	29010	27207	08728	01232	00089	00003	00000	00000
2	0.16304	17104	05749	11769	30490	15320	03001	00255	00008	00000
3	0.06365	19934	03208	14143	01919	27053	21337	05462	00559	00021
4	0.02181	12600	13824	00094	15649	00156	20131	25764	08521	01035
5	0.00690	06017	14519	05926	03421	11767	03163	12594	28149	
6	0.00208	02452	09555	12105	01024	07766	06307	07500		
7	0.00061	00905	04937	11202	07554	00092	10081			
8	0.00017	00314	02205	07326	10511	03236				
9	0.00005	00104	00896	03932	08809					
10	0.00001	00034	00342	01861						
11	0.00000	00011	00125							
12	0.00000	00003								

Lifetimes

$$d^3\Pi_g \rightleftharpoons a^3\Pi_u \qquad 185 \pm 70 \text{ nsec } (73.135)$$

$$D^1\Sigma_u^+ \rightleftharpoons X^1\Sigma_g^+ \qquad 14.6 \pm 1.5 \text{ nsec } (69.108)$$

Oscillator strengths

$$A^1\Pi_u \rightleftharpoons X^1\Sigma_g^+ \qquad f_{(0,0)} = 0.8931 \qquad (70.127)$$

$$e^3\Pi_g \rightarrow a^3\Pi_u \qquad f_{(0,0)} = 0.0995 \qquad (70.127)$$

$$D^1\Sigma_u^+ \rightleftharpoons X^1\Sigma_g^+ \qquad f_{(0,0)} = 0.1206 \qquad (70.127)$$

$$C^1\Pi_g \rightarrow A^1\Pi_u \qquad f_{(0,0)} = 0.2016 \qquad (70.127)$$

$$d^3\Pi_g \rightleftharpoons a^3\Pi_u \qquad f_{(0,0)} = 0.017 \pm .007 \qquad (73.135)$$

Potential energy curves (69.109, 63.41):

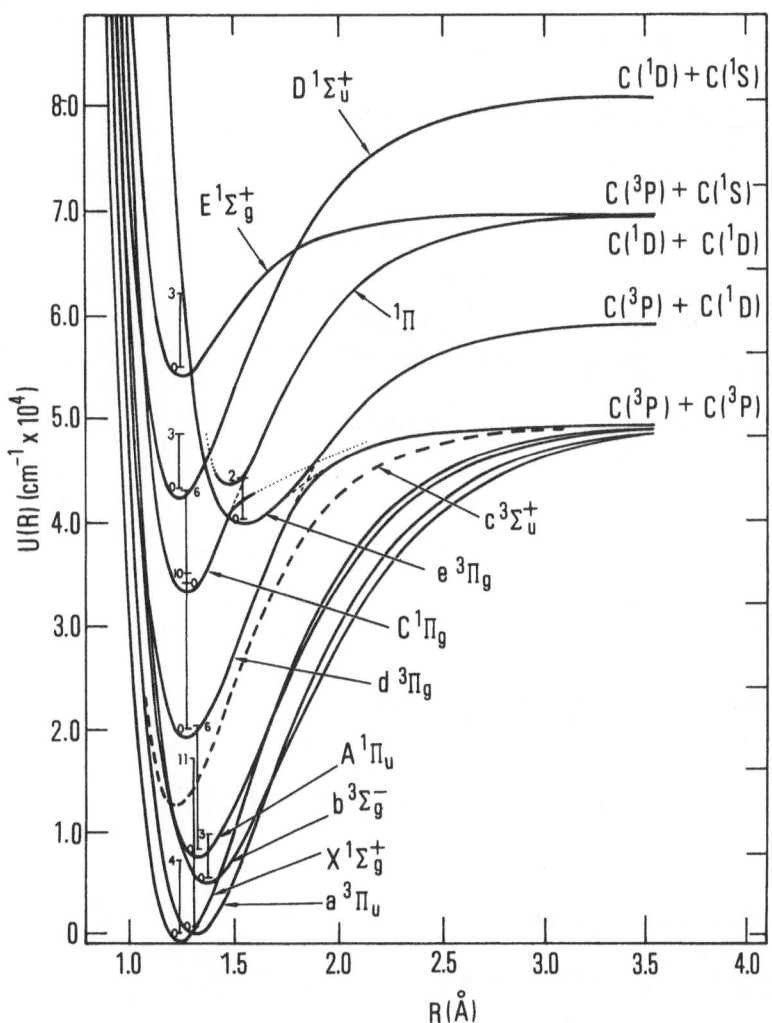

Isotope effects (69.107)

C_2

BIBLIOGRAPHY

(10. 1) Bands at High Pressure,
         F. Croze,
         C.R. Acad. Sci. 150, 1672-3

(27. 2) Swan System,
         R. C. Johnson,
         Trans. Roy. Soc. A 226, 157-230

(29. 3) Bands at High Pressure,
         R. C. Johnson and R. K. Asundi,
         Proc. Roy. Soc. A 124, 668-88

(30. 4) Deslandres-d'Azambuja System,
         R. C. Johnson,
         Nature 125, 89-90

(39. 5) Mulliken System,
         O. G. Landsverk,
         Phys. Rev. 56, 769-77

(40. 6) Deslandres-d'Azambuja System,
         G. Herzberg and R. B. Sutton,
         Can. J. Res. A 18, 74-82

(48. 7) Transition Probabilities, Swan System,
         R. B. King,
         Astrophys. J. 108, 429-33

(49. 8) Fox-Herzberg System,
         J. G. Phillips,
         Astrophys. J. 110, 73-89

(50. 9) Deslandres-d'Azambuka System,
         J. G. Phillips,
         Astrophys. J. 112, 131-5

(51. 10) Freymark System,
         H. Freymark,
         Ann. Physik 8, 221-39

(52. 11) Mulliken and Swan Systems, Absorption in Flames and Explosions,
         R. G. W. Norrish, G. Porter, and B. A. Thrush,
         Nature 169, 582-3

(52. 12) Flame Spectra,
          R. A. Durie,
          Proc. Roy. Soc. A 211, 110-21

(53. 13) Transition Probabilities, Swan System,
          A. A. Wyller,
          Mem. Soc. Roy. Sci., Liege 13, 137-44

(54. 14) Swan System,
          G. Lukas and L. Herman,
          C. R. Acad. Sci. 239, 640-2

(54. 15) Electron Affinity,
          R. E. Honig,
          J. Chem. Phys. 22, 126-31

(56. 16) Swan System,
          O. Scari,
          Acta Phys. Acad. Sci. Hung. 6, 73-85

(56. 17) Transition Probabilities, Swan System,
          R. W. Nicholls,
          Proc. Phys. Soc. A 69, 741-53

(57. 18) Transition Probabilities, Swan System,
          J. G. Phillips,
          Astrophys. J. 125, 153-62

(58. 19) Preliminary to (63. 40),
          E. A. Ballik and D. A. Ramsay,
          J. Chem. Phys. 29, 1418-9

(58. 20) Flame Spectra
          T. P. Clark,
          NACA Report, T. N. 4266

(59. 21) Preliminary to (63. 41),
          E. A. Ballik and D. A. Ramsay,
          J. Chem. Phys. 31, 1128

(59. 22) Absorption and Fluorescence in Rare Gas Matricies,
          M. MacCarty, Jr., and G. W. Robinson,
          J. Chim. Phys. 56, 723-31

(60. 23) Theory, Transition Probabilities,
          E. Clementi,
          Astrophys. J. 132, 898-904

C_2

(60.24) Thermochemistry,
R. L. Altman,
J. Chem. Phys. 32, 615-6

(60.25) Calculation of the States of Low Energy,
E. Clementi and K. S. Pitzer,
J. Chem. Phys. 32, 656-62

(60.26) Stellar Absorption; Isotope Abundance,
A. M. MacKellar,
J. Roy. Astron. Soc. Can. 54, 97-109

(61.27) Abundance in Stellar Atmospheres; Partition Function,
E. Clementi,
Astrophys. J. 133, 303-8

(61.28) Flame Emission,
S. L. N. G. Krishnamachari and H. P. Broida,
J. Chem. Phys. 34, 1709-11

(61.29) Growth Curves, Cosmic Relation of $^{12}C/^{13}C$,
J. L. Climenhaga,
Publ. Dominion Astrophys. Obs., Victoria 11, 307-37

(61.30) Absorption in a Rare Gas Matrix,
L. J. Schoen,
5th Intern. Symp. Free Radical, Sweden No. 61, 1-4

(61.31) Calculation of Transition Probabilities,
W. R. Jarmain,
U. S. Dept. Com. Off. Tech. Serv. AD 262-593

(62.32) General,
L. Wallace,
Astrophys. J. 136, 1158

(62.33) Perturbations in the $e^3\Pi_g$ State. Potential Energy Curves,
J. H. Jaffe, S. Kimel, and M. A. Hirshfeld,
Can. J. Phys. 40, 113-21

(62.34) Potential Energy Curves of the $d^3\Pi_g$ State,
N. L. Singh and D. C. Jain,
Can. J. Phys. 40, 530-3

(62.35) Heat of Sublimination and Dissociation Energy,
L. Brewer, W. T. Hicks, and O. N. Krikorian,
J. Chem. Phys. 36, 182-8

(62.36)   General,
          R. Moccia,
          J. Chem. Phys. 37, 910-1

(62.37)   Theory, Potential Energy Curves,
          S. M. Read and J. T. Vanderslice,
          J. Chem. Phys. 36, 366-9

(62.38)   Theory, Wave Functions,
          D. C. Jain and P. Sah,
          Proc. Phys. Soc. 80, 525-33

(62.39)   Internuclear Distances,
          F. Rogowski,
          Z. Naturforsch. A 17, 933

(63.40)   Ballik-Ramsay System; Perturbations,
          E. A. Ballik and D. A. Ramsay,
          Astrophys. J. 137, 61-83

(63.41)   Phillips System; Perturbations,
          E. A. Ballik and D. A. Ramsay,
          Astrophys. J. 137, 84-101

(63.42)   Swan and Deslandres-d'Azambuja Systems; Perturbations,
          J. H. Callomon and A. G. Gilby,
          Can. J. Phys. 41, 995-1004

(63.43)   Detailed Study of the Swan System,
          S. P. Davis and J. G. Phillips,
          Berkeley Analyses of Molecular Spectra, I

(64.44)   Transition Probabilities in the Swan and Fox-Herzberg Systems,
          D. C. Jain,
          J. Quant. Spectrosc. Radiative Transfer 4, 427-40

(64.45)   Franck-Condon Factors for the Swan and Phillips Systems,
          F. S. Ortenberg,
          Optics Spectrosc. 16, 398-400

(64.46)   Absorption in a Matrix,
          R. L. Barger and H. P. Broida,
          NASA Report AD 62 78 14

(65.47)   Spectra of C_2 in a Matrix,
          R. L. Barger and H. P. Broida,
          J. Chem. Phys. 43, 2371-6

C_2

(65.48) Absorption in Flames, Intensity Distribution in the Rotational Structure,
R. Bleekrode and W. C. Nieuwpoort,
J. Chem. Phys. 43, 3680-7

(65.49) G. D. Brabson,
"Spectroscopic Investigation of High Temperature Species Isolated
in Inert Gas Matrices,"
U. C. Berkely, Ph.D. Thesis

(65.50) Calculation of the Franck-Condon Factors for the Swan and Phillips
Bands,
R. J. Spindler,
J. Quant. Spectrosc. Radiative Transfer 5, 165-204

(65.51) Excitation by Shock Waves; Transition Probabilities,
A. G. Sviridov, N. N. Sobolev, and V. M. Sutovskii,
J. Quant. Spectrosc. Radiative Transfer 5, 525-43

(65.52) Origin of the Swan Bands in Comet Emission,
L. J. Stief and V. J. DeCarlo,
Nature 205, 1197

(65.53) Tail Sequences of the Swan System,
E. D. Bugrim, A. I. Lyutyi, V. S. Rossikhin, and I. L. Tsikora,
Optics Spectrosc. 19, 292-3

(65.54) Swan System,
J. E. Mentall and R. W. Nicholls,
Proc. Phys. Soc. 86, 873-6

(65.55) P. F. Fougere,
"Configuration Interaction in Diatomic Molecules Application to C_2
and BN,"
Boston University, Thesis

(66.56) Phillips Bands in Stellar Absorption,
P. Solomon and W. Stein,
Astrophys. J. 144, 825-6

(66.57) Isotope Effect and Transition Probabilities. Swan and Fox-Herzberg
Systems,
M. Halmann and I. Laulicht,
Astrophys. J. Suppl. 12, 307-12

(66.58) Emission in Flames. Intensity Distribution in the Rotational
Structure of the Swan Bands,
R. Bleekrode,
J. Chem. Phys. 45, 3153-4

(66. 59) "Ab initio" Calculations; Molecular Constants,
P. F. Fougere and R. K. Nesbet,
J. Chem. Phys. **44**, 285-98

(66. 60) Swan System,
J. A. Harrington, A. P. Modica, and D. R. Libby,
J. Chem. Phys. **44**, 3380-7

(66. 61) Matrix Spectra,
W. Weltner, Jr. and D. MacLeod, Jr.,
J. Chem. Phys. **45**, 3096-105

(66. 62) Shock Wave Excitation of the Swan System. f-values,
A. R. Fairbairn,
J. Quant. Spectrosc. Radiative Transfer **6**, 325-36

(66. 63) Absorption and Growth Curves,
A. R. Fairbairn,
J. Quant. Spectrosc. Radiative Transfer **6**, 705-15

(66. 64) Swan System Excited by Shcok Waves. f-values,
A. G. Sviridov, N. N. Sobolev, M. Z. Novgorodov, and
G. A. Arutyunova,
J. Quant. Spectrosc. Radiative Transfer **6**, 875-92

(66. 65) Emission and Absorption in Flames,
R. Bleekrode,
Thesis, Amsterdam

(66. 66) W. M. Vaidya,
"A Comparison of Cometary Spectra and Spectra of Hydrocarbon
Flames,"
Mem. Soc. Roy. Sci., Liege **12**, 433-5

(66. 67) J. G. Phillips,
"Introductory Report on Recent Laboratory Work Related to Comets,"
Mem. Soc. Roy. Sci., Liege **12**, 393-407

(66. 68) C. Arpigny,
"The Rotational and Vibrational Temperatures of C_2 in Comets,"
Mem. Soc. Roy. Sci., Liege **12**, 165-83

(66. 69) D. E. Bugrim, A. I. Lyutyi, V. S. Rossikhin, and I. L. Tsikora,
"Excitation Characteristics of the Swann Bands of C_2 in Streams of
Vapors of Metals and Organic Compounds,"
Opt. Spectrosc. **20**, 320-4

C_2

(66. 70)   C. Arpigny,
           "The C_2 Swan Bands in Comets,"
           Astrophys. J. <u>144</u>, 424-7

(66. 71)   M. I. Savadatti and H. P. Broida,
           "Spectral Study of Flames of Carbon Vapor at Low Pressure,"
           J. Chem. Phys. <u>45</u>, 2390-6

(66. 72)   E. D. Bugrim, A. I. Lyutyi, V. S. Rossikhin, and I. L. Tsikora,
           "Vibrational Relaxation of a C_2 Molecule in the Excited State,"
           Dokl. Akad. Nauk <u>169</u>, 858-60

(66. 73)   Y. A. Pdastinin and G. G. Baula,
           "Absorption Cross Section of the Electronic Systems of the Bands
           for Diatomic Molecules of N_2, O_2, N_2^+, NO, C_2, CN Under High
           Temperatures,"
           Issled. Fiz. Gaz., Akad. Nauk 41-61

(67. 74)   Lifetime of the $d^3\Pi_g$ State. f-values for the Swan System,
           E. H. Fink and K. H. Welge,
           J. Chem. Phys. <u>46</u>, 4315-8

(67. 75)   Excitation Mechanism of the "High Pressure" Bands.   A and A'
           Perturbations,
           C. Kunz, P. Harteck, and S. Dondes,
           J. Chem. Phys. <u>46</u>, 4157-8

(67. 76)   Flash Photolysis in Matrices.   Phillips and Mulliken Systems,
           D. E. Mulligen, M. E. Jacox, and L. Abouaf-Marguin,
           J. Chem. Phys. <u>46</u>, 4562-70

(67. 77)   Low Energy States by LCAO-MO-SCF Method,
           G. Verhaegen, W. G. Richards, and C. M. Moser,
           J. Chem. Phys. <u>46</u>, 160-4

(67. 78)   Oscillator Strengths, Mulliken System,
           S. R. Paglia,
           J. Molec. Spectrosc. <u>24</u>, 302-13

(67. 79)   Dissociation Energy Vibrational Extrapolation of $d^3\Pi_g$,
           G. Messerle and L. Krauss,
           Z. Naturforsch. A <u>22</u>, 1744-7

(67. 80)   Perturbations of the Deslandres-d'Azambuja System,
           G. Messerle and L. Krauss,
           Z. Naturforsch. A <u>22</u>, 2015-23

(67.81) Dissociation Energy of $C {}^1\Pi_g$,
G. Messerle and L. Krauss,
Z. Naturforsch. A <u>22</u>, 2023-6

(67.82) J. E. Mental and R. W. Nicholls,
"Spectroscopic Temperature Measurements on Laser-Produced Flames,"
J. Chem. Phys. <u>46</u>, 2881-5

(67.83) R. Bleekrode,
"Absorption and Emission Spectroscopy of C_2, CH, and OH in Low
Pressure Oxyacetylene Flames,"
University of Amsterdam, Thesis

(67.84) G. Herzberg,
"Molecular Absorption Spectra in Flash Discharges,"
Proc. Int. Conf. Spectrosc., Bombay

(67.85) B. P. Levitt and N. Wright,
"C_2 Emission During the Pyrolysis of Acetone,"
Trans. Faraday Soc. <u>63</u>, 282-8

(67.86) J. R. De La Vega and H. F. Hameka,
"Calculation of Magnetic Susceptibilities of Diatomic Molecules,"
Physica <u>35</u>, 313-22

(67.87) R. L. LeRoy,
"Relaxation Effects in Electronically Excited CH and C_2 in Low
Pressure Hydrocarbon Flames,"
Can. J. Chem. <u>45</u>, 2853-5

(67.88) G. Ndaalio and J. M. Deckers,
"Energy Distribution in Electronically Excited CH and C_2 in Low
Pressure Hydrocarbon Flames,"
Can. J. Chem. <u>45</u>, 2441-9

(68.89) Bands at High Pressure, Fine Structure,
H. Meinel and G. Messerle,
Astrophys. J. <u>154</u>, 381-4

(68.90) P. Mayer and C. R. O'Dell,
"Emission-Band Ratios in Comet Rudnicki (1966e),"
Astrophys. J. <u>153</u>, 951-62

(68.91) Theoretical,
M. A. Marchetti and S. R. LaPaglia,
J. Chem. Phys. <u>48</u>, 434-9

(68.92)   Bands at High Pressure, Isotope Effect,
          R. K. Dhumwad and N. A. Narasimham,
          Can. J. Phys. **46**, 1254-5

(68.93)   Spectra Attributed to C_2^-
          G. Herzberg and A. Lagerqvist,
          Can. J. Phys. **46**, 2363-73

(68.94)   Milliken System, Dipole Moment,
          M. A. Marchetti and S. R. LaPaglia,
          Can. J. Phys. **48**, 434-9

(68.95)   Flame Spectra. Population Inversion in $d^3\Pi_g$; Mechanism of
          Excitation,
          A. W. Naegeli and H. P. Palmer,
          Can. J. Phys. **48**, 2372-3

(68.96)   Bands at High Pressure. Excitation Mechanism,
          D. W. Naegeli and H. B. Palmer,
          J. Molec. Spectrosc. **26**, 152-4

(68.97)   Perturbations in the Swan System,
          J. G. Phillips,
          J. Molec. Spectrosc. **28**, 233-42

(68.98)   Shock Waves, Transition Moments, f-Values,
          J. O. Arnold,
          J. Quant. Radiative Transfer **8**, 1781-94

(68.99)   Comet Emission. Mechanism of Formation and Excitation,
          R. J. Fallon,
          Nature **217**, 1240-1

(68.100) Mulliken and Fox-Herzberg Systems, Excitation of the High
          Rotational Levels,
          G. Messerle,
          Z. Naturforsch. A **23**, 470

(68.101) L. N. Glusko and V. I. Tverdokhlebov,
          "Activation Energy for the Formation of the Excited Radicals CH,
          C_2 and OH in Acetylene-Air-Flames,"
          Dokl. Akad. Nauk **179**, 1370-2

(68.102) P. Connes, J. Connes, R. Bouigue, M. Querci, J. Chauville, and
          F. Querci,
          "On the Infrared Spectra of Type Mod C Stars Betweet 4000 and
          9000 cm^{-1} (2.5 to 1.1μ),"
          Ann. Astrophys. **31**, 485-92

(68. 103) Theoretical,
R. F. W. Bader and A. D. Bandrauk,
J. Chem. Phys. $\underline{49}$, 1653-65

(68. 104) K. S. Kini and M. I. Savadatti,
"High Pressure Bands of Carbon,"
J. Karnatak University $\underline{13}$, 54-60

(68. 105) C. T. Bowman and D. J. Seery,
"Chemiluminescence in the High-Temperature Oxidation of Methane,"
Combustion and Flame $\underline{12}$, 611-4

(68. 106) Y. A. Plastinin,
"Optical Absorption Cross-Sections of Two-Atom Molecules,"
Fiz. Gaz. Ioniz. Khim. Reag. Gazov. 89-125

(69. 107) Deslandres-d'Azambuja, Isotope Effect,
H. Cisak, K. Dabrowska, and M. Rytel,
Acta Phys. Polon. $\underline{36}$, 497

(69. 108) Transition Probabilities and Lifetimes, Swan and Mulliken Systems,
W. H. Smith,
Astrophys. J. $\underline{156}$, 791-4

(69. 109) Absorption in the Far Ultraviolet New State Designation,
G. Herzberg, A. Lagerqvist, and C. Malmberg,
Can. J. Phys. $\underline{47}$, 2735-43

(69. 110) Spectra Attributed to C_2^- in Matrices,
D. E. Milligan and M. E. Jacox,
J. Chem. Phys. $\underline{51}$, 1952-5

(69. 111) Swan System. Mechanism of Excitation in Flames,
W. C. Nieuwpoort and R. Bleekrode,
J. Chem. Phys. $\underline{51}$, 2051-5

(69. 112) Extension of the Swan System,
K. S. Kini and M. I. Savadatti,
J. Phys. B $\underline{2}$, 307-8

(69. 113) Rotational Population in Plasmas at Low Pressure,
H. Meinal and L. Krauss,
J. Quant. Spectrosc. Radiative Transfer $\underline{9}$, 443-60

(69. 114) M. R. Katti and H. D. Sharma,
"Evaluation of r-Centroids for the Band Systems of a Few Diatomic
Molecules Using the Hulbert-Hirschfelder Potential Function,"
Ind. J. Pure Appl. Phys. $\underline{7}$, 282-3

C_2

(69.115)   H. Cisak, K. Dabrowska, and M. Rytel,
"The Deslandres-d'Azambuka Bands of Isotopic C_2 Molecule, "
Acta Phys. Polon. 36, 497

(69.116)   W. H. Smith,
"Transition Probabilities for the Swan and Mulliken C_2 Bands, "
Astrophys. J. 156, 791-4

(70.117)   A. G. Aozhkova, V. I. Tverdokhlebov, and N. N. Chirkin,
"Spectroscopic and Probe Studies of a Low-Pressure Acetylene-Air
Flame in Counter Currents, "
Primen. Plazma. Spekt., Mater Vses. Simp. (1968) 16-18

(70.118)   L. M. Glusko, L. A. Kovalenko, A. C. Razhkova, and
I. V. Tverdokhlebov,
Zh. Prikl. Spektrosk. 13, 907-9

(70.119)   M. Querci and F. Querci,
"$^{14}N/^{12}C$, $^{12}C/^{13}C$ and $^{14}N/^{15}N$ in the Carbon Star UU Aur, "
Astron. Astrophys. 9, 1-14

(70.120)   W. L. Gebel,
"Spectrophotometry of Comets 1967n, 1968b, and 1968c, "
Astrophys. J. 161, 765-77

(70.121)   A. G. Sviridov,
"Experimental Determination of a Matrix Element of the Dipole
Moment of Electronic Transitions of the System of Bands for the
C_2 Molecule, "
Tr. Fiz. Inst. Akad. Nauk 51, 124-93

(70.122)   D. C. Jain,
"A Study of Some Potential Energy Functions for Diatomic Molecules,'
Int. J. Quant. Chem. 4, 579-86

(70.123)   E. M. Bulewicz, P. J. Padley, and R. E. Smith,
"Spectroscopic Studies of C_2, CH and OH Radicals in Low Pressure
Acetylene + Oxygen Flames, "
Proc. Roy. Soc. A 315, 129-48

(70.124)   H. Meinel,
"Spectroscopic Investigation of the Two CH-and C_2-Groups in
Various Rotational Distribution, "
Z. Naturforsch. 25a, 302-4

(70.125) S. A. Yantovskii, N. A. Shvartsman, and M. P. Porsov,
"Spectrographic Investigation of Flames of Preliminarily Mixed
Propane — Air Mixtures in the Process of Inhibition. II. Influence
of Inhibitors on the Intensity of the Emission of C_2 and CH Radicals,"
Kinet. Katal. 11, 1109-14

(70.126) I. R. Marenin and H. R. Johnson,
"New Molecular Constants for the Phillips System of C_2,"
J. Quant. Spectrosc. Radiative Transfer 10, 305-9

(70.127) A. N. Singh and D. K. Rai,
"Transition Probabilities for Electronic Spectra of Li_2 and C_2,"
Proc. Nat. Acad. Sci. A 40, 101-3

(71.128) I. D. Singh and R. C. Maheshwari,
"Isotope Effect on Franck-Condon Factors and r-Centroids for the
New ($^1\Sigma_u^+$ - $^1\Sigma_g^+$) System of Isotopic C_2 Molecule,"
Ind. J. Pure Appl. Phys. 9, 296-7

(71.129) J. Peeters, J. F. Lambert, P. Hertoghe, and A. Van Tiggelsen,
"Mechanisms of C_2 and CH Formation in a Hydrogen-Oxygen Flame
Containing Hydrocarbon Traces,"
Proc. 13th Int. Symp. Combust (1970) 321-32

(71.130) R. P. Frosch,
"C_2 and C_2^- Spectra Produced by the X Irradiation of Acetylene in
Rate-Gas Matrices,"
J. Chem. Phys. 54, 2660-6

(72.131) J. Barsuhn,
"An SCF-MO-CI Calculation of the C_2 Molecule Using an Atomic
Basis of Contracted Gaussian Lobe Functions,"
Z. Naturforsch. 27a, 1031-41

(72.132) F. Gosse, N. Sadeghi, and J. C. Pebay-Peyroula,
"Study of Mechanisms Producing C_2($A^3\Pi_g$) Molecules in a CO
Afterglow,"
Chem. Phys. Letters 13, 557-60

(72.133) W. Von Niessen,
"Density Localization of Atomic and Molecular Orbitals,"
Theoret. Chim. Acta 27, 9-23

(72.134) G. Schatz and M. Kaufman,
"Chemiluminescence Excited by Atomic Fluorine,"
J. Phys. Chem. 76, 3586-90

(73. 135)   N. Grevesse and A. J. Sauval,
"A Study of Molecular Lines in the Solar Photospheric Spectrum,"
Astron. Astrophys. 27, 29-43

(73. 136)   K. R. Bodanl, G. C. Joshi, and M. C. Pande,
"C_2 in Eta Aquilae Spectrum,"
Bull. Astron. Institute Czech. 24, 169-70

(74. 137)   G. C. Lie and E. Clementi,
"Study of the Electronic Structure of Molecules. XXII. Correlation
Energy Corrections as a Function of the Hartree-Fock Type Density
and its Application to the Homonuclear Diatomic Molecules of the
Second Row Atoms,"
J. Chem. Phys. 60, 1288-96

(74. 138)   K. H. Becker, D. Haaks, and T. Tatarczyk,
"Measurements of C_2 Radicals in Flames With a Tunable Dye-Laser,"
Z. Naturforsch. 29a, 829-30

(74. 139)   C. J. Hsu, H. B. Palmer, and C. F. Aten,
"C_2 Band Emission From Diffusion Flames of Alkali Metals and
Halogenated Methanes,"
Combustion and Flame 22, 133-5

(74. 140)   L. L. Danylewych and R. W. Nicholls,
"Intensity Measurements on the C_2 ($d^3\Pi_g$ - $a^3\Pi_u$) Swan Band System.
I. Intercept and Partial Band Methods,"
Proc. Roy. Soc. Lond. A 339, 197-212

(74. 141)   L. L. Danylewych and R. W. Nicholls,
"Intensity Measurements on the C_2 ($d^3\Pi_g$ - $^3\Pi_u$) Swan Band System.
II. Interpretation of Band Intensity Measurement from Synthetic
Spectra,"
Proc. Roy. Soc. Lond. A 339, 213-22

Ca_2

Methods of Production and Experimental Technique

Emission from a hollow cathode discharge.

Absorption in a King furnace, $T > 2300^{\circ} K$.

BAND SYSTEMS

	System	Transition	Sources	Wavelength Limits	Degrading	Band Head, $v_{0,0}$	Remarks	Bibliography
	I	$^1\Sigma - X^1\Sigma_g^+$	Absorption	6500-5900		16395.6(0,0)		(68.5, 67.4)
	II	$A^1\Sigma_u^+ \leftarrow X^1\Sigma_g^+$	Absorption	5300-4300		19401.0(v,0)		(71.7)
	III	Continuum		5100-3980				(31.1)

Ca$_2$

I. $^1\Sigma^+ - X\,^1\Sigma_g^+$ System

Band heads, λ (67.4):

v', v''	λ
0, 0	6099.2
0, 1	6180.2
0, 2	6259.1
1, 0	6014.5
1, 1	6093.4
2, 1	6013.1

II. $A\,^1\Sigma_u^+ \leftarrow X\,^1\Sigma_g^+$ System

Band heads of the v',0 progression, λ (71.7):

v''	v	v + 1	v + 2	v + 3	v + 4	v + 5	v + 6
λ	5154.4	5119.9	5086.2	5053.4	5021.5	4990.3	4960.0

SPECTROSCOPIC CONSTANTS

$_2$

State	T_e	ω_e	$x_e\omega_e$	B_e	$\alpha_e \times 10^3$	$D_e \times 10^6$	r_e	Remarks	Bibliography
$^1\Sigma$	16395.59								(71.7)
$X^1\Sigma_g^+$	0.0	65.0	1.11	0.046			4.28		(71.7)

Dissociation energy = 0.12 ± 0.1 eV, 2.69 kcal/mole, 940 cm^{-1} (71.7).

BIBLIOGRAPHY

(31. 1)   H. Hamada,
          Philos. Mag. 12, 50-67

(40. 2)   R. Sanford,
          Publ. Astron. Soc. Pacific 52, 325-6

(42. 3)   A. S. King,
          Publ. Astron. Soc. Pacific 54, 6

(67. 4) $^1\Sigma$ - X$^1\Sigma_g^+$ System, Vibrational Analysis,
          G. V. Kovalenok and V. A. Sokolov,
          "Vibrational Analysis of the Orange Bands of the Diatomic Calcium
          Molecule, "
          Izv. Sib. Otoi 4, 118-20

(68. 5) $^1\Sigma$ - X$^1\Sigma_g^+$ System, Rotational Analysis,
          G. V. Kovalenok and V. A. Sokolov,
          "Rotational Analysis of the Orange Bands of the Diatomic Calcium
          Molecule, "
          Izv. Sib. Otoi 11, 27-30

(70. 6)   R. H. Ewing and A. M. Mellor,
          "Further Calculations of Equilibrium Dimer Properties, "
          J. Chem. Phys. 53, 2983-4

(71. 7)   W. J. Balfour and R. F. Whitlock,
          "Ca$_2$: A Van der Waals Molecule, "
          J. Chem. Soc. 19, 1231

Methods of Production and Experimental Technique

Absorption.

Emission from normal and hollow cathode discharges.

Fluorescence when excited by Cu, Zn, or Al sparks.

Cd_2

BAND SYSTEMS

System	Absorption (33.10, 29.2, 29.1)	Emission (35.14, 34.12, 32.6, 31.4, 31.3, 29.1)	Fluorescence (34.12, 32.7)
I		Continuum: 5400-4058 Bands: 4463|4439|4416|4392|4369|4343|	Continuum: 5000-3800 max. at 4000
II	Broad line: 3261 Band: 3178	Continuum: 4044-3178 max. at 3939 and 3261 min. at 3370 Bands: 4044-3836, 3186-3148	Broad line: 3261 Band: 3178
III	Continuum: 2800-2212 Band: 2212	6 Bands: 2870-2780 Continuum: 3000-2191 max. at 2980 and 2288 min. at 2450 Band: 2214	Continuum: 3050-2260 max. at 2288 Bands: 3050-2700, 2560-2288
IV	Band: 2114	Bands: 2140-2110 max. at 2114	Bands: 2140-2110 max. at 2125

Spectroscopic Constants

Dissociation energy = 0.07 ± 0.01 eV, 1.63 kcal/mole, 940 cm^{-1}.

Perturbations and General Information

Potential energy curves (35.18).

Cd_2

Spectroscopic Constants

Dissociation energy = 0.07 ± 0.01 eV, 1.63 kcal/mole, 940 cm^{-1}.

Perturbations and General Information

Potential energy curves (35.18).

BIBLIOGRAPHY

(29. 1)   Absorption and Emission in Ultraviolet,
           J. G. Winans,
           Philos. Mag. 7, 555-66

(29. 2)   Absorption and Emission in Ultraviolet,
           J. H. Walter and S. Barratt,
           Proc. Roy. Soc. A 122, 201-10

(31. 3)   Absorption and Emission in Ultraviolet and Visible,
           H. Hamada,
           Nature 127, 555

(31. 4)   Absorption and Emission in Ultraviolet and Visible,
           H. Hamada,
           Philos. Mag. 12, 50-67

(32. 5)   Fluorescence,
           L. Sosnowski,
           C. R. Acad. Sci. 195, 224-6

(32. 6)   Emission in Ultraviolet and Visible,
           J. K. Robertson,
           Philos. Mag. 14, 795-806

(32. 7)   Emission and Fluorescence,
           S. W. Cram and J. G. Winans,
           Phys. Rev. 41, 388-9

(33. 8)   C. Kalinowska,
           Acta Phys. Polon. 2, 111-7

(33. 9)   R. Siksna,
           Acta Phys. Polon. 2, 253-65

(33. 10)   Absorption Coefficient,
           H. Kuhn and S. Arrhenius,
           Z. Physik 82, 716-22

(34. 11)   F. Swietoslawska,
           Acta Phys. Polon. 3, 261-70

(34. 12)   Emission and Fluorescence, Potential Energy Curves,
           S. W. Cram,
           Phys. Rev. 46, 205-9

(35. 13) Emission at 2212Å,
          J. K. Robertson,
          Nature 135, 308-9

(35. 14) Emission at 2212Å,
          L. Vegard,
          Nature 132, 682

(35. 15) Discussion,
          T. S. Subbaraya,
          Proc. Ind. Acad. Sci. A 1, 484-8

(35. 16) Theory of Van der Waals Molecules,
          F. Finkelnburg,
          Z. Physik 96, 714-9

(35. 17) Potential Energy Curves,
          F. Finkelnburg,
          Z. Physik 96, 714-9

(37. 18) A. Kotecki,
          Acta Phys. Polon. 6, 75-88

(37. 19) A. Kotecki,
          Acta Phys. Polon. 6, 144-9

(37. 20) A. Kotecki,
          Verhandl. Dtsch. Phys. Ges. 18, 53

(38. 21) T. Zamlynski,
          Acta Phys. Polon. 7, 24-33

(38. 22) Emission at 3261Å and 2288Å,
          R. Planiol,
          Ann. Physique 9, 177-235

Ce$_2$

Spectroscopic Constants

Dissociation energy = 2.83 ± 0.01 eV, 65.3 kcal/mole, 22840 cm^{-1} (72.4).

Ce_2

BIBLIOGRAPHY

(69. 1)   Dissociation Energy,
G. Balducci, G. DeMaria, and M. Guido,
J. Chem. Phys. 50, 5424-5

(69. 2)   Dissociation Energy,
K. A. Gingerich,
J. Chem. Soc. D 9-10

(69. 3)   Dissociation Energy,
K. A. Gingerich and H. S. Finkbeiner,
J. Chem. Soc. D 901-2

(72. 4)   A. Kant and S. S. Lin,
"Dissociation Energies of Homonuclear Diatomic Rare Earth Molecules,
Monatshefte für Chemie 103, 757-63

Cl_2

Methods of Production and Experimental Technique

Absorption.

Emission in many different discharges, flames, thermoluminescence, and "active" nitrogen flames.

BAND SYSTEMS

System	Transition	Sources	Wavelength Limits	Degrading	Band Head, $\nu_{0,0}$	Remarks	Bibliography
I	$A^3\Pi\left(0_u^+\right) \rightleftharpoons X^1\Sigma_g^+$	Discharge Absorption	6500-5400 6000-4780	R	17651.5		(63.50, 62.49, 36.22)
II	(a)	Absorption Discharge Discharge Discharge Discharge Absorption	<4785 3063-2715 2565-2097 1997-1855 <3400 <1950	Continuum Continuum Continuum Continuum Continuum Continuum	- (58000) (67700) (75000) - -	(b) λ max. 3300 - - - - <1560	(59.46, 47.38, 47.37, 38.28, 37.26, 37.24, 30.14)
III	$? \rightarrow A^3\Pi\left(0_u^+\right)$	Discharge	2600-2390	R	40119		(59.45)
IV	$? \rightarrow A^3\Pi\left(0_u^+\right)$	Discharge	2365-2239	V	43632		(59.44)
V	?	Discharge	1870-1070			Rydberg transitions	(59.46)

(a) Attributed to $^{1,3}\Sigma_u^+ \rightarrow {}^1\Sigma_g^+$ transitions, discussed in (59.46).

(b) Analysis is difficult.

Cl_2

I. $\underline{A^3\Pi\left(0_u^+\right) \rightleftarrows X^1\Sigma_g^+ \text{ System}}$

Bands in emission, λ (37.25, 36.22, 28.8); isotope effect (63.50, 43.36); numbering of v' (62.49).

v', v''	0	1
15	4932.8	5070.6
16	4911.7	5048.8
17	4892.7	5030.0
18	4875.7	5013.4
19	4859.6	4995.6

II. <u>Continuum Spectra</u>

<u>3063-1855Å System</u>

Three groups of continuum bands due to transitions to the unstable lower states:

$$\left(^2P_{3/2} + {}^2P_{3/2}\right), \quad \left(^2P_{3/2} + {}^2P_{1/2}\right), \quad \left(^2P_{1/2} + {}^2P_{1/2}\right)$$

<u><3400Å System</u>

Principal maxima (37.24): $\lambda|3063|2564$
Secondary maxima: $\lambda|2957|2881|2819|2758|2714|2432$

When produced using "active" nitrogen two series of diffuse bands are observed. These bands are interpreted as arising from a $^1\Sigma_u^+ \rightarrow {}^1\Pi_g$ transition with a pronounced minimum in the $^1\Pi_g$ state.

<u><1950Å System</u>

Corresponds to a transition from the ground state to the $^3\Sigma_g^+$ state, which dissociates to two $^2P_{1/2}$ atoms (47.37).

III.   2600-2390Å System

Band heads, λ (59.45):

v', v''	0	1	2
0	2492.6	2519.8	2523.5
1	2477.6	2493.5	
2	2462.6	2477.9	

IV.   2365-2239Å System

Band heads, λ (59.44):

v', v''	0	1	2
0	2291.9	2305.1	
1	2278.7		
2		2277.6	2290.4

V.   1870-1070Å Systems

Many Rydberg transitions.   Molecular orbital discussion (59.46).

Cl_2

SPECTROSCOPIC CONSTANTS

State	T_e	ω_e	$x_e\omega_e$	B_e	$\alpha_e \times 10^3$	$D_e \times 10^6$	r_e	Remarks	Bibliography
?	61433	261.5[a]	0.812[b]	-	-	-	-		(59.44)
?	57916	246.6[a]	0.615[b]	-	-	-	-		(59.45)
$A^3\Pi(0_u^+)$	17801.2	261.9	5.45	0.1680	3.7	-	2.396		(63.50, 62.49)
$X^1\Sigma_g^+$	0	559.71	2.70	0.24407	1.53	-	1.9878		(63.50, 62.49)

[a] ω_o ; [b] $x_o\omega_o$

Dissociation energy = 2.4795 ± 0.0003 eV, 57.2 kcal/mole, 19999 cm^{-1} (63.50).

Perturbations and General Information

Ionization potential = 92590 ± 80 cm^{-1}, 11.48 eV (57.42).

Potential energy curves (47.16).

Franck-Condon factors — RKRV potential for ^{35}Cl$_2$ (71.58)

$$- \text{B}^3\Pi\left(0_u^+\right) - \text{X}^1\Sigma_g^+$$

v', v''	0		1		2		3		4	5	6
0	2.	-9	6.	-8	7.	-7	6.	-6	0.00003	0.00016	0.00061
1	3.	-8	6.	-7	7.	-6	0.00005		0.00027	0.00112	0.00363
2	2.	-7	3.	-6	0.00003		0.00023		0.00110	0.00391	0.01089
3	6.	-7	0.00001		0.00012		0.00072		0.00300	0.00930	0.02202
4	2.	-6	0.00004		0.00032		0.00170		0.00631	0.01698	0.03380
5	5.	-6	0.00009		0.00069		0.00333		0.01030[a]	0.02425[a]	0.04075[a]
6	0.00001		0.00018		0.00129		0.00560		0.01623	0.03227	0.04337
7	0.00002		0.00032		0.00213		0.00836		0.02134	0.03607	0.03842
8	0.00004		0.00053		0.00319		0.01129		0.03611	0.03611	0.02935
9	0.00006		0.00079		0.00439		0.01406		0.02778	0.03287	0.01924
10	0.00009		0.00109		0.00563		0.01635		0.02835	0.02749	0.01054

[a] $q_{v', v''}$ for J' = 25, J'' = 26. All others rotationless.

BIBLIOGRAPHY

(22. 1)   Ultraviolet Emission,
          E. von Angerer,
          Z. Physik 11, 167-9

(24. 2)   Ultraviolet Emission,
          E. B. Ludlam and W. West,
          Proc. Roy. Soc. Edinburgh 44, 185-96

(26. 3)   Ultraviolet Absorption Coefficient,
          D. S. Villars,
          J. Am. Chem. Soc. 48, 1874-6

(26. 4) $^3\Pi \leftarrow \,^1\Sigma$ System,
          G. Nakamura,
          Mem. Coll. Sci. Univ. Kyoto A 9, 315-67

(26. 5) $^3\Pi \leftarrow \,^1\Sigma$ System,
          H. Kuhn,
          Z. Physik 39, 77-91

(27. 6)   Absorption,
          G. Kornfeld and W. Steiner,
          Z. Physik 45, 325-30

(27. 7)   R. Mecke,
          Z. Physik 42, 390-425

(28. 8)   Thermoluminescence,
          C. Kondratjew and A. Leipunsky,
          Z. Physik 50, 366-71

(29. 9)   Emission in Flames,
          E. B. Ludlam,
          Nature 123, 86

(29. 10) $^3\Pi \leftarrow \,^1\Sigma$ System,
          A. Elliott,
          Proc. Roy. Soc. A 123, 629-44

(29. 11) Absorption in Solution,
          A. E. Gillam and R. A. Morton,
          Proc. Roy. Soc. A 124, 604-6

(30. 12)   A. Elliott,
Nature 126, 133

(30. 13) $^3\Pi \leftarrow {}^1\Sigma$ System, Vibrational, Rotational Analysis,
A. Elliott,
Proc. Roy. Soc. A 127, 638-57

(30. 14)   Ultraviolet Absorption,
H. Cordes and H. Sponer,
Z. Physik 63, 334-44

(32. 15)   G. E. Gibson and N. S. Bayliss,
Phys. Rev. 41, 388

(32. 16)   Potential Energy Curves,
H. Cordes and H. Sponer,
Z. Physik 79, 170-85

(33. 17)   Continuum Absorption 4300Å-2500Å,
G. E. Gibson and N. S. Bayless,
Phys. Rev. 44, 188-92

(33. 18)   Continuum Absorption Theory,
G. E. Gibson, O. K. Rice, and N. S. Bayliss,
Phys. Rev. 44, 193-200

(34. 19)   Ultraviolet Emission,
A. Campetti,
Atti Accad. Sci. Torino 70, 618-31

(34. 20)   Theory,
R. S. Mulliken,
Phys. Rev. 46, 549-71

(35. 21)   Absorption Coefficient, 5320-5040Å,
F. W. Jones and W. Sponer,
Trans. Faraday Soc. 31, 811

(36. 22)   Thermochemiluminescence,
Y. Uchida,
Sci. Papers Inst. Phys. Chem. Res. 30, 121-37

(36. 23) $^3\Pi \leftarrow {}^1\Sigma$ System,
C. F. Goodeve and B. A. Stephens,
Trans. Faraday Soc. 32, 1517-8

(37. 24)   Ultraviolet Emission,
           A. E. Elliott and W. H. B. Cameron,
           Proc. Roy. Soc. A 158, 681-91

(37. 25) $^3\Pi \rightarrow {}^1\Sigma$ System,
           F. Kitagawa,
           Rev. Phys. Chem. 11, 25-38

(37. 26)   Absorption Coefficient, 5450Å-4000Å,
           R. G. Aickin and N. S. Bayliss,
           Trans. Faraday Soc. 33, 1333-8

(37. 27)   Theory,
           C. H. D. Clark,
           Trans. Faraday Soc. 33, 1398-401

(38. 28)   Excitation by "Active" Nitrogen,
           W. H. B. Cameron and A. Elliott,
           Proc. Roy. Soc. A 169, 463-9

(40. 29)   Theory of Electronic States,
           R. S. Mulliken,
           J. Chem. Phys. 8, 382-95

(40. 30)   Theory,
           G. B. B. M. Sutherland,
           J. Chem. Phys. 8, 161-4

(40. 31)   Theory of Electronic States,
           R. S. Mulliken,
           Phys. Rev. 57, 500-15

(40. 32)   Theory,
           C. H. D. Clark,
           Trans. Faraday Soc. 36, 370-6

(41. 33)   Theory,
           H. M. Hulburt and J. O. Hirschfelder,
           J. Chem. Phys. 9, 61-9

(42. 34)   Theory,
           F. London,
           J. Phys. Chem. 46, 305-16

(42. 35)   Generalities,
           H. Zeise,
           Z. Elektrochem. 48, 476-509

(43. 36) Isotope Effect,
E. F. Schrader,
Phys. Rev. <u>64</u>, 57-9

(47. 37) Potential Energy Curves,
R. K. Asundi and P. Venkateswarlu,
Ind. J. Phys. <u>21</u>, 76

(47. 38) Ultraviolet Emission,
P. Venkateswarlu,
Proc. Ind. Acad. Sci. A <u>26</u>, 22-30

(48. 39) Continuum Study,
A. Herczog and K. Wieland,
Helv. Phys. Acta <u>21</u>, 436-44

(49. 40) Absorption as f(T),
A. Herczog and K. Wieland,
Helv. Phys. Acta <u>22</u>, 552-4

(57. 41) Ultraviolet Emission,
P. B. V. Haranath and P. T. Rao,
Ind. J. Phys. <u>31</u>, 175-6

(57. 42) Photoionization,
K. Watanabe,
J. Chem. Phys. <u>26</u>, 542-7

(58. 43) Ultraviolet Emission,
P. B. V. Haranath and P. T. Rao,
J. Molec. Spectrosc. <u>2</u>, 428-63

(59. 44) Ultraviolet Emission,
B. N. Khanna,
Proc. Ind. Acad. Sci. A <u>49</u>, 293-301

(59. 45) Ultraviolet Emission,
P. Venkateswarlu and B. N. Khanna,
Proc. Ind. Acad. Sci. A <u>49</u>, 117-27

(59. 46) Ultraviolet Absorption,
J. Lee and A. D. Walsh,
Trans. Faraday Soc. <u>55</u>, 1281-92

(62. 47) Ultraviolet Fluorescence,
Y. V. Rao and P. Venkateswarlu,
J. Molec. Spectrosc. <u>9</u>, 173-90

Cl_2

(62.48)   Electron Diffraction,
          S. Shibata,
          J. Phys. Soc., Suppl. B II 17, 34-6

(62.49)   Visible Absorption,
          W. G. Richards and R. F. Barrow,
          Proc. Chem. Soc. 297

(63.50)   Visible Absorption,
          A. E. Douglas, C. K. Moeller, and B. P. Stoicheff,
          Can. J. Phys. 41, 1174-92

(64.51)   Absorption from $A^3\Pi(0_u^+)$,
          R. G. W. Norrish,
          J. Chim. Phys. 61, 1359-71

(65.52)   High Temperature Absorption Coefficient,
          T. A. Jacobs and R. R. Geidt,
          J. Quant. Spectrosc. Radiative Transfer 5, 457-63

(65.53)   Potential Energy Curve of the A State,
          R. Colin, J. Drowart, and G. Verhaegen,
          Trans. Faraday Soc. 61, 1364-71

(65.54)   Potential Energy Curves,
          F. Jenc,
          "The Reduced Potential Curves of Heavy Diatomic Molecules. I.
          The Reduced Potential Curves of Halogens and Interhalogens,"
          Coll. Czech. Chem. Comm. 30, 3772-84

(67.55)   Potential Energy Curves,
          F. Jenc,
          "Ground-State Reduced Potential Curve (RPC) and the Anomaly of Cl_2,"
          J. Chem. Phys. 47, 4910-6

(70.56)   Theory,
          H. F. Schaefer,
          "New Approach to Electronic Structure Calculations for Diatomic
          Molecules: Application to F_2 and Cl_2,"
          J. Chem. Phys. 52, 6241-7

(70.57)   W. Holzer, W. F. Murphy, and H. J. Bernstein,
          "Resonance Fluorescence of Iodine, Bromine, and Chlorine Gases
          Obtained With Argon-Ion Laser Excitation,"
          J. Chem. Phys. 52, 469-70

(71.58) Franck-Condon Factors, A - X,
J. A. Coxon,
"Franck-Condon Factors and r-Centroids for Halogen Molecules. I.
The $B^3\Pi(0_u^+) - X^1\Sigma_g^+$ System of $^{35}Cl_2$,"
J. Quant. Spectrosc. Radiat. Transfer 11, 1355-64

(71.59) Photoelectron Spectrum,
A. B Cornford, D. C. Frost, C. A. McDowell, J. L. Ragle, and
I. A. Stenhouse,
"Photoelectron Spectra of the Halogens,"
J. Chem. Phys. 54, 2651-7

(72.60) Vacuum Ultraviolet Absorption,
A. A. Zakharyan and V. M. Novikov,
"Optical Absorption of Molecular Chlorine in the Vacuum Ultraviolet,"
Sov. J. Opt. Tech. 39, 119-120

(70.61) Rotational Raman Spectra,
P. J. Hendra and C. J. Vear,
"The Pure Rotational Laser Raman Spectrum of Chlorine,"
Spectrochim. Acta 28A, 1949-61

(72.62) Ionization Cross Section,
R. E. Center and A. Mandl,
"Ionization Cross Sections of F_2 and Cl_2 by Electron Impact,"
J. Chem. Phys. 57, 4104-6

(72.63) Chemiluminescence,
G. Schatz and M. Kaufman,
"Chemiluminescence Excited by Atomic Fluorine,"
J. Phys. Chem. 76, 3586-90

(72.64) E - B System,
K. Weiland, J. G. Tellinghuisen, and A. Nobs,
"The Band Systems E - B (4000-4360Å) and F - X (2530-2740Å) of
$^{127}I_2$ and $^{129}I_2$, and the Corresponding System E - B of Br_2 and
Cl_2,"
J. Molec. Spectrosc. 41, 69-83

(73.65) P. T. T. Wong and E. Whalley,
"Intensity of the Pressure-Induced Fundamental Infrared Band of
Gaseous Chlorine,"
Can. J. Phys. 51, 696-7

(73.66) Theory,
K. K. Yee and T. J. Stone,
"Analysis of RKR Long-Range Potentials of the $B^3\Pi(0_u^+)$ States of
Br_2 and Cl_2,"
Molec. Phys. 26, 1169-76

Cl_2

(73. 67)   Potential Energy Curves,
           M. S. Child and R. B. Bernstein,
           "Diatomic Interhalogens: Systematics and Implications of Spectro-
           scopic Interatomic Potentials and Curve Crossings,"
           J. Chem. Phys. 59, 5916-25

(74. 68)   R. C. Oldenborg, J. L. Gole, and R. N. Zare,
           "Chemiluminescent Spectra of Alkali-Halogen Reactions,"
           J. Chem. Phys. 60, 4032-42

(74. 69)   M. Brith, O. Schnepp, and P. Stephens,
           "Magnetic Circular Dichroism Spectra of the Halogen Molecules I_2,
           Br_2, and Cl_2. Resolution of Overlapping $0_u^+(^3\Pi)$ and $^1\Pi$ Bands,"
           Chem. Phys. Letters 26, 549-52

(74. 70)   R. J. LeRoy,
           "Long-Range Potential Coefficients From RKR Turning Points: C_6
           and C_8 for $B(^3\Pi_{0_u^+})$-State Cl_2, Br_2, and I_2,"
           Can. J. Phys. 52, 246-56

Co_2

Spectroscopic Constants

Dissociation energy = 1.69 ± 0.26 eV, 39 kcal/mole, 13600 cm^{-1} (64.1).

Co_2

BIBLIOGRAPHY

(64. 1)   A. Kant and B. Strauss,
          "Dissociation Energies of Diatomic Molecules of the Transition
          Elements. II. Titanium, Chromium, Manganese, and Cobalt,"
          J. Chem. Phys. 41, 3806-8

Methods of Production and Experimental Technique

Single band at 4600Å, attributed to Cr$_2$, has been observed by photolysis of Cr(CO)$_6$ (74.4).

Spectroscopic Constants

Dissociation energy = 1.56 ± 0.31 eV, 36.0 kcal/mole, 12600 cm^{-1} (62.1, 64.2, 68.3).

Cr$_2$

BIBLIOGRAPHY

(62.1) M. Ackerman, F. E. Stafford, and G. Verhaegen,
        J. Chem. Phys. 36, 1560-2

(64.2) A. Kant and B. Strauss,
        "Dissociation Energies of Diatomic Molecules of the Transition
        Elements. II. Titanium, Chromium, Manganese, and Cobalt,"
        J. Chem. Phys. 41, 3806-8

(68.3) A. Kant and B. Strauss,
        "Dissociation Energy of Cr$_2$,"
        J. Chem. Phys. 45, 3161-2

(74.4) Y. M. Efremov, A. N. Samoilova, and L. V. Gurvich,
        "The λ = 4600-Å Band in a Spectrum Produced by Pulsed Photolysis
        of Chromium Carbonyl,"
        Opt. Spectrosc. 36, 381-2

$$Cs_2$$

Methods of Production and Experimental Technique

Heat pipe.

Absorption.

Laser Fluorescence.

BAND SYSTEMS

	System	Transition	Sources	Wavelength Limits	Degrading	Band Head, $\nu_{0,0}$	Remarks	Bibliography
7667Å 6250Å	I	$^3\Pi_u \to X^1\Sigma_g^+$	Heat pipe	11900				(71.23, 68.15)
	II	$^1\Sigma_u^+ \to X^1\Sigma_g^+$	Heat pipe	11300				(71.23, 68.15)
	III	$A \leftarrow X^1\Sigma_g^+$	Absorption	11300-8800	R			(38.13, 34.7)
	IV	$B \leftarrow X^1\Sigma_g^+$	Absorption	8350-6950	R			(69.18, 38.13, 34.7)
	V	$C \rightleftharpoons X^1\Sigma_g^+$	Absorption, fluorescence	6443-6340	R, V		Spectra not fully analyzed	(68.16, 36.11)
	VI	$D \leftarrow X$	Absorption	6360-6260	R, V		Spectra not fully analyzed	(68.16, 36.11)
	VII	$E \leftarrow X$	Absorption	5230-4450			55 bands of which 16 are provisionally classified	(34.7)
	VIII			Ultraviolet and visible				(35.9, 32.4, 32.3)

IV. $\quad$ B $\leftarrow$ X$^1\Sigma_g^+$ System (7667Å)

Bands of greatest intensity (38. 13):

v', v''	1, 3	0, 2	0, 1	0, 0	1, 0	3, 1	2, 0
λ	7720. 4	7716. 0	7691. 2	7666. 6	7646. 6	7631. 4	7626. 7
(Intensity)	2	4	5	3	5	2	4

V. $\quad$ C $\rightleftharpoons$ X$^1\Sigma_g^+$ System (6250Å)

Band heads (Q). There are 3 bands degrading R and 1 degrading V (68. 16):

v', v''	0	1	2	3	4
0	6265. 4	6281. 3		6313. 1	
1	6255. 1	6271. 0	6286. 9	6302. 8	6318. 6
2	6244. 8	6260. 7	6276. 6	6292. 5	6308. 3
3		6250. 4	6266. 3	6282. 1	6298. 0
4			6255. 8		6287. 6

VI. $\quad$ D $\leftarrow$ X$^1\Sigma_g^+$ System

Observed bands (v' numbering uncertain):

v', v''	9, 17	7, 13	8, 13	7, 12	5, 9	8, 10
λ	6359. 8	6317. 5	6307. 1	6301. 7	6274. 9	6259. 6
(Intensity)	7	10	7	9	9	8

VIII. $\quad$ Ultraviolet and Visible Bands

    a. Diffuse bands, λ|7185|7128

    b. Bands degrading V, λ|7078|7075|7072

    c. System at ~ 5600, intensified by Ar at 300°C

    d. Bands degrading R, λ|3959|3953|3947|3941

    e. Bands degrading V, λ|3920

<u>Perturbations and General Information</u>

Maximum photoionization cross-section = 21.6 Mb at 3214Å (73.24).

Recombination rate constant $3Cs \rightarrow Cs_2 + Cs$ (71.22):

$$k \approx 3 \times 10^{-30} \ cm^6/sec \ at \ 600°K$$

Radiative lifetimes (70.19):

λ_{exc}(Å)	Upper State	Lifetime (nsec)
4879.86		27.0 ± 2
4764.86	?, above	12.0 ± 1.5
4726.86	D state	12.0 ± 1.5
4579.35		6.4 ± 0.8

Cs_2

SPECTROSCOPIC CONSTANTS

State	T_e	ω_e	$x_e\omega_e$	B_e	$\alpha_e \times 10^3$	$D_e \times 10^6$	r_e	Remarks	Bibliography
E	(20500)	(31)	-	-	-	-	-	-	(34.7)
D	(16175.8)	(27.34)	(0.0733)	-	-	-	-	-	(36.11)
C	15948.6	29.7030	0.05756	-	-	-	-	$y_e\omega_e = 1.608 \times 10^{-3}$	(68.16)
B	13043.88	34.3293	0.07996	-	-	-	-	$y_e\omega_e = -1.511 \times 10^{-4}$	(69.18)
A	(10000)	-	-	-	-	-	-	-	(38.13, 34.7)
$^1\Sigma_u^+$	(8850)	-	-	-	-	-	-	-	(71.23, 68.15)
$^3\Pi$	(8403)	-	-	-	-	-	-	-	(71.23, 68.15)
$X^1\Sigma_g^+$	0	42.0267	0.08348	-	-	-	-	$y_e\omega_e = -2.361 \times 10^{-5}$	(69.18, 68.16)

Dissociation energy = 0.40 ± 0.01 eV, 9.13 kcal/mole, 3197 cm^{-1} (69.18).

BIBLIOGRAPHY

(23. 1)   Preliminary Search,
            J. C. MacLennan and D. S. Ainslie,
            Proc. Roy. Soc. A 103, 304-14

(28. 2)   Preliminary Search,
            J. M. Walter and S. Barratt,
            Proc. Roy. Soc. A 119, 257-75

(32. 3)   Broadening of Lines,
            S. Datta and B. Chakravarty,
            Ind. J. Phys. 7, 273-82

(32. 4)   Van der Waals Bands,
            H. Kuhn,
            Z. Physik 76, 782-92

(32. 5)   Absorption,
            R. Rompe,
            Z. Physik 74, 175-86

(34. 6)   E. Matuyama,
            Nature 133, 567-8

(34. 7)   All Systems,
            F. W. Loomis and P. Kusch,
            Phys. Rev. 46, 292-301

(34. 8)   Absorption,
            E. Matuyama,
            Sci. Rept. Tohoku Univ. 23, 322-33

(35. 9)   Broadening of Lines,
            N. Tsi-Ze and C. Shin-Piaw,
            J. Phys. Radium 6, 203-8

(36. 10)   7400-7230Å Bands, Possibly RbCs,
            P. Kusch,
            Phys. Rev. 49, 218-22

(36. 11)   C, D ← X Systems,
            P. Kusch and F. W. Loomis,
            Phys. Rev. 49, 217-8

(37. 12)   Tabulation,
C. H. D. Clark and C. W. Scaife,
Trans. Faraday Soc. 33, 1394-8

(38. 13)   A, B ← X Systems,
W. Finkelnburg and O. T. Hahn,
Z. Physik 39, 98-100

(66. 14)   Absorption Cross-Sections, D, C ← X System,
M. Lapp and L. P. Harris,
"Absorption Cross Sections of Alkali-Vapor Molecules: I. Cs$_2$ in
the Visible, II. K$_2$ in the Red,"
J. Quant. Spectrosc. Radiat. Transfer 6, 169-79

(68. 15)   1.13 μm and 1.19 μm Observed,
D. S. Bayley, E. C. Eberlin, and J. H. Simpson,
"Absorption Spectrum of Diatomic Cesium Molecules,"
J. Chem. Phys. 49, 2863-4

(68. 16)   C - X System,
P. Kusch and M. M. Hessel,
"An Analysis of the 6250-Å Band System of Cs$_2$,"
J. Molec. Spectrosc. 25, 205-23

(68. 17)   Cross-Section Measurements,
D. M. Creek and G. V. Marr,
"Some Ultraviolet Cross-Section Measurements on Molecular
Alkali-Metal Vapours,"
J. Quant. Spectrosc. Radiat. Transfer 8, 1431-6

(69. 18)   B - X System,
P. Kusch and M. M. Hessel,
"A Study of the 7667-Å Band System of Cs$_2$; The Magnetic Rotation
Spectra of the 7667 and 6250-Å Systems,"
J. Molec. Spectrosc. 32, 181-201

(70. 19)   Lifetime Measurements,
G. Baumgartner, W. Demtroder, and M. Stock,
"Lifetime-Measurements of Alkali-Molecules Excited by Different
Laserlines,"
Z. Physik 232, 462-72

(70. 20)   Two-Photon Ionization,
D. Popescu,
"Two-Photon Ionization of the Cs$_2$ Molecule,"
Rev. Roum. Phys. 15, 859-63

(71.21) Reaction in Molecular Beams,
Y. T. Lee, R. J. Gordon, and D. R. Herschbach,
"Molecular Beam Kinetics: Reactions of H and D Atoms With
Diatomic Alkali Molecules,"
J. Chem. Phys. 54, 2410-23

(71.22) Recombination Rates,
R. J. Gordon, Y. T. Lee, and D. R. Herschbach,
"Supersonic Molecular Beams of Alkali Dimers,"
J. Chem. Phys. 54, 2393-2409

(71.23) 1.13 μm and 1.19 μm Bands,
P. P. Sorokin and J. R. Lankard,
"Emission Spectra of Alkali-Metal Molecules Observed With a Heat-
Pipe Discharge Tube,"
J. Chem. Phys. 55, 3810-3

(71.24) Photoionization,
D. Popescu, M. L. Pascu, C. B. Collins, B. W. Johnson, and
I. Popescu,
"Use of Space-Charge-Amplification Techniques in the Absorption
Spectroscopy of Cs and Cs$_2$,"
Phys. Rev. A 8, 1666-72

(74.25) R. C. Oldenborg, J. L. Gole, and R. N. Zare,
"Chemiluminescent Spectra of Alkali-Halogen Reactions,"
J. Chem. Phys. 60, 4032-42

(74.26) R. W. Molof, T. M. Miller, H. L. Schwartz, B. Bederson, and
J. T. Park,
"Measurements of the Average Electric Dipole Polarizabilities of
the Alkali Dimers,"
J. Chem. Phys. 61, 1816-22

Cu_2

Methods of Production and Experimental Technique

Absorption.

Emission: thermal, discharges, exploding wires, "active" nitrogen.

BAND SYSTEMS

	System	Transition	Sources	Wavelength Limits	Degrading	Band Head, $\nu_{0,0}$	Remarks	Bibliography
	I	$A\,^1\Pi_u \leftarrow X\,^1\Sigma_g^+$	Absorption	5750-4850	R	20401.79		(71.8, 65.5, 54.2)
	II	$B\,^1\Sigma_u^+ \leftarrow X\,^1\Sigma_g^+$	Absorption	4720-4490	R	21748.48		(71.7, 65.5, 54.2)
	III			Ultraviolet			3 or 4 bands not analyzed	(71.7, 65.5, 54.2)

I. $A\,^1\Pi_u \leftarrow X\,^1\Sigma_g^+$ System

Band heads for $^{63}Cu_2$, λ (71.8):

v', v''	1, 0	0, 0	0, 1
λ	4856.3	4901.5	4966.0

II. $B\,^1\Sigma_u^+ \leftarrow X\,^1\Sigma_g^+$ System

Band heads for $^{63}Cu_2$, isotope shifts, λ (71.7):

v', v''	0	1	2
0	4598.1	4654.7	4712.3
1	4547.4	4603.0	4659.7
2	4498.8		4608.1

III. Ultraviolet Systems

Bands observed in the regions: $2900 > \lambda > 2700$, $2560 > \lambda > 2490$, $2460 > \lambda > 2330$

SPECTROSCOPIC CONSTANTS

State	T$_e$	ω_e	x$_e\omega_e$	B$_e$	$\alpha_e \times 10^3$	D$_e \times 10^8$	r$_e$	Remarks	Bibliography
B$^1\Sigma_u^+$	21760.4	245.2	1.90						(71.7, 65.5, 54.2)
A$^1\Pi_u$	20434.2	191.9	0.348	0.08185	0.62	3.81	2.559		(71.8, 54.2)
X$^1\Sigma_g^+$	0	268.9	1.63	0.10867	0.56	5.11	2.22		(71.7, 65.5, 54.2)

Dissociation energy = 2.05 ± 0.09 eV, 47.3 kcal/mole, 16550 cm^{-1} (60.4).

Perturbations and General Information

Franck-Condon factors, $B^1\Sigma_u^+ - X^1\Sigma_g^+$, RKRV potentials (71.9):

v', v''	0	1	2	3
0	0.237	0.283	0.261	0.104
1	0.251	0.051	0.078	
2	0.227	0.010	0.156	
3	0.149			

Potential energy curves (71.9):

State	v	$U+T_e(cm^{-1})$	$r_{max}(\text{Å})$	$r_{min}(\text{Å})$
$X^1\Sigma_g^+$	0	132.3	2.286	2.158
	1	396.9	2.338	2.117
	2	659.8	2.376	2.091
	3	919.6	2.409	2.070
	4	1177.7	2.439	2.053
	5	1433.7	2.467	2.038
	6	1687.7	2.492	2.024
	7	1939.1	2.517	2.013
	8	2188.8	2.541	2.002
	9	2436.9	2.563	1.992
	10	2682.7	2.585	1.983
	11	2926.6	2.607	1.975
	12	3169.5	2.628	1.967
	13	3409.5	2.649	1.959
	14	3647.5	2.670	1.952
	15	3883.4	2.689	1.945
	16	4117.3	2.708	1.939
	17	4349.5	2.728	1.933
	18	4579.3	2.747	1.927
	19	4807.5	2.767	1.921
	20	5033.7	2.786	1.916
	21	5257.9	2.806	1.911
	22	5479.9	2.826	1.906
	23	5700.5	2.842	1.901
$B^1\Sigma_u^+$	0	21869.3	2.396	2.263
	1	22109.0	2.453	2.221
	2	22344.6	2.495	2.193
	3	22575.9	2.531	2.171
	4	22802.7	2.563	2.152

BIBLIOGRAPHY

(25. 1)   B - X System, Assigned to CuI,
R. S. Mulliken,
Phys. Rev. 26, 1-32

(54. 2)   A, B - X Systems, Vibrational Analysis,
B. Kleman and S. Lindkvist,
Arkiv Fysik 8, 333-9

(59. 3)   Ultraviolet Bands,
J. Ruamps,
Ann. Phys. 4, 1111-57

(60. 4)   Dissociation Energy,
M. Ackerman, F. E. Stafford, and J. Drowart,
J. Chem. Phys. 33, 1784-9

(65. 5)   B - X System, Rotational Analysis,
N. Aslund, R. F. Barrow, W. G. Richards, and D. N. Travis,
Arkiv Fysik 30, 171-85

(71. 6)   G. V. Kovalenok,
"Identification of the Resolved Vibrational Structure of Diatomic
Molecules,"
Izv. Tomsk. Politekh. Inst. 184, 40-3

(71. 7)   B - X System,
D. S. Pesic and S. Weniger,
"Study of the Vibrational Spectrum of the B - X Transition in the
^{63}Cu^{65}Cu Molecule,"
C. R. Acad. Sci. B 272, 46-9

(71. 8)   A - X System,
D. S. Pesic and S. Weniger,
"Rotational Analysis of the A - X System of ^{63}Cu$_2$,"
C. R. Acad. Sci. B 273, 602-4

(71. 9)   Franck-Condon Factors, B - X System,
T. V. R. Rao and S. V. J. Lakshman,
"RKRV Curves, r-Centroids and Franck-Condon Factors for Bands
of the ^{63}Cu$_2$ Molecule,"
J. Quant. Spectrosc. Radiative Trans. 11, 1157-61

(71. 10)   Theory,
P. Joyes and M. Leleyter,
"Ab Initio Study of Cu$_2$ and Cu$_2^+$,"
J. Phys. B 6, 150-4

Dy$_2$

Spectroscopic Constants

Dissociation energy = 0.78 ± 0.17 eV, 18 kcal/mole, 6291 cm^{-1} (71.1).

Dy$_2$

BIBLIOGRAPHY

(71. 1)   S. Lin and A. Kant,
           "Dissociation Energies of Diatomic Rare Earth Molecules Dy$_2$, Ho$_2$,
           Er$_2$, Tm$_2$ and Yb$_2$,"
           TR No. AMMRC TR 71-34, Army Materials and Mechanics Research
           Center, Watertown, Mass., Sept 1971

(72. 2)   A. Kant and S. Lin,
           "Dissociation Energies of Homonuclear Diatomic Rare Earth Molecules,"
           Monatshefte für Chemie 103, 757-63

Er$_2$

Spectroscopic Constants

Dissociation energy = 0.78 ± 0.17 eV, 18 kcal/mole, 6291 cm^{-1} (71.1).

BIBLIOGRAPHY

(71. 1) S. Lin and A. Kant,
"Dissociation Energies of Diatomic Rare Earth Molecules Dy_2, Ho_2, Er_2 Tm_2 and Yb_2,"
TR No. AMMRC TR 71-34, Army Materials and Mechanics Research Center, Watertown, Mass., Sept 1971

(72. 2) A. Kant and S. Lin,
"Dissociation Energies of Momonuclear Diatomic Rare Earth Molecules, Monatshefte für Chemie 103, 757-63

Eu$_2$

Spectroscopic Constants

Dissociation energy = 0.43 ± 0.17 eV, 10 kcal/mole, 3500 cm^{-1} (72.1).

Eu$_2$

(72. 1)   A. Kant and S. Lin,
          "Dissociation Energies of the Homonuclear Diatomic Molecules of
          the Rare Earths,"
          Monatshefte für Chemie 103, 757-63

$$F_2$$

Methods of Production and Experimental Technique

Absorption.

Emission in a discharge.

BAND SYSTEMS

	System	Transition	Sources	Wavelength Limits	Degrading	Characteristic Bands, λ	Remarks	Bibliography
	I	$A\,^1\Pi_u - X\,^1\Sigma_g^+$	Absorption	4400-2200	Continuum	-	Peaks, $\lambda = 2845\text{Å}$	(56.25, 37.5, 31.3)
	II	$C\,^1\Sigma^+ \rightarrow B\,^1\Pi$	Discharge	7510-4220	R	6517.2(0, 5) 5398.0(0, 2) 5388.5(1, 3)	Perturbed	(68.39, 66.36, 29.2)
	III	$C\,^1\Sigma^+ \rightarrow B'\,^1\Pi$	Discharge	2 bands	R	5852.6(0, ?) 5495.7(1, ?)	3 heads	(68.39)
	IV		Absorption	Far ultraviolet	R V	1935 952	Rydberg series	(59.29)

F_2

II. $\underline{C^1\Sigma^+ \to B^1\Pi \text{ System}}$

Band heads, λ (Intensity) (68.39, 32.4, 29.2, 24.1):

v', v''	0	1	2	3	4	5	6
0			5398(10)[a]	5731.5(25)	6102.8(70)	6517.2(100)[b]	6982.5(4)
1		4823(1)[a]	5092.3(3)[a]	5388.5(10)	5715.6(4)[a]		6480.4(25)

[a] No rotational analysis; [b] partial rotational analysis

III. $\underline{C^1\Sigma^+ \to B'^1\Pi \text{ System}}$

Two bands have been attributed to this system (68.39):

(v', v'')	(0, ?)	(1, ?)
λ	5852.6	5495.7
(Intensity)	8	1

IV. Rydberg Series

Band series have been reported for a number of systems (59.29):

System	Degrades	Intensity	λ	System	Degrades	Intensity	λ
V	R	2	1035.8	B_2''		7	946.6
			1031.1				937.0
			1026.6				927.7
			1021.2				918.8
			1017.4				910.3
			1012.4	C_2'	R	10	874.2
			1008.2				867.2
			1003.5				861.2
			998.4	B_3'		12	859.8
V'	R	5	969.1				850.2
			962.8	B_3''		15	855.6
B_2'		7	952.3				848.7
			942.7	B_4'		18	830.7
			933.7	B_4''		18	828.3
			924.8	B_5'		18	814.7
				B_5''		18	813.8
				B_6'		18	807.0

SPECTROSCOPIC CONSTANTS

State	T_e	ω_e	$x_e\omega_e$	B_e	$\alpha_e \times 10^3$	$D_e \times 10^6$	r_e	Remarks	Bibliography
$C\,^1\Sigma^+$ [a]	$x+(20900)$ [b]	(1110)	–	0.804	8.0	–	1.485		$(68.39,\ 29.2)$
$B'\,^1\Pi$ [a]	–	–	–	(1.005)	–	–	(1.329)		(68.39)
$B\,^1\Pi$ [a]	x	(1100)	–	(1.047)	(12.0)	–	(1.302)		$(68.39,\ 29.2)$
$A\,^1\Pi_u$	–	–	–	–	–	–	–		$(56.25,\ 37.5,\ 31.3)$
$X\,^1\Sigma_g^+$	0	891.85 [c]	–	0.8828 [d]	–	–	1.417 [e]		(51.18)

[a] Possibly a Rydberg series; [b] position uncertain because of perturbations; [c] $\Delta G_{1/2}$; [d] B_o; [e] r_o

Dissociation energy = 1.61 ± 0.05 eV, 37.1 kcal/mole, 12976 cm^{-1} (72.45).

F_2

Perturbations and General Information

Intense perturbations are noted in the positions and intensities of the $C^1\Sigma$ state (32.4).

Potential curve of the repulsive $A^1\Pi_u$ state (57.27).

Molecular orbital wavefunctions of the $X^1\Sigma_g^+$ state (66.35, 64.34).

Calculated energies of the $X^1\Sigma_g^+$ state using Hartree-Fock method (64.33).

BIBLIOGRAPHY

(24. 1)   Measurements,
          H. G. Gale and G. S. Monk,
          Astrophys. J. 59, 125-32

(29. 2)   Vibrational Analysis,
          H. G. Gale and G. S. Monk,
          Astrophys. J. 69, 77-102

(31. 3)   Continuum Absorption,
          H. Von Wartenberg, G. Sprenger, and J. Yaylor,
          Bodenstein-Festband, 61-8

(32. 4)   Perturbations,
          J. Aars,
          Z. Physik 79, 122-38

(37. 5)   Continuum Absorption, Dissociation Energy,
          M. Bodenstein, H. Jockusch, and S. H. Chong,
          Z. Anorg. Allg. Chem. 231, 24-33

(38. 6)   Electron Diffraction,
          L. O. Brockway,
          J. Am. Chem. Soc. 60, 1348-9

(40. 7)   Force Constants,
          G. B. B. M. Sutherland,
          J. Chem. Soc. 8, 161-4

(41. 8)   Electron Diffraction,
          M. T. Rogers, V. Schomaker, and D. P. Stevenson,
          J. Am. Chem. Soc. 63, 2610-11

(41. 9)   Electron Diffraction,
          V. Schomaker and D. P. Stevenson,
          J. Am. Chem. Soc. 63, 37-40

(42. 10) Dissociation Energy,
          A. L. Wahrhaftig,
          J. Chem Phys. 10, 248

(47. 11) Dissociation Energy,
          H. Schmitz and H. J. Schumacher,
          Z. Naturforsch. A 2, 362

F_2

(47. 12) Dissociation Energy,
A. G. Gaydon,
"Dissociation Energies and Spectra of Diatomic Molecules,"
Chapman and Hall, London

(49. 13) Dissociation Energy,
A. D. Caunt and R. F. Barrow,
Nature 164, 753-4

(50. 14) Raman Spectrum,
D. Andrychuk,
J. Chem Phys. 18, 233

(50. 15) Dissociation Energy,
M. W. Nathans,
J. Chem. Phys. 18, 1122

(50. 16) Dissociation Energy,
M. G. Evans, E. Warhurst, and E. Whittle,
J. Chem. Soc. 1524-34

(50. 17) Dissociation Energy,
A. D. Caunt and R. F. Barrow,
Trans. Faraday Soc. 46, 154-6

(51. 18) Raman Spectrum,
D. Andrychuk,
Can. J. Phys. 29, 151-8

(51. 19) Dissociation Energy,
R. N. Doescher,
J. Chem Phys. 19, 1070-1

(52. 20) Dissociation Energy,
R. N. Doescher,
J. Chem. Phys. 20, 330-4

(52. 21) Dissociation Energy,
H. Wise,
J. Chem. Phys. 20, 927

(52. 22) Dissociation Energy,
E. Wicke,
J. Phys. Chem. 56, 358-60

(53. 23) Dissociation Energy,
P. W. Gilles and J. L. Margrave,
J. Chem. Phys. 21, 381-2

(53.24) Dissociation Energy,
R. F. Barrow and J. L. Caunt,
Proc. Roy. Soc. A 219, 120-40

(56.25) Absorption 2100-8000Å,
R. K. Stevenberg and R. C. Vogel,
J. Am. Chem. Soc. 78, 901-4

(56.26) Dissociation Energy,
K. L. Wray and D. F. Hornig,
J. Chem. Phys. 24, 1271-2

(57.27) Potential Curve,
A. L. G. Rees,
J. Chem. Phys. 26, 1567-71

(58.28) Dissociation Energy, Electron Affinity,
J. G. Stamper and R. F. Barrow,
Trans. Faraday Soc. 54, 1592-4

(59.29) Vacuum Ultraviolet Absorption,
R. P. Iczkowski and J. L. Margrave,
J. Chem. Phys. 30, 403-5

(61.30) Valence-Band Calculation of $X^1\Sigma_g^+$
K. Hijikata,
J. Chem. Phys. 34, 221-31

(61.31) K. Hijikata,
J. Chem. Phys. 34, 231-9

(62.32) SCF-MO Calculation of $X^1\Sigma_g^+$,
S. Fraga and B. J. Ransil,
J. Chem. Phys. 36, 1127-42

(64.33) Hartree-Fock Energy,
R. M. Stevens and W. N. Lipscomb,
J. Chem. Phys. 41, 3710-6

(64.34) SCF-MO Calculation for $X^1\Sigma_g^+$,
A. C. Wahl,
J. Chem. Phys. 41, 2600-13

(66.35) Extended Hartree-Fock Calculation for $X^1\Sigma_g^+$,
G. Das and A. C. Wahl,
J. Chem. Phys. 44, 87-96

F_2

(66. 36) Emission Spectrum,
          W. Stricker,
          "Band Emission of F_2 and F_2^+ in the Spectral Region Between 4500 and
          8500Å,"
          Z. Naturforsch. 21a, 1518-9

(67. 37) J. R. De La Vega and H. F. Hameka,
          "Calculation of Magnetic Susceptibilities of Diatomic Molecules,"
          Physica 35, 313-22

(67. 38) Magnetic Moment,
          J. R. De La Vega and H. F. Hameka,
          "Calculation of Magnetic Susceptibilities of Diatomic Molecules.   VIII.
          Anisotropies and Rotational Magnetic Moments,"
          J. Chem. Phys. 47, 1834-6

(68. 39) Emission Spectrum,
          T. L. Porter,
          J. Chem. Phys. 48, 2071-83

(68. 40) Vacuum Absorption, Dissociation Energy,
          W. Stricker and L. Krauss,
          Z. Naturforsch. A 23, 486-91

(69. 41) Oscillator Strength,
          R. P. Main and A. Schadee,
          "On the Oscillator Strengths of MgO and F_2,"
          J. Quant. Spectrosc. Radiat. Transfer 9, 713-4

(69. 42) Theory,
          R. Janoschek and H. Preuss,
          "Molecular Properties of F_2 by the SCF-MO-LC(LCGO)-Method,"
          Z. Naturforsch. 24a, 674-5

(70. 43) Theory,
          H. F. Schaefer III,
          "New Approach to Electronic Structure Calculations for Diatomic
          Molecules: Application to F_2 and Cl_2,"
          J. Chem. Phys. 52, 6241-7

(72. 44) Dissociation Energy Discussion,
          N. Cohen,
          Aerospace Report No. TR-0073(3430)-9

(72. 45) Dissociation Energy,
          J. Blauer and W. Solomon,
          "Shock Tube Calorimeter for the Dissociation Energy of Fluorine,"
          J. Chem. Phys. 57, 3587-9

180

(72. 46)   Theory,
A. Hernandez and E. V. Ludena,
"Loge Localization Analysis of Diatomic Molecules: Li_2 and F_2,"
J. Chem. Phys. 57, 5350-3

(72. 47)   Ionization Cross-Section,
R. E. Center and A. Mandl,
"Ionization Cross Sections of F_2 and Cl_2 by Electron Impact,"
J. Chem. Phys. 57, 4104-6

(72. 48)   Theory,
G. L. Bendazzoli, F. Bernardi, and A. Geremia,
"A. M. O. Calculations for Some First Row Diatomic Molecules,"
Theoret. Chim. Acta 27, 63-8

(72. 49)   Theory,
W. Von Niessen,
"Density Localization of Atomic and Molecular Orbitals,"
Theoret. Chim. Acta 27, 9-23

(73. 50)   Dissociation Energy,
A. A. Dronin and L. N. Gorokhov,
"Form of Curve for the Dissociative Ionization and the Dissociation
Energy of the F_2 Molecule,"
High Temp. 10, 674-6

(73. 51)   Potential Energy Curves,
M. S. Child and R. B. Bernstein,
"Diatomic Interhalogens: Systematics and Implications of Spectro-
scopic Interatomic Potentials and Curve Crossings,"
J. Chem. Phys. 59, 5916-25

(73. 52)   Theory,
E. Kasseckert,
"Ab Initio Calculations for the F_2 Molecule,"
Z. Naturforsch. 28a, 704-8

(74. 53)   Theory,
G. C. Lie and E. Clementi,
"Study of the Electronic Structure of Molecules: XXII. Correlation
Energy Corrections as a Function of the Hartree-Fock Type Density
and its Application to the Homonuclear Diatomic Molecules of the
Second Row Atoms,"
J. Chem. Phys. 60, 1288-96

Fe$_2$

Fe$_2$

Spectroscopic Constants

Dissociation energy = 1.30 ± 0.22 eV, 30 kcal/mole, 10500 cm^{-1} (69.1).

BIBLIOGRAPHY

(69. 1)   S. Lin and A. Kant,
          "Dissociation Energy of Fe$_2$,"
          J. Phys. Chem. 73, 2450-1

Ga$_2$

Methods of Production and Experimental Technique

Emission from a King furnace.

Band Systems

The spectra of gallium has been observed in emission and consists primarily of red-degraded bands in the region 4600-5500Å (65.3).

Spectroscopic Constants

Dissociation energy = 1.39 ± 0.22 eV, 32 kcal/mole, 11200 cm^{-1} (58.2, 57.1).

Ga$_2$

BIBLIOGRAPHY

(57. 1)   J. Drowart and R. E. Honig,
          Bull. Soc. Chim. Belg. 66, 411-2

(58. 2)   W. A. Chupka, J. Berkowitz, C. F. Giese, and M. G. Inghram,
          J. Phys. Chem. 62, 611-4

(65. 3)   D. E. S. Ginter,
          "Electronic Spectra of the Homonuclear Molecules of the Group III
          Metals,"
          Thesis, Vanderbilt University

Gd$_2$

Spectroscopic Constants

Dissociation energy = 1.78 ± 0.35 eV, 41 kcal/mole, 14340 cm^{-1} (72.1).

Gd$_2$

BIBLIOGRAPHY

(72. 1)   A. Kant and S. Lin,
           "Dissociation Energies of the Homonuclear Diatomic Molecules of
           the Rare Earths,"
           Monatshefte für Chemie 103, 757-63

Ge$_2$

Spectroscopic Constants

Dissociation energy = 2.81 ± 0.12 eV, 64.9 kcal/mole, 22700 cm^{-1} (66.2, 66.1).

BIBLIOGRAPHY

(66. 1)   A. Kant,
           J. Chem. Phys. **44**, 2450-6

(66. 2)   A. Kant and B. H. Strauss,
           J. Chem. Phys. **45**, 822-6

H$_2$

Methods of Production and Experimental Technique

Absorption.

Emission from discharge into H$_2$ gas.

SINGLET SYSTEMS   BAND SYSTEMS

	System	Transition	Sources	Electronic Configuration	Degrades	Band Head, $\nu_{0,0}$	Remarks	Bibliography
Lyman	I	$B\,^1\Sigma_u^+ \rightleftharpoons X\,^1\Sigma_g^+$	Absorption, emission	$2p\sigma \leftarrow 1s\sigma$	R	90203.55	Perturbed	(73.92, 68.30, 68.26, 64.23, 61.17, 60.16, 59.15)
Werner	II	$C\,^1\Pi \rightleftharpoons X\,^1\Sigma_g^+$	Absorption, emission	$2p\pi \leftarrow 1s\sigma$	R	99081.72	Perturbed	(64.23, 62.21, 62.20, 58.14, 33.1)
	III	$B'\,^1\Sigma_u^+ \leftarrow X\,^1\Sigma_g^+$	Absorption	$2p\sigma \leftarrow 1s\sigma$	R	110478.54	Perturbed	(68.28, 64.24, 61.18, 34.5)
Hopfield-Beutler	IV	$D\,^1\Pi_u^- \leftarrow X\,^1\Sigma_g^+$	Absorption	$2p\pi \leftarrow 1s\sigma$	R	112871.74	Perturbed, predissociated	(68.28, 64.24)
	V	$B''\,^1\Sigma_u^+ \leftarrow X\,^1\Sigma_g^+$	Absorption	$2p\sigma \leftarrow 1s\sigma$	R	116882.00	Perturbed	(68.28, 64.24, 61.18)
	VI	$D'\,^1\Pi_u^- \leftarrow X\,^1\Sigma_g^+$	Absorption	$2p\pi \leftarrow 1s\sigma$	R	117834.65	Perturbed	(68.28, 64.24)
	VII	$D''\,^1\Pi_u^- \leftarrow X\,^1\Sigma_g^+$	Absorption	$2p\pi \leftarrow 1s\sigma$	R	120172.21		(68.28, 62.22)
	VIII	$E\,^1\Sigma_g^+ \rightarrow B\,^1\Sigma_g^+$	Emission	$2s\sigma \rightarrow 2p\sigma$	V	8961.2		(36.7)
	IX	$F\,^1\Sigma_g^+ \rightarrow B\,^1\Sigma_g^+$	Emission	$(2p\sigma)^2 \rightarrow 2p\sigma$	R	~14000	Perturbed	(49.13)

H_2

SINGLET SYSTEMS   BAND SYSTEMS

System	Transition	Sources	Electronic Configuration	Degrades	Band Head, $\nu_{0,0}$	Remarks	Bibliography
X	$Q \rightarrow B^1\Sigma_g^+$	Emission	$\rightarrow 2p\sigma$	R	21151		(34.4, 34.3, 34.2)
XI	$K \rightarrow B^1\Sigma_g^+$	Emission	$\rightarrow 2p\sigma$	R	21425.4		(34.5)
XII	$G^1\Sigma_g^+ \rightarrow B^1\Sigma_g^+$	Emission	$3d\sigma \rightarrow 2p\sigma$	V	(21609)	Perturbed	(34.5)
XIII	$I^1\Pi_g \rightarrow B^1\Sigma_g^+$	Emission	$3d\pi \rightarrow 2p\sigma$	V	21813		(34.5)
XIV	$J^1\Delta_g \rightarrow B^1\Sigma_g^+$	Emission	$3d\delta \rightarrow 2p\sigma$	V	22150		(34.5)
XV	$H^1\Sigma_g^+ \rightarrow B^1\Sigma_g^+$	Emission	$3s\sigma \rightarrow 2p\sigma$	V	22754.1		(34.5)
XVI	$L^1\Sigma_g^+ \rightarrow B^1\Sigma_g^+$	Emission	$\rightarrow 2p\sigma$		23054.8		(34.5)
XVII	$M^1\Sigma_g^+ \rightarrow B^1\Sigma_g^+$	Emission	$\rightarrow 2p\sigma$		23190		(34.5)
XVIII	$N^1\Sigma_g^+ \rightarrow B^1\Sigma_g^+$	Emission	$\rightarrow 2p\sigma$		24896		(34.5)
XIX	$T^1\Sigma_g^+ \rightarrow B^1\Sigma_g^+$	Emission	$\rightarrow 2p\sigma$		27130		(34.5)
XX	$P^1\Sigma_g^+ \rightarrow B^1\Sigma_g^+$	Emission	$4d\sigma \rightarrow 2p\sigma$		27148		(34.5)

System	Transition	Sources	Electronic Configuration	Degrades	Band Head, $\nu_{0,0}$	Remarks	Bibliography
XXI	$R^1\Pi_g \rightarrow B^1\Sigma_g^+$	Emission	$4d\pi \rightarrow 2p\sigma$		~27400	Perturbed	(34.5)
XXII	$S^1\Delta_g \rightarrow B^1\Sigma_g^+$	Emission	$4d\delta \rightarrow 2p\sigma$		27460		(34.5)
XXIII	$O^1\Sigma_g^+ \rightarrow B^1\Sigma_g^+$	Emission	$4s\sigma \rightarrow 2p\sigma$		27487		(34.5)
XXIV	$K \rightarrow C^1\Pi_u^-$	Emission	$\rightarrow 2p\pi$	R	12541.2		(34.5)
XXV	$G^1\Sigma_g^+ \rightarrow C^1\Pi_u^-$	Emission	$3d\sigma \rightarrow 2p\pi$	R	12725	Perturbed	(34.5)
XXVI	$I^1\Pi_g \rightarrow C^1\Pi_u^-$	Emission	$3d\pi \rightarrow 2p\pi$	R	12925		(34.5)
XXVII	$J^1\Delta_g \rightarrow C^1\Pi_u^-$	Emission	$3d\delta \rightarrow 2p\pi$	R	13264		(34.5)
XXVIII	$H^1\Sigma_g^+ \rightarrow C^1\Pi_u^-$	Emission	$3s\sigma \rightarrow 2p\pi$	R	13866.6		(34.5)
XXIX	$P^1\Sigma_g^+ \rightarrow C^1\Pi_u^-$	Emission	$4d\sigma \rightarrow 2p\pi$		18260		(34.5)
XXX	$E^1\Pi_g \rightarrow C^1\Pi_u^-$	Emission	$4d\pi \rightarrow 2p\pi$		~18400	Perturbed	(34.5)
XXXI	$D^1\Pi_u^- \rightarrow E^1\Sigma_g^+$	Emission	$2p\pi \rightarrow 2s\sigma$	R	13713.3		(38.11, 37.10)

	System	Transition	Sources	Electronic Configuration	Degrades	Band Head, $\nu_{0,0}$	Remarks	Bibliography
Fulcher (α)	I'	$e\,{}^3\Sigma_u^+ \to a\,{}^3\Sigma_g^+$	Emission	$3p\sigma \to 2s\sigma$	R	11605.7		(34.5)
	II'	$d\,{}^3\Pi_u \to a\,{}^3\Sigma_g^+$	Emission	$3p\pi \to 2s\sigma$	R	16619.0	Perturbed	(35.6)
	III'	$f\,{}^3\Sigma_u^+ \to a\,{}^3\Sigma_g^+$	Emission	$4p\sigma \to 2s\sigma$	R	20526.0		(34.5)
(β)	IV'	$k\,{}^3\Pi_u \to a\,{}^3\Sigma_g^+$	Emission	$4p\pi \to 2s\sigma$	R	22271.5		(34.5)
	V'	$m\,{}^3\Sigma_u^+ \to a\,{}^3\Sigma_g^+$	Emission	$4f\sigma \to 2s\sigma$		23295.1		(34.4, 34.3, 34.2)
(γ)	VI'	$n\,{}^3\Pi_u \to a\,{}^3\Sigma_g^+$	Emission	$5p\pi \to 2s\sigma$	R	24847.5		(34.5)
	VII'	$t\,{}^3\Sigma_u \to a\,{}^3\Sigma_g^+$	Emission	$5f\sigma \to 2s\sigma$	R	25343		(34.4)
(δ)	VIII'	$u\,{}^3\Pi_u \to a\,{}^3\Sigma_g^+$	Emission	$6p\pi \to 2s\sigma$	R	26232.5		(34.5)
	IX'	$a\,{}^3\Sigma_g^+ \to b\,{}^3\Sigma_u^+$	Emission	$2s\sigma \to 2p\sigma$		Continuum $20000 < \nu < 62500$		(44.12)
	X'	$g\,{}^3\Sigma_g^+ \to c\,{}^3\Pi_u$	Emission	$3d\sigma \to 2p\pi$		16926		(34.5, 34.4, 34.3, 34.2)

System	Transition	Sources	Electronic Configuration	Degrades	Band Head, $\nu_{0,0}$	Remarks	Bibliography
XI'	$h\,{}^3\Sigma_g^+ \to c\,{}^3\Pi_u$	Emission	$3s\sigma \to 2p\pi$	R	16990	Predissociated	(34.4, 34.3, 34.2)
XII'	$i\,{}^3\Pi_g \to c\,{}^3\Pi_u$	Emission	$3d\pi \to 2p\pi$		17162		(34.5)
XIII'	$j\,{}^3\Delta_g \to c\,{}^3\Pi_u$	Emission	$3d\delta \to 2p\pi$		17355	Preionized	(34.5)
XIV'	$p\,{}^3\Sigma_g^+ \to c\,{}^3\Pi_u$	Emission	$4d\sigma \to 2p\pi$		22588	Perturbed	(34.5)
XV'	$s\,{}^3\Delta_g \to c\,{}^3\Pi_u$	Emission	$4d\delta \to 2p\pi$		22626	Perturbed	(34.5)
XVI'	$r\,{}^3\Pi_g \to c\,{}^3\Pi_u$	Emission	$4d\pi \to 2p\pi$		22699	Perturbed	(34.5)
XVII'	$v \to c\,{}^3\Pi_u$	Emission	$\to 2p\pi$	R	22487		(34.5)
XVIII'	$q \to c\,{}^3\Pi_u$	Emission	$\to 2p\pi$	R	25220		(34.4, 34.3, 34.2)

H_2

SINGLET SYSTEMS

I. $\quad$ $B\,^1\Sigma_u^+ \rightleftharpoons X\,^1\Sigma_g^+$ System (Lyman)

Band heads, λ (59.15):

v',v''	0	1	2	3	4	5	6	7	8
0	1108.6			1275.2	1334.2	1394.5	1455.7	1517.1	1578.0
1	1092.6		(1198.5)	1254.1	1311.1	1369.3	1428.2	1487.3	1545.8
2	1077.5		(1180.3)	1234.3	1289.4	1345.7	1402.6	1459.5	1515.8
3	1063.3	(1112.4)		1215.5	1269.0	1323.5	1378.5	1433.4	1487.7
4	1049.7			1197.9	1249.8	1302.6	1355.8	1408.9	1461.3
5	1036.9			1181.2	1231.6	1282.9	1334.4	1385.9	1436.5
6	1024.7	(1070.3)				1264.2	1314.3	1364.2	1413.2
7	1013.1					1246.7	1295.3	1343.8	1391.3
8	1002.1						1268.6	1324.5	1370.6
9	991.6						1260.4	1306.3	1351.1
10	981.7						1244.4	1289.0	1332.7
11	972.2							1272.8	1315.3
12	963.2							1257.4	1298.9
13	954.6							1242.8	1283.3

II. $\quad$ $C\,^1\Pi_u^- \rightleftharpoons X\,^1\Sigma_g^+$ System (Werner)

Band heads, λ (33.1):

v',v''	1	2	3	4	5	6	7	8
0		1098.9	1145.4	1192.7				
1	1028.4		1115.8	1160.8	1206.2			
2		1047.3	1089.4	1132.2	1175.4	1218.5		
3				1106.6		1188.9	1229.6	
4					1123.1		1201.3	1239.2

Absorption series, $v'' = 0$ (58.14):

$$C'\Pi_u^+ \leftarrow X^1\Sigma_g^+ \text{ System}$$

v'	λ	v'	λ	v'	λ
0	1009.3	6	901.2	12	848.5
1	986.3	7	889.0	13	845.0
2	965.6	8	878.2		
3		9			
4	930.0	10	860.5		
5	914.8	11	853.8		

$$C^1\Pi_u^- \leftarrow X^1\Sigma_g^+ \text{ System}$$

v'	λ	v'	λ	v'	λ
0	1009.3	5	914.8	10	860.5
1	986.3	6	901.2	11	853.8
2	965.6	7	889.0	12	848.6
3	946.9	8	878.1	13	845.0
4	930.0	9	868.7		

III. $\quad \underline{B'^1\Sigma_u^+ \leftarrow X^1\Sigma_g^+ \text{ System}}$

Band heads, λ (68.28):

(v', v'')	(0, 0)	(1, 0)	(2, 0)	(3, 0)	(4, 0)	(5, 0)	(6, 0)
λ	905.2	890.0	876.6	865.0	855.4	848.5	845.1

IV. $\quad \underline{D^1\Pi_u^- \leftarrow X^1\Sigma_g^+ \text{ System}}$

Band heads, λ for $v'' = 0$ (68.28):

v'	λ	v'	λ	v'	λ
0	886.0	6	804.2	12	760.7
1	868.8	7	794.7	13	756.5
2	853.3	8	786.1	14	753.1
3	839.2	9	778.4	15	750.7
4	826.3	10	771.6		
5	814.7	11	765.7		

$D\,{}^1\Pi_u^+ \leftarrow X\,{}^1\Sigma_g^+$ System

Band heads, λ for $v'' = 0$ (68.28):

v'	λ	v'	λ	v'	λ
0	886.0	6	804.2	12	760.7
1	868.8	7	794.7	13	756.5
2	853.3	8	786.1	14	753.1
3	839.2	9	778.4	15	750.7
4	826.4	10	771.6		
5	814.7	11	765.7		

V. $B''\,{}^1\Sigma_u^+ \leftarrow X\,{}^1\Sigma_g^+$ System

Band heads, λ (68.28):

(v', v'')	(0, 0)	(1, 0)	(2, 0)	(3, 0)	(4, 0)	(5, 0)	(6, 0)
λ	855.6	840.7	827.3	815.3	804.4	794.7	786.1

VI. $D'\,{}^1\Pi_u^- \leftarrow X\,{}^1\Sigma_g^+$ System

Band heads, λ (68.28):

(v', v'')	(0, 0)	(1, 0)	(2, 0)	(3, 0)
λ	848.65	833.06	818.94	806.02

$D'\,{}^1\Pi_u^+ \leftarrow X\,{}^1\Sigma_g^+$ System

Band heads, λ (68.28):

(v', v'')	(0, 0)	(1, 0)	(2, 0)	(3, 0)
λ	848.64	833.06	818.89	806.00

VII. $D''\,{}^1\Pi_u^- \leftarrow X\,{}^1\Sigma_g^+$ System

Band heads, λ (68.28):

(v', v'')	(0, 0)	(1, 0)	(2, 0)
λ	832.14	817.18	803.56

$D''{}^1\Pi_u^+ \leftarrow X^1\Sigma_g^+$ System

Band heads, λ (68.28):

(v', v'')	$(0,0)$	$(1,0)$	$(2,0)$
λ	832.09	817.21	803.67

XI. $K \rightarrow B^1\Sigma_g^+$ System

Band heads, λ (34.5):

v', v''	0	1	2	3	4	5
0	4662.7	4968.2	5305.9	5681.8		6515.9
1	4223.9	4472.2	4744.7	5043.1		5735.3
2	3871.8	4080.0	4305.1	4549.2	4814.9	5707.3

XII. $G^1\Sigma_g^+ \rightarrow B^1\Sigma_g^+$ System

Band heads, λ (34.5):

v', v''	0	1	2	3	4	5
0	4629.4	4930.2	5262.6	5632.2		
1	4196.9	4442.7	4710.8	5004.8	5328.2	5685.7
2	3861.0	4068.0	4291.8	4534.3	4798.2	5086.5
3	3617.4	3798.5	3993.0	4202.2	4427.9	4672.0

XIII. $I^1\Pi_a \rightarrow B^1\Sigma_g^+$ System

Band heads, λ (34.5):

v', v''	0	1	2	3	4	5	6
0	4559.5	4851.3		5529.4			
1	4163.0	4404.9	4668.3	4956.9	5274.0	5624.0	6012.1
2	3841.6	4046.6					5363.9
3		3785.3	3978.2				

H_2

$I \, ^1\Pi_b \to B \, ^1\Sigma_g^+$ System

Band heads, λ (34.5):

v', v''	0	1	2	3	4	5	6
0	4580.6	4874.7	5198.9	5559.0			
1	4178.9	4422.2	4687.6	4978.1	5297.5	5650.4	
2	3865.2	4072.3	4296.3	4539.1	4803.3	5091.6	5407.2
3	3617.3	3798.2	3992.3				

XIV. $J \, ^1\Delta_g \to B \, ^1\Sigma_g^+$ System

Band heads, λ (34.5):

v', v''	0	1	2	3	4
0	4503.3	4786.1	5097.6	5442.2	
1	4098.5	4335.0	4585.3	4854.4	5165.8

XV. $H \, ^1\Sigma_g^+ \to B \, ^1\Sigma_g^+$ System

Band heads, λ (34.5):

v', v''	0	1	2	3	4	5	6
0	4383.5	4652.2	4947.1				
1	3985.2	4206.1		4706.5	4991.5		5647.8
2	3685.3	3873.4	4075.8	4293.9	4512.6		

XVI. $L \, ^1\Sigma_g^+ \to B \, ^1\Sigma_g^+$ System

Band heads, λ (34.5):

v', v''	0	1	2	3	4
0	4333.9	4596.4	4884.2	5202.4	5550.9
1					
2	3741.4	3935.5	4666.6	4370.2	4614.9

XVII. $\underline{M\,{}^1\Sigma_g^+ \to B\,{}^1\Sigma_g^+ \text{ System}}$

Band heads, λ (34.5):

v', v''	0	1	2	3	4	5
0	4307.6	4566.8		5162.9	5507.8	5890.7
1	3941.0	4157.0		4645.1		
2	3634.9	3817.8	4014.3			4701.2

XVIII. $\underline{N\,{}^1\Sigma_g^+ \to B\,{}^1\Sigma_g^+ \text{ System}}$

Band heads, λ (34.5):

v', v''	0	1	2	3	4	5	6
0	4010.7	4234.6	4477.7	4742.3	5031.7		5686.0
1	3717.1	3908.5	4114.7	4334.1	4577.9	4839.3	5124.0

XX. $\underline{P\,{}^1\Sigma_g^+ \to B\,{}^1\Sigma_g^+ \text{ System}}$

Band heads, λ (34.5):

(v', v'')	(0, 0)	(0, 1)	(0, 2)	(0, 3)	(0, 4)	(0, 5)
λ	3675.4	3862.6	4072.8	4280.6	4515.1	4769.2

XXI. $\underline{R\,{}^1\Pi_a \to B\,{}^1\Sigma_g^+ \text{ System}}$

Band heads, λ (34.5):

(v', v'')	(0, 1)	(0, 2)	(0, 4)
λ	3640.6	3824.1	4462.5

$\underline{R\,{}^1\Pi_b \to B\,{}^1\Sigma_g^+ \text{ System}}$

Band heads, λ (34.5):

v', v''	0	1	2	3	4	5
0	3656.7	3841.6	4040.2	4264.0	4485.5	4736.0
1	3391.1		3718.4	3898.9	4092.2	4299.6

XXII. $S\,^1\Delta_a \rightarrow B\,^1\Sigma_g^+$ System

Band heads, λ (34.5):

(v', v'')	(0, 0)	(0, 1)	(0, 2)	(0, 3)	(0, 4)	(0, 5)
λ	3636. 1	3817. 5	4012. 5	4198. 5	2387. 9	4623. 6

$S\,^1\Delta_b \rightarrow B\,^1\Sigma_g^+$ System

(v', v'')	(0, 0)	(0, 1)	(0, 2)	(0, 3)	(0, 4)	(0, 5)
λ	3634. 4	3816. 4	4011. 9	4222. 3	4421. 8	4695. 3

TRIPLET SYSTEMS

II'. $d\,^3\Pi_u \rightarrow a\,^3\Sigma_g^+$ System (Fulcher α)

Band heads, λ (35.6):

v', v''	0	1	2	3	4
0	6020. 1	7097. 2			
1	5304. 5	6123. 3	7170. 5		
2	4769. 4	5421. 5	6226. 7	7242. 7	
3	4355. 4	4892. 6	5538. 9	6328. 7	7311. 5
4		4481. 5	5017. 8	5657. 4	6430. 0
5		4153. 9	4610. 6	5145. 1	5776. 3

IV'. $k\,^3\Pi_u \rightarrow a\,^3\Sigma_g^+$ System (β)

Band heads, λ (34.5):

v', v''	0	1	2	3	4
0	4491. 8	5065. 3			
1	4086. 1	4555. 4	5110. 6		
2	3765. 5	4160. 4	4618. 7	5155. 4	5789. 0
3		3846. 4	4235. 0	4681. 6	5198. 6
4			3928. 3	4309. 8	4744. 1
5				4011. 1	4384. 8

VI'. $n^3\Pi_u \rightarrow a^3\Sigma_g^+$ System (γ)

Band heads, λ (34.5):

v', v''	0	1	2	3	4
0	4025.9	4480.7			
1		3934.8	4519.0		
2			4134.4	4556.6	
3				4185.0	4593.3

SPECTROSCOPIC CONSTANTS

TRIPLET STATES

State	T_o	ω_e [a]	$\omega_e x_e$ [b]	B_e [c]	α [d]	D_o	r_e	Remarks	Configuration
$u\,^3\Pi_u$	121316.7			~29.3[f]			1.07[g]		$1s\sigma\ 6p\pi$
$t\,^3\Sigma_u$	120427	~2661.4	~121.9	~31.5[f]			1.03[g]		$1s\sigma\ 5f\sigma$
q	120017	2172.6[e]		~30[f]			~1.06[g]		
$n\,^3\Pi_u$	119931.7	2322	62.9	29.03	1.3		1.057		$1s\sigma\ 5p\pi$
$m\,^3\Sigma_u^+$	118379.3	2457.1[e]		~36[f]			~0.96[g]		$1s\sigma\ 4f\sigma$
$r\,^3\Pi_g$	117492	2170[e]						(h)	$1s\sigma\ 4d\pi$
$s\,^3\Delta_g$	117419	2170[e]						(h)	$1s\sigma\ 4d\delta$
$p\,^3\Sigma_g^+$	117381	2147.7[e]						(h)	$1s\sigma\ 4d\sigma$
$k\,^3\Pi_u$	117356	2336	60	29.40	1.58		1.067		$1s\sigma\ 4p\pi$
v	117280	~2339	~57	~29.1[f]			1.07[g]		
$f\,^3\Sigma_u^+$	115610	2140.1[e]		29.61	2.18	~2.2	1.063		$1s\sigma\ 4p\sigma$
$j\,^3\Delta_g$	112148	2265	58						$1s\sigma\ 3s\delta$

SPECTROSCOPIC CONSTANTS

TRIPLET STATES

State	T_o	ω_e (a)	$x_e\omega_e$ (b)	B_e (c)	α (d)	D_o	r_e	Remarks	Configuration
$i\,^3\Pi_g$	111955	2268	~75						$1s\sigma\ 3d\pi$
$h\,^3\Sigma_g^+$	111783	2395.2	64.2	30.6	1.26	2.0	1.04		$1s\sigma\ 3s\sigma$
$g\,^3\Sigma_g^+$	111719	2265.5	89						$1s\sigma\ 3d\sigma$
$d\,^3\Pi_u$	111703	2371.58	66.27	30.364	1.545	2.0	1.0496		$1s\sigma\ 3p\pi$
$e\,^3\Sigma_u^+$	106690	2195.8	65.80	27.30	1.515	1.65	1.107		$1s\sigma\ 3p\sigma$
$a\,^3\Sigma_g^+$	95084	2664.83	71.65	34.216	1.671	2.21	0.9887		$1s\sigma\ 2s\sigma$
$c\,^3\Pi_u$	94793	2465.0	61.4	31.07	1.425	1.9	1.038		$1s\sigma\ 2p\pi$
$b\,^3\Sigma_u^+$	Unstable								$1s\sigma\ 2p\sigma$

SPECTROSCOPIC CONSTANTS

SINGLET STATES

State	T_σ	ω_e (a)	$x_e\omega_e$ (b)	B_e (c)	α (d)	D_o	r_e	Remarks	Configuration
D''$^1\Pi_u^-$	120172.21	2323.56	62.15	30.76	1.45	21232.48	1.043		1sσ 2pπ
D'$^1\Pi_u^-$	117834.65	2330.19	63.26	29.89	1.11	21102.27	1.058		1sσ 2pπ
O$^1\Sigma_g^+$	117680			~32 (f)			~1.0 (g)		1sσ 4sσ
S$^1\Delta_g$	117656			~28.8 (f)			~1.08 (g)		1sσ 4dδ
R$^1\Pi_g$	~117600	2142 (e)		~30 (f)			~1.1 (g)		1sσ 4dπ
P$^1\Sigma_g^+$	117342			~30 (f)			~1.1 (g)		1sσ 4dσ
T$^1\Sigma_g^+$	117330			~19			~1.3		
B''$^1\Sigma_u^+$	116882.00	2197.50	68.14	27.13	1.30	16723.31	1.111		1sσ 2pσ
N$^1\Sigma_g^+$	115090	1983.3 (e)		~17.5 (f)			~1.38 (g)		
M$^1\Sigma_g^+$	113390	2176 (e)		~13 (f)			~1.6 (g)		
L$^1\Sigma_g^+$	113250	1835 (e)		~10 (f)			~1.8 (g)		

SPECTROSCOPIC CONSTANTS

SINGLET STATES

State	T_o	ω_e (a)	$x_e\omega_e$ (b)	B_e (c)	α (d)	D_o	r_e	Remarks	Configuration
$H\ ^1\Sigma_g^+$	112949	2538	124	~29.5 (f)			~1.07 (g)		$1s\sigma\ 3s\sigma$
$D\ ^1\Pi_u^-$	112871.74	2361.59	69.15	30.81	0.102	20733.57	1.042		$1s\sigma\ 2p\pi$
$J\ ^1\Delta_g$	112345	2220 (e)		~28.8 (f)			~1.07 (g)		$1s\sigma\ 3d\delta$
$I\ ^1\Pi_g$	112007	2265.2	78.47	29.79			1.060		$1s\sigma\ 3d\pi$
$G\ ^1\Sigma_g^+$	111805.3	2404.3	88.8	28.4 (f)			~1.08 (g)		$1s\sigma\ 3d\sigma$
K	111621.5	2293	30	~11 (f)			~1.7 (g)		
Q	111347.2	742 (e)		16.3 (f)			~1.43 (g)		
$B'\ ^1\Sigma_u^+$	110478.54	2074.93	113.39	26.78	0.380	7893.55	1.118		$1s\sigma\ 2p\sigma$
$F\ ^1\Sigma_g^+$	~100400	~1000 (e)		~6.24 (f)		(16938)	~2.32 (g)	(h)	$(2p\sigma)^2$
$E\ ^1\Sigma_g^+$	99157.3	2588.9		32.68	1.82		1.012		$1s\sigma\ 2s\sigma$
$C\ ^1\Pi_u^-$	99081.72	2431.22	60.88	31.50	1.83	19290.37	1.031		$1s\sigma\ 2p\pi$

SPECTROSCOPIC CONSTANTS

SINGLET STATES

State	T_o	ω_e (a)	$x_e\omega_e$ (b)	B_e (c)	α (d)	D_o	r_e	Remarks	Configuration
$B\ ^1\Sigma_u^+$	90203.55	1357.39	20.42	20.035	1.2312	28168.54	1.29253		$1s\sigma\ 2p\sigma$
$X\ ^1\Sigma_g^+$	0	4401.21	121.34	60.8530	3.0622	36113.05	0.74116		$(1s\sigma)^2$

(a) Actual value of $Y_{10}(\sim\omega_e)$; (b) Actual value of $-Y_{20}(\sim x_e\omega_e)$; (c) Actual value of $Y_{01}(\sim B_e)$;

(d) Actual value of $-Y_{11}(\sim\alpha)$; (e) $\Delta G_{1/2}$; (f) B_o; (g) r_o; (h) Observed in perturbation.

Dissociation energy = 4.47733 eV, 103.251 kcal/mole, 36118.62 cm^{-1}.

Perturbations and General Information

Strong heterogeneous perturbations are observed in the $B^1\Sigma_u^+$, $B'^1\Sigma_g^+$, $D^1\Pi_u^-$, and $C^1\Pi_u^-$ states (73.93, 64.24, 61.17).

Homogeneous perturbations of the $C^1\Pi_u$ (v = 3) state by the $B^1\Sigma_u^+$ (v = 14) state are observed (73.110).

Weak heterogeneous perturbations are observed in the $B'^1\Sigma_g^+$ and $D'^1\Pi_u$ state (64.24).

Weak homogeneous perturbations are observed in the $B'^1\Sigma_g^+$ and $B''^1\Sigma_g^+$ state (64.24).

Strong vibrational and rotational perturbations are seen in the $F^1\Sigma_g^+$ state.

Strong vibrational perturbations are observed in the $G^1\Sigma_g^+$ state (37.9).

The $R^1\Pi_g^+$ and $d^3\Pi_u^+$ states are perturbed (73.106).

The $p^3\Sigma_g^+$, $p^3\Pi_g$, and $s^3\Delta_g$ states are perturbed.

The $h^3\Sigma_g^+$ state is predissociated at level v = 3 (36.8).

The $D^1\Pi_u$ state is predissociated at level v = 3 by $B'^1\Sigma_u^+$ state (72.48, 61.19).

The $j^3\Delta_g$ state is preionized at the v = 3 level.

The E and F states have been shown to represent a single state, the E, $F^1\Sigma_g^+$ state with a double minimum (74.134, 74.125, 71.41, 69.33).

Isotope shifts (68.26).

Lasing has been observed from the Werner and Lyman bands (74.115, 74.114, 73.95, 73.91, 70.37).

H$_2$

Franck-Condon factors for the B$^1\Sigma_u^+$ - X$^1\Sigma_g^+$ system (69.32):

v', v''	0	1	2	3	4	5	6	7	8
0	0.0037	0.0273	0.0903	0.1799	0.2403	0.2248	0.1474	0.0654	0.0181
1	0.0133	0.0689	0.1384	0.1220	0.0269	0.0139	0.1367	0.2294	0.1754
2	0.0273	0.0984	0.1066	0.0186	0.0235	0.1081	0.0559	0.0065	0.1613
3	0.0427	0.1036	0.0471	0.0047	0.0787	0.0392	0.0126	0.1041	0.0272
4	0.0563	0.0878	0.0080	0.0394	0.0571	0.0007	0.0715	0.0213	0.0438
5	0.0663	0.0624	0.0007	0.0599	0.0129	0.0342	0.0422	0.0112	0.0676
6	0.0723	0.0371	0.0130	0.0509	0.0005	0.0534	0.0022	0.0517	0.0073
7	0.0740	0.0175	0.0294	0.0278	0.0160	0.0352	0.0101	0.0394	0.0110
8	0.0725	0.0055	0.0402	0.0084	0.0337	0.0096	0.0334	0.0074	0.0392
9	0.0688	0.0005	0.0426	0.0003	0.0289	0.0000	0.0379	0.0014	0.0338
10	0.0634	0.0005	0.0382	0.0024	0.0315	0.0067	0.0239	0.0163	0.0106

Franck-Condon factors for the I$^1\Pi_g$ - B$^1\Sigma_g^+$ system (69.32):

v', v''	0	1	2	3	4	5	6	7	8
0	0.4663	0.3126	0.1397	0.0528	0.0185	0.0064	0.0023	0.0008	0.0003
1	0.3902	0.0248	0.1893	0.1849	0.1117	0.0551	0.0247	0.0106	0.0046
2	0.1263	0.3604	0.0264	0.0482	0.1303	0.1254	0.0845	0.0478	0.0248
3	0.0167	0.2551	0.2285	0.0988	0.0008	0.0589	0.0981	0.0895	0.0630
4	0.0005	0.0461	0.3440	0.1319	0.1267	0.0099	0.0160	0.0592	0.0746
5	0.0000	0.0009	0.0716	0.4042	0.0905	0.1156	0.0284	0.0015	0.0296
6	0.0000	0.0001	0.0002	0.0775	0.4576	0.0941	0.0856	0.0390	0.0002
7		0.0000	0.0004	0.0009	0.0539	0.4970	0.1494	0.0465	0.0440
8			0.0000	0.0005	0.0094	0.0114	0.4678	0.2763	0.0086
9			0.0000	0.0002	0.0000	0.0222	0.0109	0.2883	0.4363
10				0.0000	0.0003	0.0034	0.0126	0.1093	0.0371

Franck-Condon factors for the $d^3\Pi_u - a^3\Sigma_g^+$ system (69.32):

v',v''	0	1	2	3	4	5	6	7
0	0.93058	0.06863	0.00078	0.00000	0.00000			
1	0.06414	0.79842	0.13516	0.00226	0.00001	0.00000		
2	0.00485	0.11677	0.67463	0.19937	0.00433	0.00003	0.00000	
3	0.00039	0.01423	0.15683	0.56046	0.26113	0.00687	0.00008	0.00000
4	0.00003	0.00171	0.02725	0.18385	0.45689	0.32038	0.00972	0.00017
5	0.00000	0.00020	0.00448	0.04251	0.19799	0.36464	0.37715	0.01269
6	0.00000	0.00002	0.00073	0.00910	0.05825	0.20006	0.28411	0.43155
7		0.00000	0.00011	0.00193	0.01566	0.07249	0.19146	0.21539
8		0.00000	0.00002	0.00041	0.00417	0.02378	0.08332	0.17411
9			0.00000	0.00009	0.00113	0.00772	0.03259	0.08913
10			0.00000	0.00002	0.00032	0.00257	0.01258	0.04071

Franck-Condon factors for the E, $F^1\Sigma_g^+ \to B^1\Sigma_u^+$ System (74.125):

v', v''	0	1	2	3	4	5	7	9
0	0.341	0.313	0.187	0.910-1	0.403-1	0.166-1	0.266-2	0.428-3
0*	0.825-4	0.207-2	0.180-1	0.843-1	0.216	0.314	0.994-1	0.451-3
1	0.424	0.497-3	0.794-1	0.174	0.138	0.818-1	0.260-1	0.102-1
1*	0.615-3	0.842-2	0.482-1	0.130	0.132	0.143-1	0.290	0.597-1
2	0.186	0.333	0.190-1	0.439-2	0.122-1	0.103	0.226-1	0.557-1
2*	0.874-4	0.232-1	0.877-1	0.958-1	0.338-1	0.373-1	0.405-2	0.317
3	0.129-1	0.129	0.193-1	0.172	0.120-2	0.549-1	0.177-2	0.112
3*	0.743-2	0.618-1	0.932-1	0.496-1	0.189-3	0.836-1	0.481-1	0.532-1
4	0.291-4	0.870-2	0.242-1	0.212-1	0.382-1	0.105	0.501-1	0.167-2
4*	0.184-1	0.802-4	0.199	0.132-1	0.579-1	0.169-2	0.130	0.343-1
5								
5*	0.227-2	0.905-1	0.137	0.244-1	0.496-1	0.381-2	0.296-1	0.199-2
6								
6*	0.637-2	0.196-1	0.254-1	0.786-1	0.102	0.642-2	0.114-1	0.356-1
7								
7*	0.699-3	0.107-2	0.509-1	0.495-3	0.115	0.325-1	0.562-1	0.300-6
8								
8*	0.326-3	0.661-2	0.175-2	0.347-1	0.184-5	0.778-1	0.166-2	0.360-1

v' with an asterisk refers to the outer minimum. v' without an asterisk refers to the inner minimum.

Franck-Condon factors followed by a factor of ten.

Franck-Condon factors for the $C^1\Pi_u \rightarrow E$, $F^1\Sigma_g^+$ system (74.125):

v', v''	0	0*	1	1*	2
0	0.993	0.672-6	0.706-2	0.101-9	0.232-3
1	0.678-2	0.122-5	0.981	0.323-4	0.500-2
2	0.555-3	0.146-4	0.662-2	0.145-3	0.794
3	0.256-4	0.201-3	0.283-2	0.133-2	0.725-2
4	0.738-5	0.203-2	0.483-4	0.929-2	0.326-1
5	0.190-6	0.143-1	0.536-3	0.435-1	0.594-2
6	0.503-6	0.673-1	0.191-3	0.118	0.229-3
7	0.578-7	0.200	0.190-4	0.142	0.115-1
8	0.259-6	0.347	0.158-3	0.185-1	0.172-2
9	0.119-6	0.292	0.136-3	0.121	0.835-2
10	0.205-7	0.748-1	0.222-3	0.413	0.120-2

v', v''	2*	3	3*	4*	5*
0	0.255-4	0.548-6	0.921-5	0.370-4	0.181-5
1	0.244-2	0.105-2	0.284-4	0.245-2	0.108-2
2	0.146-2	0.459-5	0.131-1	0.164	0.974-2
3	0.482-2	0.463	0.169-1	0.444-1	0.206
4	0.248-1	0.159-3	0.516-1	0.425-1	0.674-1
5	0.745-1	0.523-2	0.909-1	0.710-1	0.311-1
6	0.111	0.343-1	0.598-1	0.152-1	0.470-2
7	0.354-1	0.727-3	0.284-3	0.151-1	0.388-1
8	0.237-1	0.212-1	0.645-1	0.364-1	0.615-3
9	0.847-1	0.896-3	0.691-2	0.704-2	0.345-1
10	0.335-1	0.204-1	0.510-1	0.301-1	0.112-2

v'' with an asterisk refers to the outer minimum.   v'' without an asterisk refers to the inner minimum.

Franck-Condon factors followed by a factor of ten.

Oscillator strengths (74.127, 74.126):

System	Band	$f_{v',0} \times 10^3$
$B\,^1\Sigma_u^+ \rightleftarrows X\,^1\Sigma_g^+$	(0,0)	1.75
	(1,0)	5.19
	(2,0)	11.5
	(3,0)	17.6
	(4,0)	24.5
	(5,0)	25.8
$C\,^1\Pi_u \rightleftarrows X\,^1\Sigma_g^+$	(1,0)	59.2
	(2,0)	64.2
	(3,0)	44.2
	(4,0)	31.7
	(5,0)	22.4
	(6,0)	17.0
$B'\,^1\Sigma_u^+ \leftarrow X\,^1\Sigma_g^+$	(1,0)	2.84
	(3,0)	4.84
$D\,^1\Pi_u^- \leftarrow X\,^1\Sigma_g^+$	(0,0)	6.14
	(2,0)	10.9

Radiative lifetimes:

$d\,^3\Pi_u$ state $\quad \tau_1 = 29.4 \pm 3.2$ nsec $\quad$ (73.75, 73.74)

$a\,^3\Sigma_g^+$ state $\quad \tau_0 = 11.9 \pm 1.2$ nsec $\quad$ (72.65)

$\tau_1 = 10.8 \pm 1.1$ nsec

$\tau_{2,3} = 10 \pm 2$ nsec

$B'\,^1\Sigma_u^+$ state $\quad \tau_{8-11} = 1 \pm 0.2$ nsec $\quad$ (72.65)

$c\,^3\Pi_a$ state $\quad \tau_0 = 3.15$ msec $\quad$ (69.35)

BIBLIOGRAPHY

References in this bibliography are only those dated from 1968. All those publications from which information is taken directly are listed because of the extensive bibliography given in B. Rosen's compendium (see Reference in front of book). An excellent paper by T. E. Sharp (71.41) also gives extensive spectroscopic data and bibliography.

(33. 1)   C. R. Jeppesen,
          Phys. Rev. 44, 165-84

(34. 2)   O. W. Richardson and T. B. Rymer,
          Proc. Roy. Soc. A 147, 24-47

(34. 3)   O. W. Richardson and T. B. Rymer,
          Proc. Roy. Soc. A 147, 251-72

(34. 4)   O. W. Richardson and T. B. Rymer,
          Proc. Roy. Soc. A 147, 272-92

(34. 5)   O. W. Richardson,
          Molecular Hydrogen and Its Spectrum,
          Yale University Press, New Haven

(35. 6)   G. H. Dieke and R. W. Blue,
          Phys. Rev. 47, 261-72

(36. 7)   G. H. Dieke,
          Phys. Rev. 50, 797-805

(36. 8)   H. Beutler and H. O. Jünger,
          Z. Physik 101, 285-303

(37. 9)   Perturbations and Isotope Effect,
          G. H. Dieke and M. N. Lewis,
          Phys. Rev. 52, 100-25

(37. 10) O. W. Richardson,
          Proc. Roy. Soc. A 160, 487-507

(38. 11) O. W. Richardson,
          Proc. Roy. Soc. A 164, 316-45

(44. 12) Continuum,
          A. S. Coolidge,
          Phys. Rev. 65, 236-46

(49. 13) Doubly Excited Levels,
          G. H. Dieke,
          Phys. Rev. 76, 50-7

(58. 14) H_2 Emission Spectra,
          G. H. Dieke,
          J. Molec. Spectrosc. 2, 494-517

(59. 15) Lyman Bands in Absorption.   Study of the X State,
          G. Herzberg and L. L. Howe,
          Can. J. Phys. 37, 636-59

(60. 16) Ground State Dissociation Energy,
          G. Herzberg and A. Monfils,
          J. Molec. Spectrosc. 5, 482-98

(61. 17) Perturbations and Absorption in the B and C States,
          A. Monfils,
          Bull. Cl. Sci. Acad. Roy. 47, 585-98

(61. 18) B' and B'' States,
          A. Monfils,
          Bull. Cl. Sci. Acad. Roy. 47, 599-606

(61. 19) D and D' States,
          A. Monfils,
          Bull. Cl. Sci. Acad. Roy. 47, 816-23

(62. 20) Perturbations and Absorption in the B and C States,
          A. Monfils,
          Bull. Cl. Sci. Acad. Roy. 48, 460-81

(62. 21) Λ Doubling in the C State,
          A. Monfils,
          Bull. Cl. Sci. Acad. Roy. 48, 482-9

(62. 22) D'' State,
          A. Monfils and B. Rosen,
          J. Quant. Spectrosc. Radiative Trans. 2, 321-5

(64. 23) B and C States in Absorption,
          V. G. Ryabova and L. V. Gurvich,
          High Temp. 2, 749-50

(64. 24) B', B'', D, and D' States,
          V. G. Ryabova and L. V. Gurvich,
          Teplofiz. Vysokikh Temp. 2, 834-5

(68. 25) Potential Functions for B, B', B'', C, D, D', and D'' States,
          A. Monfils,
          Bull. Cl. Sci. Acad. Roy. 54, 44-80

(68.26)   Lyman Bands,
          P. G. Wilkinson,
          Can. J. Phys. 46, 1225-36

(68.27)   Potential Functions of the a, B, and C States,
          W. Kolos and L. Wolniewicz,
          J. Chem. Phys. 48, 3672-80

(68.28)   B, B', B'', C, D, D', and D'' States,
          A. Monfils,
          J. Molec. Spectrosc. 25, 513-43

(68.29)   G. N. Haddad, K. H. Lokan, A. J. D. Farmer, and J. H. Carver,
          "An Experimental Determination of the Oscillator Strengths for
          Some Transitions in the Lyman Bands of Molecular Hydrogen,"
          J. Quant. Spectrosc. Radiative Trans. 8, 1193-1200

(68.30)   D. A. Dahlberg, D. K. Anderson, and I. E. Dayton,
          "Vacuum Ultraviolet Emission Produced by Proton and H-Atom
          Impact on H_2,"
          Phys. Rev. 170, 127-30

(68.31)   F. J. Comes and H. O. Wellern,
          "The Spectroscopy of the Hydrogen Molecule Near Its Ionization
          Limit,"
          Z. Naturforsch. 23a, 881-7

(69.32)   R. J. Spindler, Jr.,
          "Franck-Condon Factors for Band Systems of Molecular Hydrogen I,"
          J. Quant. Spectrosc. Radiative Trans. 9, 597-626

(69.33)   W. Kolos and L. Wolniewicz,
          "Theoretical Investigation of the Lowest Double-Minimum State E,
          F $^1\Sigma_g^+$ of the Hydrogen Molecule,"
          J. Chem. Phys. 50, 3228-40

(69.34)   F. P. Lossing and G. P. Semeluk,
          "Threshold Ionization Efficiency Curves for Monoenergetic Electron
          Impact on H_2, D_2, CH_4 and CD_4,"
          Int. J. Mass Spect. Ion Phys. 2, 408-13

(69.35)   R. C. Bigelow,
          "A Laser Study: Absolute Spectral Intensities From the Transition
          a $^3\Sigma_g$ to b $^3\Sigma_u$ in Molecular Deuterium and Hydrogen Excited by
          Pulse Discharge,"
          Thesis, University of Colorado

(70.36) K. W. Chow,
"I. Vacuum-Ultraviolet Emission and Absorption Study of Rate-Gas Molecules: Determination of Potential Curves of the $A^1\Sigma_u^+$ and $X^1\Sigma_g^+$ States of He_2
II. Electronic Excitation Transfer Between Atomic and Molecular Hydrogen,"
Thesis, Yale University

(70.37) R. T. Hodgson,
"Vacuum-Ultraviolet Laser Action Observed in the Lyman Bands of Molecular Hydrogen,"
Phys. Rev. Letters 25, 494-7

(70.38) A. Weingartshofer, H. Ehrhardt, V. Hermann, and F. Linder,
"Measurements of Absolute Cross Sections for (e, H_2) Collision Processes. Formation and Decay of H_2 Resonances,"
Phys. Rev. A 2, 294-304

(71.39) R. W. Reno,
"A Measurement of the Hyperfine Structure of the $^3\Pi_u$ State of H_2,"
Thesis, Yale University

(71.40) K. H. Becker and E. H. Fink,
"Relative Line Intensities in the Lyman Bands of HD and H_2,"
Z. Naturforsch. 26a, 319-20

(71.41) T. E. Sharp,
"Potential-Energy Curves for Molecular Hydrogen and Its Ions,"
Atomic Data 2, 119-69

(72.42) R. W. Carlson and W. R. Fenner,
"Absolute Raman Scattering Cross-Section of Molecular Hydrogen,"
Astrophys. J. 178, 551-6

(72.43) H. Yokoyama,
"Heitler-London Description of H_2 and LiH With Variable Orbital Exponents,"
Int. J. Quant. Chem. 6, 1121-47

(72.44) R. F. Heidner III,
"Quantitative Vacuum Ultraviolet Absorption Studies of Vibrationally-Excited Hydrogen,"
Thesis, University of California, Los Angeles

(72.45) M. Ratner,
"SPPA Method for Two-Particle Green's Function of Minimum-Basis H_2,"
Int. J. Quant. Chem. 6, 1165-72

H_2

(72.46)   G. Mainfray, C. Manus, and I. Tugov,
           "Multiphoton Dissociation, Predissociation, and Autoionization of
           the Hydrogen Molecule,"
           JETP Letters 16, 19-23

(72.47)   G. J. Prangsma, L. J. M. Borsboom, H. F. P. Knaap,
           C. J. N. Van Den Meijdenberg and J. J. M. Beenakker,
           "Rotational Relaxation in Ortho Hydrogen Between 170 and 300 K,"
           Physica 61, 527-38

(72.48)   F. Fiquet-Fayard and O. Gallais,
           "Predissociation of H_2 and D_2 ($D^1\Pi_u$): Comparison of Calculated
           and Experimental Line-Widths,"
           Chem. Phys. Letters 16, 18-19

(72.49)   B. Ritchie,
           "Semiclassical Theory for Low-Energy Molecular Collisions.
           $H^+ - H_2$ Vibrational Excitation,"
           Phys. Rev. A 6, 1902-7

(72.50)   B. Ritchie,
           "Theory of Small-Energy-Transfer Collisions in Dominant Long-
           Ranged Forces: $H^+ - H_2$ and $e^- - H_2$ Vibrational Excitation,"
           Phys. Rev. A 6, 1456-60

(72.51)   J. L. Koster and P. J. A. Ruttink,
           "Non-Empirical Approximate Calculations for the Ground States of
           H_2 and H_3 Including Complete Configuration Interaction,"
           Chem. Phys. Letters 17, 419-21

(72.52)   R. Jost,
           "Calculation of the Intermediate Coupling Between the Hund's
           Cases b and d, for the 3d Rydberg States of H_2,"
           Chem. Phys. Letters 17, 393-6

(72.53)   J. M. Schulman and D. N. Kaufman,
           "Perturbation Calculation of the Nuclear Spin-Spin Coupling
           Constant in HD Based on the Bare-Nucleus Potential,"
           J. Chem. Phys. 57, 2328-32

(72.54)   B. M. Hopkins and H. L. Chen,
           "Vibrational Relaxation of Hydrogen,"
           J. Chem. Phys. 57, 3161-6

(72.55)   A. Cohn,
           "Photoionization of Excited H_2 Molecules,"
           J. Chem. Phys. 57, 2456-8

(72.56)   R. A. White and E. F. Hayes,
          "Quantum Mechanical Studies of the Vibrational Excitation of H$_2$
          by Li$^+$,"
          J. Chem. Phys. 57, 2985-93

(72.57)   C. F. Bender, H. F. Schaefer, and P. A. Kollman,
          "The Long Range Intermolecular Potential of H$_2$-H$_2$,"
          Molec. Phys. 24, 235-9

(72.58)   E. A. G. Armour,
          "An Application of a Transcorrelated Procedure to the Calculation
          of the Energy of the Hydrogen Molecule,"
          Molec. Phys. 24, 181-203

(72.59)   N. K. Berezhetskaya, N. B. Delone, and T. T. Urazbaev,
          "Temperature Dependence of Multiphoton Ionization of the Hydrogen
          Molecule,"
          JETP Letters 15, 478-80

(72.60)   D. L. Akins,
          "Collisional Broadening Cross-Sections for HD*(B$^1\Sigma_u^+$)-HD(X$^1\Sigma_g^+$)
          and HD*(B$^1\Sigma_u^+$)-He From Peak-Absorbancy Measurements,"
          J. Quant. Spectrosc. Radiative Trans. 12, 1357-65

(72.61)   M. Misakian and J. C. Zorn,
          "Dissociative Excitation of Molecular Hydrogen by Electron Impact,"
          Phys. Rev. A 6, 2180-96

(72.62)   M. Lu Van, G. Mainfray, C. Manus, and I. Tugov,
          "Multiphoton Ionization and Dissociation of Molecular Hydrogen at
          1.06 μm,"
          Phys. Rev. Letters 29, 1134-7

(72.63)   D. G. Truhlar,
          "Vibrational Matrix Elements of the Quadrupole Moment Functions
          of H$_2$, N$_2$ and CO,"
          Int. J. Quant. Chem. 6, 975-88

(72.64)   M. G. Dondi, U. Valbusa, and G. Scoles,
          "Energy Dependence of the Differential Collision Cross Section
          for Hydrogen at Thermal Energies,"
          Chem. Phys. Letters 17, 137-141

(72.65)   W. H. Smith and R. Chevalier,
          "Radiative-Lifetime Studies of the Emission Continua of the
          Hydrogen and Deuterium Molecules,"
          Astrophys. J. 177, 835-9

H_2

(72.66) J. D. Doll and W. H. Miller,
"Classical S-Matrix for Vibrational Excitation of H_2 by Collision With He in Three Dimensions,"
J. Chem. Phys. 57, 5019-26

(72.67) G. Sperber,
"Analysis of Reduced Density Matrices in the Co-ordinate Representation. III. Electron Density and Correlation in the Ground States of H_2 and the H_6 Ring System Within Some Approximations of the Simple LCAO Type,"
Int. J. Quant. Chem. 6, 881-98

(72.68) A. K. Chandra and R. Sundar,
"Vibrational Force Constant and Electron Relaxation in H_2 and Li_2,"
Chem. Phys. Letters 14, 577-82

(73.69) P. J. Hicks, J. Comer, and F. H. Read,
"Autoionizing Transitions in N_2 and H_2 Produced by Electron Impact,"
J. Phys. B: Atom. Molec. Phys. 6, L65-L69

(73.70) L. Y. Nelson, G. J. Mullaney, and S. R. Byron,
"Superfluorescence in N_2 and H_2 Electron-Beam-Stabilized Discharge,"
Appl. Phys. Letters 22, 79-80

(73.71) R. V. Reid, Jr. and M. L. Vaida,
"Electric Field Gradient and Magnetic Spin-Spin Interactions in Isotopes of the Hydrogen Molecule,"
Phys. Rev. A 7, 1841-9

(73.72) J. S. Margolis,
"Measurement of Some 1-0 H_2 Quadrupole Transition Strengths,"
J. Molec. Spectrosc. 48, 409-10

(73.73) S. P. Reddy and K. S. Chang,
"Collision-Induced Fundamental Band of H_2 in H_2-He and H_2-Ne Mixtures at Different Temperatures,"
J. Molec. Spectrosc. 47, 22-38

(73.74) T. A. Miller and R. S. Freund,
"Fine Structure and Hyperfine Structure of Ortho-H_2 d(3p) $^3\Pi_u$ (v = 0-3) via Microwave Optical Magnetic Resonance Induced by Electrons,"
J. Chem. Phys. 58, 2345-57

(73.75)   R. S. Freund and T. A. Miller,
"Fine Structure of Para-H_2 d(3p) $^3\Pi_u$ (v = 0-3) via Microwave
Optical Magnetic Resonance Induced by Electrons,"
J. Chem. Phys. 58, 3565-73

(73.76)   T. J. Venanzi and B. Kirtman,
"Calculation of the Dynamic Polarizability of H_2 by the Distinguish-
able Electron Method,"
J. Chem. Phys. 58, 3953-8

(73.77)   L. Bouscasse, R. Phan-Tan-Luu, E. J. Vincent, and J. Metzger,
"Technique D'Optimisation de L'Energie SCF en Fonction des
Exposants de Slater,"
Int. J. Quant. Chem. 7, 357-64

(73.78)   M. Lombardi,
"Fine and Hyperfine Structure of the 2p and 3p $^3\Pi_u$ States of H_2,"
J. Chem. Phys. 58, 797-802

(73.79)   F. R. McCourt and H. Moraal,
"Reorientation Cross Sections for Molecular Hydrogen,"
J. Chem. Phys. 59, 2370-5

(73.80)   R. Shafer and R. G. Gordon,
"Quantum Scattering Theory of Rotational Relaxation and Spectral
Line Shapes in H_2-He Gas Mixtures,"
J. Chem. Phys. 58, 5422-43

(73.81)   A. L. Ford and J. C. Browne,
"Direct-Resolvent-Operator Computations on the Hydrogen-Molecule
Dynamic Polarizability, Rayleigh, and Raman Scattering,"
Phys. Rev. A 7, 418-26

(73.82)   J. W. Liu,
"Total Inelastic Cross Section for Collisions of H_2 With Fast
Charged Particles,"
Phys. Rev. A 7, 103-9

(73.83)   M. LuVan, G. Mainfray, C. Manus, and I. Tugov,
"Multiphoton Ionization of Atomic and Molecular Hydrogen at 0.53 μ,"
Phys. Rev. A 7, 91-8

(73.84)   E. A. G. Armour,
"A Calculation of the Energy and Properties of the Hydrogen
Molecule Using the Method of Moments,"
Molec. Phys. 25, 993-1010

(73. 85)   R. Browning and J. Fryar,
"Dissociative Photoionization of H_2 and D_2 Through the $1s\sigma_g$ Ionic State,"
J. Phys. B: Atom. Molec. Phys. <u>6</u>, 364-71

(73. 86)   L. Julien, M. Glass-Maujean, and J. P. Descoubes,
"On the Dissociation of the H_2 Molecule Following Electron Impact and Leading to the Obtention of Atomic Hydrogen in the n = 3 Levels,"
J. Phys. B: Atom. Molec. Phys. <u>6</u>, L196-L200

(73. 87)   J. W. Liu and V. H. Smith, Jr.,
"The Differential Cross Section for Elastic Scattering of Electrons by H_2 in the First Born Approximation,"
J. Phys. B: Atom. Molec. Phys. <u>6</u>, L275-L279

(73. 88)   P. Baltayan,
"Non-Zero Magnetic Field Level-Crossing of $4p\ ^3\Pi_u$ Hydrogen Molecular Level Excited by Electron Impact,"
Phys. Letters <u>42A</u>, 435-6

(73. 89)   A. Crowe and J. W. McConkey,
"Experimental Evidence for New Dissociation Channels in Electron-Impact Ionization of H_2,"
Phys. Rev. Letters <u>31</u>, 192-6

(73. 90)   A. Jones and J. L. J. Rosenfeld,
"Monte-Carlo Simulation of Hydrogen-Atom Recombination,"
Proc. Roy. Soc. A <u>333</u>, 419-34

(73. 91)   R. W. Dreyfus and R. T. Hodgson,
"Relativistic Electron-Beam Pumped UV Gas Lasers,"
J. Vac. Sci. Tech. <u>10</u>, 1033-6

(73. 92)   P. D. Feldman and W. G. Fastie,
"Fluorescence of Molecular Hydrogen Excited by Solar Extreme-Ultraviolet Radiation,"
Astrophys. J. <u>185</u>, L101-4

(73. 93)   W. C. Stwalley,
"Potential Energy Curve of the $B\ ^1\Sigma_u^+$ State of H_2,"
J. Chem. Phys. <u>58</u>, 536-40

(73. 94)   A. Crowe and J. W. McConkey,
"Dissociative Ionization by Electron Impact. I. Protons From H_2,"
J. Phys. B: Atom. Molec. Phys. <u>6</u>, 2088-2107

(73. 95)   V. S. Antonov, I. N. Knyazev, V. S. Letokhov, and V. G. Movshev,
"Hydrogen Laser in Vacuum Ultraviolet at Atmospheric Pressure,"
JETP Letters <u>17</u>, 545-8

(73.96) R. David, M. Faubel, and J. P. Toennies,
"Measurements of Differential Cross Sections for Vibrational
Quantum Transitions in Scattering of Li^+ on H_2,"
Chem. Phys. Letters 18, 87-90

(73.97) M. M. Audibert, C. Joffrin, and J. Ducuing,
"Vibrational Relaxation in Hydrogen – Rare-Gases Mixtures,"
Chem. Phys. Letters 19, 26-8

(73.98) A. L. Ford and J. C. Browne,
"Elastic Scattering of Electrons by H_2 in the Born Approximation,"
Chem. Phys. Letters 20, 284-90

(73.99) J. Schaefer and W. A. Lester, Jr.,
"Effect of Rotation on Vibrational Excitation of H_2 by Li^+ Impact,"
Chem. Phys. Letters 20, 575-80

(73.100) H. P. Kelly,
"The Photoionization Cross Section for H_2 From Threshold to
30 eV,"
Chem. Phys. Letters 20, 547-50

(73.101) P. Jeffers, D. Hilden, and S. H. Bauer,
"v-v Pumping in H_2: Estimate of Induced Dissociation at $300°K$,"
Chem. Phys. Letters 20, 525-7

(73.102) R. S. Freund, T. A. Miller, and B. R. Zegarski,
"Fine Structure of Para-H_2, k(4p) $^3\Pi_u$,"
Chem. Phys. Letters 23, 120-2

(73.103) G. Karl, E. Obryk, and J. D. Poll,
"The Hexadecapole Moment of the Hydrogen Molecule,"
Can. J. Phys. 51, 2216-7

(73.104) E. A. Colbourn,
"Singlet-Triplet Differences in the Lowest Π_u States of H_2,"
J. Phys. B: Atom. Molec. Phys. 6, 2618-24

(73.105) M. Koumanova and T. Rebane,
"Magnetic Properties Calculating of the Hydrogen Molecule by
Variation Method of Vector Potential and by Variation Method of
Induced Current,"
Izvest. Fiz. Inst. Aneb. 22, 27-34

(73.106) T. A. Miller and R. S. Freund,
"Fine Structure and Perturbations in the d(3p) $^3\Pi_u$ States of H_2
and D_2,"
J. Chem. Phys. 59, 4093-4104

H_2

(73. 107)   D. A. Jennings, W. Braun, and H. P. Broida,
"Vibrational Relaxation of Hydrogen by Direct Detection of
Electronic and Vibrational Energy Transfer With Alkali Metals,"
J. Chem. Phys. 59, 4305-8

(73. 108)   S. V. Rode and S. S. Vasil'ev,
"Molecular Hydrogen Excitation Processes in a Glow Discharge
at Moderate Pressures,"
Russ. J. Phys. Chem. 47, 666-9

(73. 109)   J. P. Colpa and R. E. Brown,
"Hund's Rules and the Interpretation and Common Misinterpreta-
tion of Energy Differences,"
Molec. Phys. 26, 1453-63

(73. 110)   H. Schmoranzer and J. Geiger,
"Light Emission of Electron Impact Excited Hydrogen Molecules
and the Dependence of Electronic Transition Moment on Internuclear Distance,"
J. Chem. Phys. 59, 6153-6

(73. 111)   A. Cohn and M. Marcucci,
"Stimulated Emission Cross Section of H_2 Dissociation Transition,"
J. Appl. Phys. 44, 1930-1

(73. 112)   D. G. Truhlar,
"Ab Initio Hartree-Fock Calculations of Electronic Wave Functions
for the $c^3\Pi_u$ State of H_2,"
Int. J. Quant. Chem. 7, 1175-82

(74. 113)   G. C. Lie and E. Clementi,
"Study of the Electronic Structure of Molecules. XXII. Correla-
tion Energy Corrections as a Function of the Hartree-Fock Type
Density and Its Application to the Homonuclear Diatomic Mole-
cules of the Second Row Atoms,"
J. Chem. Phys. 60, 1288-96

(74. 114)   M. Gallardo, C. A. Massone, and M. Garavaglia,
"On the Inversion Mechanism of the Molecular Hydrogen Vacuum
Ultraviolet Laser,"
IEEE J. Quant. Elect. QE-10, 525-6

(74. 115)   R. W. Dreyfus and R. T. Hodgson,
"Molecular-Hydrogen Laser: 1098-1613 Å,"
Phys. Rev. A 9, 2635-48

(74. 116)   A. U. Hazi,
"Distribution of Final States Resulting From the Autoionization of the $^1\Sigma_g$ (2pσ_u^2) States of H_2 and D_2,"
J. Chem. Phys. 60, 4358-61

(74. 117)   B. R. Lewis,
"Experimentally-Determined Oscillator Strengths for Molecular Hydrogen – III.   Rotational Variation of Band Strength,"
J. Quant. Spectrosc. Radiative Trans. 14, 723-9

(74. 118)   A. L. Ford,
"Nonadiabatic Effects in the Ground State of the Hydrogen Molecule,"
J. Molec. Spectrosc. 52, 358-62

(74. 119)   M. Lombardi,
"Fine and Hyperfine Structure of the 4p and 5p $^3\Pi_u$ States of H_2,"
J. Chem. Phys. 60, 4094-5

(74. 120)   E. H. Fink, P. Hafner, and K. H. Becker,
"Energy Exchange Between HD(B $^1\Sigma_u^+$, v', J') and Ground State H_2 and D_2 by Interaction of Electronic Transition and Dipole Moments,"
Z. Naturforsch. 29a, 194-9

(74. 121)   A. N. Jette,
"Fine-Structure of the Metastable, c$^3\Pi_u$(1s, 2p), State of Molecular Hydrogen,"
Chem. Phys. Letters 25, 590-2

(74. 122)   S. B. Elston, S. A. Lawton, and F. M. J. Pichanick,
"High-Resolution Studies of Total Electron Excitation Functions Near Threshold for the States B$^1\Sigma_u^+$ and c$^3\Pi_u$ of Molecular Hydrogen,"
Phys. Rev. A 10, 225-30

(74. 123)   T. A. Miller, R. S. Freund, and B. R. Zegarski,
"Fine and Hyperfine Structure of Ortho-H_2, k(4p) $^3\Pi_u$,"
J. Chem. Phys. 60, 3195-3202

(74. 124)   A. N. Jette,
"Spin-Other-Orbit and Spin-Spin Interactions in the Metastable, c$^3\Pi_u$(1s, 2p) State of H_2,"
J. Chem. Phys. 61, 816-9

(74. 125)   C. S. Lin,
"Theoretical Analysis of the Vibrational Structure of the Electronic Transitions Involving a State With Double Minimum: E, F $^1\Sigma_g^+$ of H_2,"
J. Chem. Phys. 60, 4660-4

H$_2$

(74.126) W. Fabian and B. R. Lewis,
"Experimentally Determined Oscillator Strengths for Molecular
Hydrogen – I. The Lyman and Werner Bands Above 900 Å,"
J. Quant. Spectrosc. Radiative Trans. 14, 523-35

(74.127) B. R. Lewis,
"Experimentally-Determined Oscillator Strengths for Molecular
Hydrogen – II. The Lyman and Werner Bands Below 900 Å, the
B'-X and the D-X Bands,"
J. Quant. Spectrosc. Radiative Trans. 14, 537-46

(74.128) P. W. Gibbs, C. G. Gray, J. L. Hunt, S. P. Reddy, R. H. Tipping,
and K. S. Chang,
"New Rotational Transition in the Hydrogen Molecule,"
Phys. Rev. Letters 33, 256-8

(74.129) B. Huron, J. P. Malrieu, and P. Rancurel,
"Perturbation Calculation of Transition Moments. Application to
H$_2$ and MgO,"
Chem. Phys. 3, 277-83

(74.130) C. E. Johnson,
"Quenching of the c$^3\Pi_u$ Metastable State of H$_2$ and D$_2$ by an
Electric Field,"
Phys. Rev. A 9, 576-7

(74.131) C. Bottcher and K. Docken,
"Autoionizing States of the Hydrogen Molecule,"
J. Phys. B: Atom. Molec. Phys. 7, L5-L8

(74.132) V. Dose and W. Hett,
"Destruction of Metastable Hydrogen in Collisions With Helium,"
J. Phys. B: Atom. Molec. Phys. 7, L79-L81

(74.133) C. T. Hsieh, N. D. Foltz, and C. W. Cho,
"Production of the Stimulated Raman Lines in H$_2$ Gas With a
Focused Laser Beam,"
J. Opt. Soc. Am. 64, 202-5

(74.134) M. Jackson and R. P. McEachran,
"The Frozen-Core Approximation for Diatomic Molecules
II. The b$^3\Sigma_u^+$, a$^3\Sigma_g^+$ and E, F $^1\Sigma_g^+$ States of H$_2$,"
J. Phys. B: Atom. Molec. Phys. 7, 1782-9

(74.135) K. J. Miller and A. E. S. Green,
"Energy Levels and Potential Energy Curves for H$_2$, N$_2$, and O$_2$
With an Independent Particle Model,"
J. Chem. Phys. 60, 2617-26

(74. 136)   J. J. Ewing,
            "Calculation of Spin Orbit Relaxation Rates by Near Resonant
            E → V Energy Transfer,"
            Chem. Phys. Letters 29, 50-5

He$_2$

Methods of Production and Experimental Technique

Absorption in He discharges.

Emission – A condensed discharge in He at pressures of 20-50 Torr produces the continuum systems.

Radio frequency, microwave, and hollow cathode discharges.

SINGLET SYSTEMS BAND SYSTEMS

System	Transition	Sources	Electronic Configuration [a]	Degrading	Band Head, $v_{0,0}$	Remarks	Bibliography [b]
I	$A^1\Sigma_u^+ \rightarrow X^1\Sigma_g^+$	Emission	Hopfield Continuum		Max. intensity ~ 810Å	Continuum	(66.48, 65.41, 65.37, 62.29, 58.24)
II	$D^1\Sigma_u^+ \rightarrow X^1\Sigma_g^+$	Emission			Max. intensity ~ 680Å	Continuum	(65.41, 58.24)
III	$A^1\Sigma_u^+ \leftarrow X^1\Sigma_g^+$	Absorption	600Å Bands		Max. intensity ~ 600-625Å	Continuum	(70.68, 69.67, 68.63, 66.46, 63.32)
IV	$C^1\Sigma_g^+ \rightarrow A^1\Sigma_u^+$	Emission	$3p\sigma \rightarrow 2s\sigma$	R	10945.50		(65.36)
V	$D^1\Sigma_u^+ \rightarrow B^1\Pi_g$	Emission	$3s\sigma \rightarrow 2p\pi$	R, V	15161.81		(65.38, 28.6)
VI	$E^1\Pi_g \rightarrow A^1\Sigma_u^+$	Emission	$3p\pi \rightarrow 2s\sigma$	R	19476.4		(71.76, 29.18, 29.17, 29.11)
VII	$F^1\Sigma_u^+ \rightarrow B^1\Pi_g$	Emission	$3d\sigma \rightarrow 2p\pi$	R	15837	(c)	(65.38, 28.6, 28.5)
VIII	$F^1\Pi_u \rightarrow B^1\Pi_g$	Emission	$3d\pi \rightarrow 2d\pi$	R, V	16360	(c)	(65.38, 29.18, 29.17, 29.14)
IX	$F^1\Delta_u \rightarrow B^1\Pi_g$	Emission	$3d\delta \rightarrow 2p\pi$	V	16360	(c)	(65.38, 29.18, 29.17, 29.14)
X	$H^1\Sigma_u^+ \rightarrow C^1\Sigma_g^+$	Emission	$4s\sigma \rightarrow 3p\sigma$	V	13719.6	Perturbed N > 6	(73.84)

System	Transition	Sources	Electronic[a] Configuration	Degrading	Band Head, $\nu_{0,0}$	Remarks	Bibliography[b]
XI	$H^1\Sigma_u^+ \rightarrow B^1\Pi_g$	Emission	$4s\sigma \rightarrow 2p\pi$	R	21163	Perturbed $N > 6$	(73.84, 28.7)
XII	$I^1\Pi_g \rightarrow A^1\Sigma_u^+$	Emission	$4p\pi \rightarrow 2s\sigma$	R	24979.6		(28.6, 27.3, 25.2)
XIII	$J^1\Sigma_u^+ \rightarrow B^1\Pi_g$	Emission	$4d\sigma \rightarrow 2p\pi$		21416.11	(c)	(73.84)
XIV	$J^1\Pi_u \rightarrow B^1\Pi_u$	Emission	$4d\pi \rightarrow 2p\pi$		21523.76	(c)	(73.84)
XV	$J^1\Delta_u \rightarrow B^1\Pi_u$	Emission	$4d\delta \rightarrow 2p\pi$		21661.60	(c)	(73.84, 29.9)
XVI	$L^1\Pi_g \rightarrow A^1\Sigma_u^+$	Emission	$5p\pi \rightarrow 2s\sigma$	R	27507		(29.18, 29.11, 25.2)
XVII	$M^1\Sigma_u^+ \rightarrow B^1\Pi_g$	Emission	$5d\sigma \rightarrow 2p\pi$		(23960)	(c, d)	(29.18, 29.14)
XVIII	$M^1\Pi_u \rightarrow B^1\Pi_g$	Emission	$5d\pi \rightarrow 2p\pi$		(24000)	(c, d)	(29.18, 29.14)
XIX	$M^1\Delta_u \rightarrow B^1\Pi_g$	Emission	$5d\delta \rightarrow 2p\pi$		(24050)	(c,d)	(29.18, 29.14)
XX	$P^1\Pi_g \rightarrow A^1\Sigma_u^+$	Emission	$6p\pi \rightarrow 2s\sigma$	R	28873		(29.18, 25.2)
XXI	$R^1\Pi_g \rightarrow A^1\Sigma_u^+$	Emission	$7p\pi \rightarrow 2s\sigma$	R	29696		(29.10)

System	Transition	Sources	Electronic[a] Configuration	Degrading	Band Head, $\nu_{0,0}$	Remarks	Bibliography[b]
XXII	$S^1\Pi_g \rightarrow A^1\Sigma_u^+$	Emission	$8p\pi \rightarrow 2s\sigma$	R	30228		(29.10)

He$_2$

System	Transition	Sources	Electronic[a] Configuration	Degrading	Band Head, $\nu_{0,0}$	Remarks	Bibliography[b]
I	$b\,^3\Pi_g \to a\,^3\Sigma_u^+$	Emission	$2p\pi \to 2s\sigma$	R	4768.2		(65.40, 56.23)
II	$c\,^3\Sigma_g^+ \to a\,^3\Sigma_u^+$	Emission	$3p\sigma \to 2s\sigma$	R	10889.48		(65.36, 53.22, 32.20)
III	$d\,^3\Sigma_u^+ \to c\,^3\Sigma_g^+$	Emission	$3s\sigma \to 3p\sigma$	V	9502.66		(65.39, 32.20)
IV	$d\,^3\Sigma_u^+ \to b\,^3\Pi_g$	Emission	$3s\sigma \to 2p\pi$	R, V	15623.1		(65.39, 29.15, 29.8, 28.6, 28.4, 22.1)
V	$e\,^3\Pi_g \to a\,^3\Sigma_u^+$	Emission	$3p\pi \to 2s\sigma$	R	21507.24		(71.76, 50.21, 28.6, 27.3, 25.2)
VI	$f\,^3\Sigma_u^+ \to c\,^3\Sigma_g^+$	Emission	$3d\sigma \to 3p\sigma$	R	10658	(c)	(66.43, 32.20)
VII	$f\,^3\Pi_u \to c\,^3\Sigma_g^+$	Emission	$3d\pi \to 3p\sigma$	V	10865	(c)	(66.43, 32.20)
VIII	$f\,^3\Delta_u \to c\,^3\Sigma_g^+$	Emission	$3d\delta \to 3p\sigma$	V	11315	(c)	(66.43)
IX	$f\,^3\Sigma_u^+ \to b\,^3\Pi_g$	Emission	$3d\sigma \to 2p\pi$	R	16779	(c)	(65.39, 29.17, 29.14, 28.6, 28.5)

System	Transition	Sources	Electronic[a] Configuration	Degrading	Band Head, $\nu_{0,0}$	Remarks	Bibliography[b]
X	$f\,^3\Pi_u \to b\,^3\Pi_g$	Emission	$3d\pi \to 2p\pi$		16986	(c)	(65.39, 29.17, 29.14, 28.6, 28.5)
XI	$f\,^3\Delta_u \to b\,^3\Pi_g$	Emission	$3d\delta \to 2p\pi$	R, V	17437	(c)	(65.39, 28.6, 28.5, 22.1)
XII	$h\,^3\Sigma_u^+ \to c\,^3\Sigma_g^+$	Emission	$4s\sigma \to 3p\sigma$	V	15870.7	Perturbed for large N	(29.9)
XIII	$h\,^3\Sigma_u^+ \to b\,^3\Pi_g$	Emission	$4s\sigma \to 2p\pi$	R	21992.2	Perturbed for large N	(73.84, 29.15, 28.6, 28.5, 22.1)
XIV	$i\,^3\Pi_g \to a\,^3\Sigma_u^+$	Emission	$4p\pi \to 2s\sigma$	R	27193.0		(29.18, 28.6, 27.3, 25.2)
XV	$j\,^3\Sigma_u^+ \to b\,^3\Pi_g$	Emission	$4d\sigma \to 2p\pi$	R	22434	(c)	(73.84, 29.18, 29.16, 29.14, 28.6, 28.5)
XVI	$j\,^3\Pi_u \to b\,^3\Pi_g$	Emission	$4d\pi \to 2p\pi$	R, V	22524	(c)	(73.84, 29.18, 29.16, 29.14, 28.6, 28.5)
XVII	$j\,^3\Delta_u \to b\,^3\Pi_g$	Emission	$4d\delta \to 2p\pi$	V	22702	(c)	(73.84, 29.18, 29.16, 29.14, 28.6, 28.5)
XVIII	$j\,^3\Sigma_u^+ \to c\,^3\Sigma_g^+$	Emission	$4d\sigma \to 3p\sigma$	R	(16313)	(c)	(29.15, 29.9)

TRIPLET SYSTEMS   BAND SYSTEMS

System	Transition	Sources	Electronic Configuration[a]	Degrading	Band Head, $\nu_{0,0}$	Remarks	Bibliography[b]
XIX	$j\,^3\Pi_u \to c\,^3\Sigma_g^+$	Emission	$4d\pi \to 3p\sigma$	R	16402	(c)	(29.15, 29.9)
XX	$j\,^3\Delta_u \to c\,^3\Sigma_g^+$	Emission	$4d\delta \to 3p\sigma$	V	16581	(c)	(29.15)
XXI	$k\,^3\Sigma_u^+ \to b\,^3\Pi_g$	Emission	$5s\sigma \to 2p\pi$	R	24804		(29.16, 29.15, 27.3)
XXII	$k\,^3\Sigma_u^+ \to c\,^3\Sigma_g^+$	Emission	$5s\sigma \to 3p\sigma$	V	18683.5		(29.15, 29.11)
XXIII	$l\,^3\Pi_g \to a\,^3\Sigma_u^+$	Emission	$5p\pi \to 2s\sigma$	R	29785.3		(68.62, 29.18, 29.11, 29.10, 25.2)
XXIV	$l\,^3\Pi_g \to d\,^3\Sigma_u^+$	Emission	$5p\pi \to 3s\sigma$	R	9393.9		(73.84)
XXV	$m\,^3\Sigma_u^+ \to b\,^3\Pi_g$	Emission	$5d\sigma \to 2p\pi$	R	25019	(c)	(29.15, 29.9)
XXVI	$m\,^3\Pi_u \to b\,^3\Pi_g$	Emission	$5d\pi \to 2p\pi$	R	(25070)	(c, d)	(29.15, 29.9)
XXVII	$m\,^3\Delta_u \to b\,^3\Pi_g$	Emission	$5d\delta \to 2p\pi$	V	25152	(c)	(29.16, 29.15)
XXVIII	$m\,^3\Sigma_u^+ \to c\,^3\Sigma_g^+$	Emission	$5d\sigma \to 3p\sigma$	R	18899	(c)	(29.15)
XXIX	$m\,^3\Pi_u \to c\,^3\Sigma_g^+$	Emission	$5d\pi \to 3p\sigma$	R	(18944)	(c, d)	(29.15)

System	Transition	Sources	Electronic Configuration[a]	Degrading	Band Head, $\nu_{0,0}$	Remarks	Bibliography[b]
XXX	$m\,^3\Delta_u \to c\,^3\Sigma_g^+$	Emission	$5d\delta \to 3p\sigma$	V	(19039)	(c)	(29.15)
XXXI	$o\,^3\Sigma_u^+ \to b\,^3\Pi_g$	Emission	$6s\sigma \to 2p\pi$	R	26290.3		(29.15)
XXXII	$o\,^3\Sigma_u^+ \to c\,^3\Sigma_g^+$	Emission	$6s\sigma \to 3p\sigma$	V	20168.8		(29.15, 29.11)
XXXIII	$p\,^3\Pi_g \to a\,^3\Sigma_u^+$	Emission	$6p\pi \to 2s\sigma$	R	31179.93		(68.62, 29.18, 29.12, 27.3, 25.2)
XXXIV	$p\,^3\Pi_g \to d\,^3\Sigma_u^+$	Emission	$6p\pi \to 3s\sigma$	R	10788.6		(n.p. 95)
XXXV	$q\,^3\Sigma_u^+ \to b\,^3\Pi_g$	Emission	$6d\sigma \to 2p\pi$	R	26409	(c)	(29.16)
XXXVI	$q\,^3\Pi_u \to b\,^3\Pi_g$	Emission	$6d\pi \to 2p\pi$	R	(26466)	(c, d)	(29.14)
XXXVII	$q\,^3\Delta_u \to b\,^3\Pi_g$	Emission	$6d\delta \to 2p\pi$	V	(26483)	(c, d)	(29.15)
XXXVIII	$q\,^3\Sigma_u^+ \to c\,^3\Sigma_g^+$	Emission	$6d\sigma \to 3p\sigma$	R	20288	(c)	(29.15)
XXXIX	$q\,^3\Pi_u \to c\,^3\Sigma_g^+$	Emission	$6d\pi \to 3p\sigma$	R	(20330)	(c, d)	(29.15)
XL	$q\,^3\Delta_u \to c\,^3\Sigma_g^+$	Emission	$6d\delta \to 3p\sigma$	V	20365	(c, d)	(29.15)

System	Transition	Sources	Electronic Configuration[a]	Degrading	Band Head, $\nu_{0,0}$	Remarks	Bibliography[b]
XLI	$r\,^3\Sigma_g^+ \to a\,^3\Sigma_u^+$	Emission	$7p\sigma \to 2s\sigma$	V	31478.62		(68.62)
XLII	$r\,^3\Pi_g \to a\,^3\Sigma_u^+$	Emission	$7p\pi \to 2s\sigma$	R	32016.56	(e)	(68.62, 29.18, 29.12, 27.3, 25.2)
XLIII	$r\,^3\Pi_g \to d\,^3\Sigma_u^+$	Emission	$7p\pi \to 3s\sigma$	R	11625.3	(e)	(68.62)
XLIV	$s\,^3\Sigma_g^+ \to a\,^3\Sigma_u^+$	Emission	$8p\sigma \to 2s\sigma$	V	32211.70	(e)	(68.62)
XLV	$s\,^3\Pi_g \to a\,^3\Sigma_u^+$	Emission	$8p\pi \to 2s\sigma$	R	32556.66	(e)	(68.62, 29.12, 25.2)
XLVI	$t\,^3\Sigma_g^+ \to a\,^3\Sigma_u^+$	Emission	$9p\sigma \to 2s\sigma$	V	32693.18	(e)	(68.62)
XLVII	$t\,^3\Pi_g \to a\,^3\Sigma_u^+$	Emission	$9p\pi \to 2s\sigma$	R	32925.96	(e)	(68.62, 29.12, 25.2)
XLVIII	$u\,^3\Sigma_g^+ \to a\,^3\Sigma_u^+$	Emission	$10p\sigma \to 2s\sigma$	V	33026.5	(e)	(68.62)
XLIX	$u\,^3\Pi_g \to a\,^3\Sigma_u^+$	Emission	$10p\pi \to 2s\sigma$	R	33189.16	(e)	(68.62, 29.12, 25.2)
L	$v\,^3\Sigma_g^+ \to a\,^3\Sigma_u^+$	Emission	$11p\sigma \to 2s\sigma$	V	33266.5	(e)	(68.62)
LI	$v\,^3\Pi_g \to a\,^3\Sigma_u^+$	Emission	$11p\pi \to 2s\sigma$	R	33383.51	(e)	(68.62)

System	Transition	Sources	Electronic Configuration[a]	Degrading	Band Head, $\nu_{0,0}$	Remarks	Bibliography[b]
LII	$w\,^3\Sigma_g^+ \to a\,^3\Sigma_u^+$	Emission	$12p\sigma \to 2s\sigma$	V	33445.1	(e)	(68.62)
LIII	$w\,^3\Pi_g \to a\,^3\Sigma_u^+$	Emission	$12p\pi \to 2s\sigma$	R	33531.04	(e)	(68.62)
LIV	$y\,^3\Pi_g \to a\,^3\Sigma_u^+$	Emission	$13p\pi \to 2s\sigma$	R	33645.7	(e)	(68.62)
LV	$z\,^3\Pi_g \to a\,^3\Sigma_u^+$	Emission	$14p\pi \to 2s\sigma$	R	33736.65	(e)	(68.62)
LVI	$^3\Pi_g \to a\,^3\Sigma_u^+$	Emission	$15p\pi \to 2s\sigma$	R	33809.8	(e)	(68.62)
LVII	$^3\Pi_g \to a\,^3\Sigma_u^+$	Emission	$16p\pi \to 2s\sigma$	R	33869.8	(e)	(68.62)
LVIII	$^3\Pi_g \to a\,^3\Sigma_u^+$	Emission	$17p\pi \to 2s\sigma$	R	33919.2	(e)	(68.62)

[a] These are the United Atom Orbital Rydberg symbols. Except for the ground state $X\,^1\Sigma_g^+$ ($\sigma_g 1s^2 \sigma_u 1s^2$), the inner shell configurations ($\sigma_g 1s^2 \sigma_u 1s\, nl\lambda$) will be designated by their $nl\lambda$ value since all of these states converge to the He$_2^+$ $X\,^2\Sigma_u^+$ state.

[b] For publications before 1925, see (29.13).

[c] States associated with the $nd\lambda$ configuration are strongly affected by l-decoupling (large Δ splitting), hence the band structures do not follow the usual intensity scheme of Hund's case b. See (66.43) for discussion.

[d] Data is fragmentary and may be in error.   [e] l-decoupling affects $np\pi\ \Pi_g^+$ and $np\sigma\ \Sigma_g^+$ levels. As a result, branches using these terms become diffuse for $N \geq 7$ (68.62).

SINGLET SYSTEMS

IV. $C\,^1\Sigma_g^+ \rightarrow A\,^1\Sigma_g^+$ System

Band origins, λ (65.36):

v', v''	0	1	2	3
0	9136.2	10923.4		
1	7989.1	9322.3	11102.5	
2	7139.5	8186.0	9527.4	11302.0
3		7343.3	8404.8	9756.1

V. $D\,^1\Sigma_u^+ \rightarrow B\,^1\Pi_g$ System

Band origins, λ (65.38, 28.6):

(v', v'')	(0, 0)	(1, 1)	(2, 2)	(1, 2)	(0, 1)
λ	6595.4	6605.0	6615.1	7400.8	7426.7

VI $E\,^1\Pi_g \rightarrow A\,^1\Sigma_u^+$ System

Band origins, λ (71.76):

(v', v'')	(1, 0)	(2, 1)	(0, 0)	(1, 1)	(2, 2)	(0, 1)
λ	4733.1	4780.1	5134.3	5171.4	5208.3	5654.2

VII. $F\,^1\Sigma_u^+ \rightarrow B\,^1\Pi_u$ System

Band origins, λ (65.38, 29.18):

(v', v'')	(0, 0)	(1, 1)	(0, 1)	(1, 2)
λ	6314.3	6367.8	7072.1	7104.8

VIII. $F\,^1\Pi_u \rightarrow B\,^1\Pi_u$ System

Band origins, λ (65.38, 29.18):

(v', v'')	(0, 0)	(1, 1)	(2, 2)	(0, 1)	(1, 2)	(2, 3)
λ	6246.9	6288.5	6335.5	6987.6	7005.7	7029.9

He$_2$

IX. $\underline{F^1\Delta_u \rightarrow B^1\Pi_g \text{ System}}$

Band origins, λ (65.38, 29.18):

(v', v'')	$(0,0)$	$(1,1)$	$(2,2)$	$(1,2)$	$(0,1)$
λ	6112.5	6135.0	6158.4	6815.7	6819.4

TRIPLET SYSTEMS

II. $\underline{c^3\Sigma_g^+ \rightarrow a^3\Sigma_u^+ \text{ System}}$

Band origins, λ (65.36, 53.22):

v', v''	0	1	2	3
0	9183.6	10920.6		
1	8084.1	9401.1	11132.1	
2	7277.5	8326.4	9657.2	11389.7
3		7543.2	8619.2	9973.1
4			7875.3	8990.4

III. $\underline{d^3\Sigma_u^+ \rightarrow c^3\Sigma_g^+ \text{ System}}$

Band origins, λ (65.39):

(v', v'')	$(3,2)$	$(2,1)$	$(1,0)$	$(3,3)$	$(2,2)$	$(1,1)$
λ	8775.0	8881.0	8962.2	9854.2	10113.3	10332.7

IV. $\underline{d^3\Sigma_u^+ \rightarrow b^3\Pi_g \text{ System}}$

Band origins, λ (65.39):

(v', v'')	$(0,0)$	$(1,1)$	$(2,2)$	$(3,3)$	$(0,1)$	$(1,2)$
λ	6400.8	6418.1	6437.5	6458.3	7181.3	7167.4

V. $\underline{e^3\Pi_g \rightarrow a^3\Sigma_u^+ \text{ System}}$

Band origins, λ (71.76, 50.21):

(v', v'')	$(0,0)$	$(1,1)$	$(2,2)$	$(0,1)$	$(1,2)$
λ	4649.6	4667.2	4683.2	5056.9	5057.7

He_2

VI. $f\,^3\Sigma_u^+ \to c\,^3\Sigma_g^+$ System

Band origins, λ (66.43, 32.20):

(v', v'')	(1, 0)	(2, 1)	(2, 2)	(1, 1)	(0, 0)	(0, 1)	(1, 2)
λ	8192.0	8206.8	9248.1	9322.3	9382.6	10895.6	10689.5

VII. $f\,^3\Pi_u \to c\,^3\Sigma_g^+$ System

Band origins, λ (66.43, 32.20):

(v', v'')	(3, 2)	(2, 1)	(1, 0)	(2, 2)	(1, 1)	(0, 0)	(1, 2)
λ	8034.7	8039.2	8040.5	9035.1	9126.6	9203.9	10433.0

VIII. $f\,^3\Delta_u \to c\,^3\Sigma_g^+$ System

Band origins, λ (66.43):

(v', v'')	(2, 1)	(1, 0)	(2, 2)	(1, 1)	(0, 0)
λ	7669.3	7720.8	8571.2	8716.9	8837.8

IX. $f\,^3\Sigma_u^+ \to b\,^3\Pi_g$ System

Band origins, λ (65.39, 29.14):

(v', v'')	(0, 0)	(1, 1)	(2, 2)	(0, 1)	(1, 2)	(2, 3)
λ	5959.8	6014.0	6076.1	6630.9	6666.7	6711.4

X. $f\,^3\Pi_u \to b\,^3\Pi_g$ System

Band origins, λ (65.39, 29.14):

(v', v'')	(0, 0)	(1, 1)	(2, 2)	(0, 1)	(1, 2)	(2, 3)
λ	5887.2	5931.5	5983.0	6541.5	6564.3	6598.0

233

He$_2$

XI.   f$^3\Delta_u \rightarrow$ b$^3\Pi_g$ System

Band origins, λ (65.39, 29.14):

(v', v'')	(0, 0)	(1, 1)	(2, 2)
λ	5734.9	5755.4	5775.7

XIV.   i$^3\Pi_g \rightarrow$ a$^3\Sigma_u^+$ System

Band origins, λ (29.18, 28.6):

(v', v'')	(1, 0)	(0, 0)	(1, 1)	(0, 1)	(1, 2)
λ	3468.5	3677.4	3690.2	3927.6	3930.2

XXIII.   l$^3\Pi_g \rightarrow$ a$^3\Sigma_u^+$ System

Band origins, λ (68.62, 29.10):

(v', v'')	(1, 0)	(0, 0)	(1, 1)	(2, 2)	(0, 1)
λ	3182.8	3357.4	3368.5	3378.8	3564.7

XXXIII.   p$^3\Pi_g \rightarrow$ a$^3\Sigma_u^+$ System

Band origins, λ (68.62, 29.12):

(v', v'')	(1, 0)	(2, 1)	(0, 0)	(1, 1)	(2, 2)
λ	3047.9	3063.8	3207.2	3217.7	3227.4

XLII.   r$^3\Pi_g \rightarrow$ a$^3\Sigma_u^+$ System

Band origins, λ (68.62, 29.12):

(v', v'')	(1, 0)	(0, 0)	(1, 1)	(0, 1)
λ	2972.1	3123.3	3133.4	3302.1

234

XLV. $\quad \underline{s^3\Pi_g \rightarrow a^3\Sigma_u^+ \text{ System}}$

Band origins, λ (68.62, 29.12):

(v', v'')	(1, 0)	(0, 0)	(1, 1)	(0, 1)
λ	2925.2	3071.5	3081.3	3244.1

XLVII. $\quad \underline{t^3\Pi_g \rightarrow a^3\Sigma_u^+ \text{ System}}$

Band origins, λ (68.62, 29.12):

(v', v'')	(1, 0)	(0, 0)	(1, 1)	(0, 1)
λ	2893.9	3037.1	3046.6	3205.7

SPECTROSCOPIC CONSTANTS

SINGLET STATES

State	$T_o - A$[a]	ω_e	$x_e\omega_e$	B_e	α_e	$D_e \times 10^6$	r_e	Remarks	Configuration
S $^1\Pi_g$	30228			(7.21)	(0.22)		(1.08)		8pπ
R $^1\Pi_g$	29696			(7.22)	(0.22)		(1.07)		7pπ
P $^1\Pi_g$	28873			(7.22)	(0.22)		(1.07)		6pπ
M $^1\Delta_u$	(27551)								5dδ
M $^1\Pi_u$	(27501)								5dπ
M $^1\Sigma_u^+$	(27461)								5dσ
L $^1\Pi_g$	27507			(7.23)	(0.22)		(1.079)		5pπ
J $^1\Delta_u$	25129.4			7.097[b]					4dδ
J $^1\Pi_u$	25003.87			7.080[b]					4dπ
I $^1\Pi_g$	24979.6			(7.24)	(0.22)		(1.078)		4pπ
H $^1\Sigma_u^+$	24665.0			(7.26)	(0.23)		(1.07)		4sσ

SINGLET STATES

SPECTROSCOPIC CONSTANTS

State	$T_0 - A^{(a)}$	ω_e	$x_e\omega_e$	B_e	α_e	$D_e \times 10^6$	r_e	Remarks	Configuration
$F\,^1\Delta_u$	19862.4	1706.59	35.06	7.230	0.225		1.079		$3d\delta$
$F\,^1\Pi_u$	19509.8	1670.57	40.03	7.156	0.235		1.085		$3d\pi$
$F\,^1\Sigma_u^+$	19339.0	(1644)	(40)	7.098	0.246		1.089		$3d\sigma$
$E\,^1\Pi_g$	19476.4	1721.2	34.9	7.270	0.215		1.076	$y_e\omega_e = -0.04$ $\gamma_e = -0.0022$	$3p\pi$
$D\,^1\Sigma_u^+$	18663.3	1746.43	35.54	7.365	0.218		1.069	$\gamma_e = -0.0059$	$3s\sigma$
$C\,^1\Sigma_g^+$	10945.5	1653.43	41.04	7.052	0.215		1.093	$y_e\omega_e = 0.355$ $z_e\omega_e = -0.131$	$3p\sigma$
$B\,^1\Pi_g$	3501.5	1765.76	34.39	7.403	0.216		1.068	$y_e\omega_e = -0.0267$ $\gamma_e = -0.0015$	$2p\pi$
$A\,^1\Sigma_u^+$	0	1861.27	35.0	7.787	0.228		1.040	$y_e\omega_e = -0.105$	$2s\sigma$
$X\,^1\Sigma_g^+$	$-A^{(c)}$								

SPECTROSCOPIC CONSTANTS

TRIPLET STATES

State	$T_o - a$ [a]	ω_e	$x_e\omega_e$	B_e	α_e	$D_e \times 10^6$	r_e	Remarks	Configuration
$^3\Pi_g$	33919.2	1698	(35)	(7.211)	(0.224)		(1.080)		17pπ
$^3\Pi_g$	33869.8	1698	(35)	(7.211)	(0.224)		(1.080)		16pπ
$^3\Pi_g$	33809.8	1698	(35)	(7.211)	(0.224)		(1.080)		15pπ
$z\,^3\Pi_g$	33736.65	1698	(35)	(7.211)	(0.224)		(1.080)		14pπ
$y\,^3\Pi_g$	(33645.7)	1698	(35)	(7.211)	(0.224)		(1.080)		13pπ
$w\,^3\Pi_g$	33531.04	1698	(35)	(7.211)	(0.224)		(1.080)		12pπ
$w\,^3\Sigma_g^+$	33445.2								12pσ
$v\,^3\Pi_g$	33383.51	(1698.95)	(35)	7.21	0.22		1.080		11pπ
$v\,^3\Sigma_g^+$	33266.6								11pσ
$u\,^3\Pi_g$	33189.16	(1699.19)	(35.2)	7.21	0.22		1.080		10pπ
$u\,^3\Sigma_g^+$	33026.58								10pσ

SPECTROSCOPIC CONSTANTS

TRIPLET STATES

State	$T_o - a$ (a)	ω_e	$x_e\omega_e$	B_e	α_e	$D_e \times 10^6$	r_e	Remarks	Configuration
t $^3\Pi_g$	32925.96	(1699.65)	(35.2)	7.21	0.23		1.080		9pπ
t $^3\Sigma_g^+$	32693.1								9pσ
s $^3\Pi_g$	32556.60	(1699.80)	(35.2)	7.213	0.224		1.080		8pπ
s $^3\Sigma_g^+$	32211.70								8pσ
r $^3\Pi_g$	32016.56	(1700.5)	(35.2)	(7.216)	(0.224)		1.080		7pπ
r $^3\Sigma_g^+$	31478.6								7pσ
q $^3\Delta_u$	31255								6dδ
q $^3\Pi_u$	(31234)								6dπ
q $^3\Sigma_u^+$	31177								6dσ
p $^3\Pi_g$	31179.9	1701.18	35.3	7.220	0.224		1.079		6pπ
o $^3\Sigma_u^+$	31058.4			(7.22)	(0.23)		1.07		6sσ

SPECTROSCOPIC CONSTANTS

TRIPLET STATES

State	$T_o - a$ (a)	ω_e	$x_e\omega_e$	B_e	α_e	$D_e \times 10^6$	r_e	Remarks	Configuration
$m\,^3\Delta_u$	29920								$5d\delta$
$m\,^3\Pi_u$	(29835)								$5d\pi$
$m\,^3\Sigma_u^+$	29788								$5d\sigma$
$l\,^3\Pi_g$	29785.3	(1704.46)	(35.2)	7.226	0.222		1.079		$5p\pi$
$k\,^3\Sigma_u^+$	29573	1635.3$^{(d)}$		7.23	0.23		1.079		$5s\sigma$
$j\,^3\Delta_u$	27470	1633$^{(d)}$							$4d\delta$
$j\,^3\Pi_u$	27292	1598$^{(d)}$							$4d\pi$
$j\,^3\Sigma_u^+$	27202	1592$^{(d)}$							$4d\sigma$
$i\,^3\Pi_g$	27193.0	(1708.4)	(35.2)	7.242	0.223		1.078		$4p\pi$
$h\,^3\Sigma_u^+$	26760.4	1637.9$^{(d)}$		7.26	0.23		1.077		$4s\sigma$
$f\,^3\Delta_u$	22205.5	1706.82	35.10	7.230	0.229		1.079		$3d\delta$

He_2

SPECTROSCOPIC CONSTANTS

TRIPLET STATES

State	$T_o - a$ (a)	ω_e	$x_e\omega_e$	B_e	α_e	$D_e \times 10^6$	r_e	Remarks	Configuration
$f^3\Pi_u$	21754.0	1661.48	44.79	7.136	0.233		1.086		$3d\pi$
$f^3\Sigma_u^+$	21548.8	1635.77	44.41	7.071	0.246		1.091		$3d\sigma$
$e^3\Pi_g$	21507.2	1721.1	34.7	7.283	0.221		1.075	$y_e\omega_e = -0.02$ $\gamma_e = -0.0013$	$3p\pi$
$d^3\Sigma_u^+$	20391.3	1728.01	36.13	7.341	0.224		1.071	$y_e\omega_e = -0.127$ $\gamma_e = 0.0027$	$3s\sigma$
$c^3\Sigma_g^+$	10889.5	1583.85	52.74	7.005	0.310		1.096	$y_e\omega_e = -1.257$ $z_e\omega_e = -0.488$ $\gamma_e = 0.0016$	$3p\sigma$
$b^3\Pi_g$	4768.2	1769.07	35.02	7.447	0.219		1.063	$y_e\omega_e = -0.048$ $\gamma_e = -0.0017$	$2p\pi$
$a^3\Sigma_u^+$	0	1808.5	38.2	7.703	0.228		1.045	$y_e\omega_e = -0.3$ $\gamma_e = -0.0046$	$2s\sigma$

(a) A and a represent the energies of the $A^1\Sigma_u^+$ and $a^3\Sigma_u^+$ states ($v = 0$, $N = 0$ levels) above the $X^1\Sigma_g^+$ state. The $A^1\Sigma_u^+$ state is 2344.1 cm^{-1} above the $a^3\Sigma_u^+$ state. (b) B_o; (c) $X^1\Sigma_g^+$ state is unstable; (d) $\Delta G_{1/2}$

He$_2$

Perturbations and General Information

Potential energy curves (70.70)

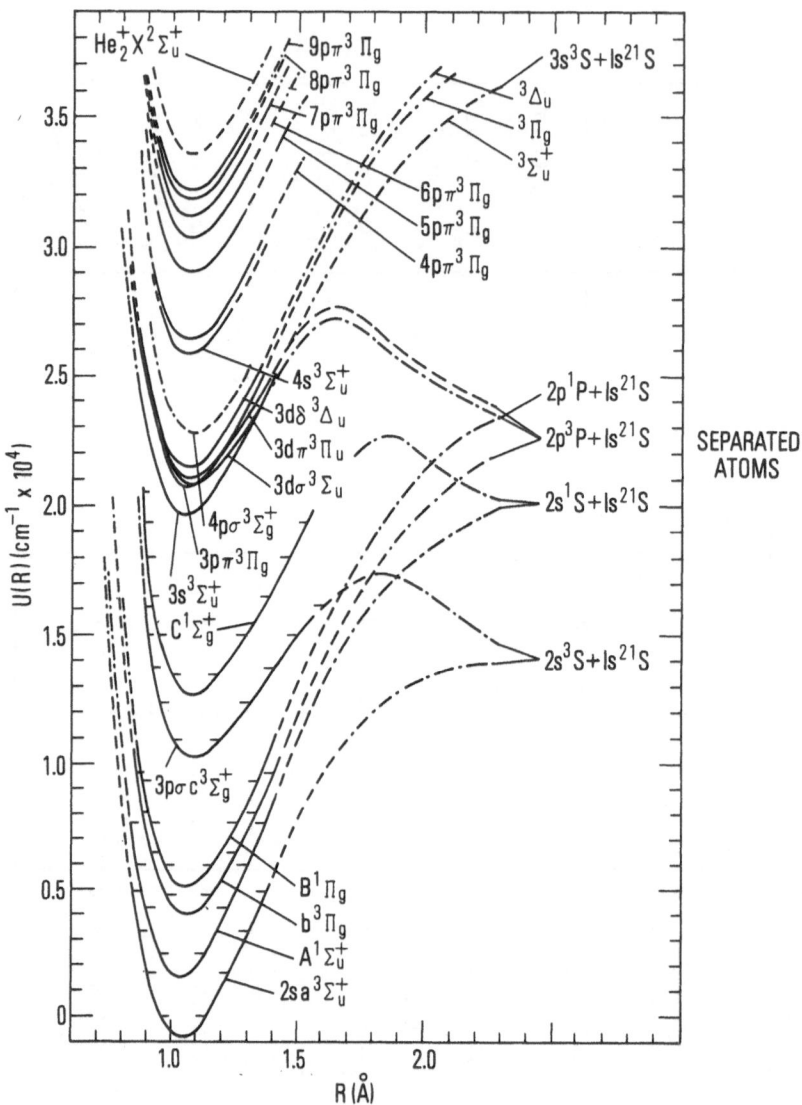

Stimulated emission has been observed on the d$^3\Sigma_u^+ \rightarrow$ b$^3\Pi_g$ transition at 6400Å (74.90).

BIBLIOGRAPHY

(22. 1)   d, f, h - b Systems,
W. E. Curtis,
Proc. Roy. Soc. A 101, 38-64

(25. 2)   E, I, L, P - A, and e, i, p, r, s, t, u - a Systems,
W. E. Curtis and R. G. Long,
Proc. Roy. Soc. A 108, 513-41

(27. 3)   I - A System, e, i, p, r - a and k - b Systems,
W. Weizel and C. Fuchtbauer,
Z. Physik 44, 431-54

(28. 4)   d, f, h - b Systems,
W. E. Curtis,
Proc. Roy. Soc. A 118, 157-69

(28. 5)   F - B System and f, j - b Systems,
W. E. Curtis and A. Harvey,
Proc. Roy. Soc. A 121, 381-401

(28. 6)   D, F - B and I - A Systems, e, i - a and d, f, h, j - b Systems,
G. H. Dieke, T. Takamine, and T. Suga,
Z. Physik 49, 865-84

(28. 7)   H - B System and j - b System,
W. Weizel,
Z. Physik 51, 328-40

(29. 8)   d - b System,
R. S. Mulliken,
Phys. Rev. 34, 1530-40

(29. 9)   J - B System and j, m - b and h, j - c Systems,
W. E. Curtis and A. Harvey,
Proc. Roy. Soc. A 125, 484-506

(29. 10)   R, S - A Systems and i - a System,
S. Imanishi,
Sci. Papers Inst. Phys. Chem. Res. 10, 193-209

(29. 11)   E - A System and i - a, k, o - c Systems,
S. Imanishi,
Sci. Papers Inst. Phys. Chem. Res. 10, 237-52

He$_2$

(29.12) p, r, s, t, u - a Systems,
S. Imanishi,
Sci. Papers Inst. Phys. Chem. Res. 11, 139-49

(29.13) History of He$_2$,
W. E. Curtis,
Trans. Faraday Soc. 25, 694-707

(29.14) F, M - B Systems and f, j, q - b Systems,
G. H. Dieke, S. Imanishi, and T. Takamine,
Z. Physik 54, 826-43

(29.15) d, h, k, m, o, q - b and h, j, m, q - c Systems,
G. H. Dieke, S. Imanishi, and T. Takamine,
Z. Physik 57, 305-24

(29.16) L - A System and j, k, m, q - b Systems,
W. Weizel,
Z. Physik 52, 175-96

(29.17) E - A and F - B System and f - b System,
W. Weizel,
Z. Physik 54, 321-40

(29.18) F, M - B and L, P - A Systems and i, p, r - a and j - b Systems,
W. Weizel and E. Pestel,
Z. Physik 56, 197-214

(30.19) A - X System,
J. J. Hopfield,
Astrophys. J. 72, 133-45

(32.20) c - a and d, f - c Systems,
W. F. Meggers and G. H. Dieke,
Bur. Stand. J. Res. 9, 121-9

(50.21) e - a System,
G. H. Dieke and E. S. Robinson,
Phys. Rev. 80, 1-5

(53.22) c - a System,
D. Cuthbertson,
C. R. Acad. Sci. 236, 1757-8

(56.23) b - a System,
G. Hepner and L. Herman,
C. R. Acad. Sci. 243, 1504-6

(58. 24) A - X System, Discussion of Hopfield Continuum,
Y. Tanaka, A. S. Jursa, and F. J. LeBlanc,
J. Opt. Soc. Am. 48, 304-8

(60. 25) He-He Interaction,
N. Moorse,
J. Chem. Phys. 33, 471-80

(61. 26) He(^{1}S) - He(^{3}S) Interaction,
G. H. Brigman, S. J. Brient, and F. A. Matsen,
J. Chem. Phys. 34, 958-60

(61. 27) LCAO-MO-SCF Calculation,
B. J. Ransil,
J. Chem. Phys. 34, 2109-18

(62. 28) Theory,
R. P. Hurst, M. Karplus, and T. P. Das,
J. Chem. Phys. 36, 2786-92

(62. 29) Hopfield Continuum Discussion,
R. E. Huffman, Y. Tanaka, and J. C. Larrabee,
J. Opt. Soc. Am. 52, 851-7

(62. 30) He-He Interaction — Theory,
D. Y. Kim,
Z. Physik 166, 359-69

(63. 31) Theory,
R. V. Miller and R. D. Present,
J. Chem. Phys. 38, 1179-82

(63. 32) A - X System, 600Å Bands,
Y. Tanaka and K. Yoshino,
J. Chem. Phys. 39, 3081-7

(63. 33) Potential Energy Curve Calculation,
R. D. Poshusta and F. A. Matsen,
Phys. Rev. 132, 307-9

(64. 34) Theory,
R. S. Mulliken,
J. Am. Chem. Soc. 86, 3183-97

(64. 35) Theory,
R. S. Mulliken,
Phys. Rev. 136, 962-5

He$_2$

(65. 36) C - A System and c - a System,
          M. L. Ginter,
          J. Chem. Phys. 42, 561-8

(65. 37) Hopfield Continuum,
          A. L. Smith and J. W. Meriwether, Jr.,
          J. Chem. Phys. 42, 2984

(65. 38) D, F - B Systems,
          M. L. Ginter,
          J. Molec. Spectrosc. 17, 224-39

(65. 39) d, f - b and d - c Systems,
          M. L. Ginter,
          J. Molec. Spectrosc. 18, 321-43

(65. 40) b - a System,
          P. Gloersen and G. H. Dieke,
          J. Molec. Spectrosc. 16, 191-204

(65. 41) R. E. Huffman, J. C. Larrabee, Y. Tanaka, and D. Chambers,
          "Helium Continuum Afterglow in the Vacuum Ultraviolet,"
          J. Opt. Soc. Am. 55, 101-2

(66. 42) Theory,
          R. S. Mulliken,
          J. Am. Chem. Soc. 88, 1849-61

(66. 43) f - c System, 1-Decoupling,
          M. L. Ginter,
          J. Chem. Phys. 45, 248-62

(66. 44) Theory, Long-Range Interactions,
          N. R. Kestner,
          J. Chem. Phys. 45, 208-12

(66. 45) Theory, Potential Energy Curves,
          N. R. Kestner and O. Sinanoglu,
          J. Chem. Phys. 45, 194-207

(66. 46) 600Å Band,
          F. H. Mies and A. L. Smith,
          J. Chem. Phys. 45, 994-1000

(66. 47) Theory,
          D. R. Scott, E. M. Greenwalt, J. C. Browne, and F. A. Matsen,
          J. Chem. Phys. 44, 2981-4

(66.48) A - X System, Hopfield Continuum,
D. Villarejo, R. R. Herm, and M. G. Inghram,
J. Opt. Soc. Am. 56, 1574-84

(67.49) Theory,
G. P. Barnett,
Can. J. Phys. 45, 137-43

(67.50) Theory,
T. L. Gilbert and A. C. Wahl,
J. Chem. Phys. 47, 3425-38

(67.51) A State Dissociation Energy,
B. K. Gupta and F. A. Matsen,
J. Chem. Phys. 47, 4860-1

(67.52) Theory,
C. J. Herbert and O. G. Ludwig,
J. Chem. Phys. 47, 3086-8

(67.53) a State, Potential Barrier,
K. H. Ludlum, L. P. Larson, and J. M. Caffrey, Jr.,
J. Chem. Phys. 46, 127-30

(67.54) Theory,
G. H. Matsumoto, C. F. Bender, and E. R. Davidson,
J. Chem. Phys. 46, 402-3

(67.55) Theory,
B. M. Morris and R. D. Present,
J. Chem. Phys. 46, 653-7

(67.56) A. L. Smith,
J. Chem. Phys. 47, 1561-2

(67.57) A. B. Callear and R. E. M. Hedges,
"Metastability of Rotationally Hot Dihelium at 77°K,"
Nature 215, 1267-8

(67.58) D. J. Klein, C. E. Rodriguez, J. C. Browne, and F. A. Matsen,
"Ground-State Potential of He$_2$,"
J. Chem. Phys. 47, 4862-3

(67.59) N. I. Zhirnov and O. P. Shadrin,
"Calculation of Franck-Condon Factors With Poschl-Teller Wave
Functions. I. The Probabilities of Some Vibrational Transitions
in the $D^1\Sigma_u^+$ - $B^1\Pi_g$ Band System of the He$_2$ Molecule,"
Opt. Spect. 24, 478-80

He$_2$

(67. 60) G. P. Barnett,
"Wave Functions for the $^1\Sigma_g^+$ Ground State of the He$_2$ Molecule, "
Can. J. Phys. <u>45</u>, 137-143

(68. 61) A. V. Phelps,
"Decay of Metastable Helium Molecules ($^3\Sigma_u^+$) and Atoms (^{3}S) in an Afterglow, "
J. Opt. Soc. Am. <u>58</u>, 1540-1

(68. 62) i, p, s, t, u - a Systems and r($^3\Sigma_g^+$, $^3\Pi_g$) - a Systems,
M. L. Ginter and D. S. Ginter,
J. Chem. Phys. <u>48</u>, 2284-91

(68. 63) 600Å Band,
A. L. Smith,
J. Chem. Phys. <u>49</u>, 4817-24

(69. 64) Theory,
R. S. Mulliken,
J. Am. Chem. Soc. <u>91</u>, 4615-21

(69. 65) Theory,
B. K. Gupta and F. A. Matsen,
J. Chem. Phys. <u>50</u>, 3797-803

(69. 66) Theory,
D. J. Klein,
J. Chem. Phys. <u>50</u>, 5151-7

(69. 67) A ← X System,
Y. Tanaka and K. Yoshino,
J. Chem. Phys. <u>50</u>, 3087-98

(70. 68) 600Å Band, a, A States, Potential Barrier,
A. L. Smith and K. W. Chow,
J. Chem. Phys. <u>52</u>, 1010-2

(70. 69) A. B. Callear and R. E. M. Hedges,
"Rotational and Vibrational Relaxation of He$_2$ a$^3\Sigma_u^+$, "
Trans. Faraday Soc. <u>60</u>, 2921-35

(70. 70) M. L. Ginter and R. Battino,
"Potential-Energy Curves for the He$_2$ Molecule, "
J. Chem. Phys. <u>52</u>, 4469-74

(70. 71) K. Chow,
Thesis, Yale University

(71.72) K. Chow and A. L. Smith,
"Repulsive Potential Curves From Molecular Continuum-Continuum Emission. III. He$_2(X^1\Sigma_g^+)$,"
J. Chem. Phys. $\underline{54}$, 1556-62

(71.73) K. Chow, A. L. Smith, and M. G. Waggoner,
"Absorption Coefficients of Helium Between 599 and 610Å; Transition Moment for He$_2$ A$^1\Sigma_u^+ \leftarrow$ X$^1\Sigma_g^+$,"
J. Chem. Phys. $\underline{55}$, 4208-13

(71.74) K. M. Sando,
"The Emission of Radiation Near 600Å by Helium,"
Molec. Phys. $\underline{21}$, 439-47

(71.75) S. Mulkamel and U. Kaldor,
"Potential of the A$^1\Sigma_u^+$ State of He$_2$,"
Molec. Phys. $\underline{22}$, 1107-17

(71.76) C. M. Brown and M. L. Ginter,
"Spectrum and Structure of the He$_2$ Molecule. VI. Characterization of the States Associated With the UAO's 3pπ and 2s,"
J. Molec. Spectrosc. $\underline{5}$, 302-16

(71.77) M. Bourene and J. LeCalve,
"Effects of Additives on the He$_2$ Afterglow Excited by an Intense Electron Pulse,"
J. Phys. $\underline{32}$, 29-33

(71.78) B. Liu,
"Dissociation Energies of He$_2^+$(X$^2\Sigma_u^+$) and He$_2$(A$^1\Sigma_u^+$),"
Phys. Rev. Letters $\underline{27}$, 1251-3

(71.79) M. L. Ginter and C. M. Brown,
"Dissociation Energies of X$^2\Sigma_u^+$(He$_2$) and A$^1\Sigma_u^+$(He$_2$),"
J. Chem. Phys. $\underline{56}$, 672-4

(72.80) D. Kunik and U. Kaldor,
"Ground State of He$_2$ by the Spin-Optimized Method,"
J. Chem. Phys. $\underline{56}$, 1741-5

(72.81) B. K. Gupta,
"On the Potential Curve of the Metastable Helium Molecule,
Molec. Phys. $\underline{23}$, 75-9

(72.82) S. L. Guberman and W. A. Goddard III,
"On the Origin of Energy Barriers in the Excited States of He$_2$,"
Chem. Phys. Letters $\underline{14}$, 460-5

(73.83) P. J. Bertoncini and A. C. Wahl,
"Ab Initio Calculation of the Helium-Helium X$^1\Sigma_g^+$ Potential at Intermediate and Large Separations. II. Changes in Intra-Atomic Correlation Energy,"
J. Chem. Phys. $\underline{58}$, 1259-61

(73.84) C. M. Brown and M. L. Ginter,
"Spectrum and Structure of the He$_2$ Molecule. Characterization of the Singlet and Triplet States Associated With the UAO's 4s, 4dσ, 4dπ, and 4dδ,"
J. Molec. Spectrosc. $\underline{46}$, 256-75

(73.85) W. B. Peatman and D. T. Wu,
"The 600Å Bands of Helium,"
Chem. Phys. $\underline{2}$, 335-41

(73.86) M. H. Mittleman and H. Tai,
"Low-Energy Atom-Atom Scattering: Corrections to the He-He Interaction,"
Phys. Rev. A $\underline{8}$, 1880-91

(73.87) L. Lenamon, J. C. Browne, and R. E. Olson,
"Theoretical Low-Energy Inelastic-Scattering Cross Sections for He(2^{3}S) + He(1^{1}S) → He(2^{3}P) + He(1^{1}S): Curve Crossing Between the c$^3\Sigma_g^+$ and b'$^3\Pi_g$ States of He$_2$,"
Phys. Rev. $\underline{8}$, 2380-6

(73.88) S. Mukamel and U. Kaldor,
"Ab Initio Calculation of the He$_2$ A$^1\Sigma_u^+$ ← X$^1\Sigma_g^+$ Absorption Spectrum,"
Molec. Phys. $\underline{26}$, 291-5

(73.89) A. B. Kunz,
"Approximation to the Method of Local Orbitals,"
J. Phys. B $\underline{6}$, L47-50

(74.90) C. B. Collins, A. J. Cunningham, S. M. Curry, B. W. Johnson, and M. Stockton,
"Stimulated Emission From the Recombining Afterglow of an Electron-Beam Discharge in Several Atmospheres of Helium,"
Appl. Phys. Letters $\underline{24}$, 245-7

(74.91) J. P. Daudey, J. P. Malrieu, and O. Rojas,
"Perturbative Ab Initio Calculations of Internolecular Energies. II. The He · · · He Problem,"
Int. J. Quant. Chem. $\underline{8}$, 17-28

(74. 92) A. P. Hickman and N. F. Lane,
"Long-Range $He_2(^3\Sigma^+_{g,u})$ Potentials and Metastability Exchange in
He*-He Collisions,"
Phys. Rev. A 10, 444-7

(74. 93) P. B. Foreman, P. K. Rol, and K. P. Coffin,
"The Repulsive $^1\Sigma^+_g$ He_2 Potential Obtained From Total Cross
Sections,"
J. Chem. Phys. 61, 1658-65

(74. 94) W. Lichten, M. V. McCusker, and T. L. Vierima,
"Fine Structure of the Metastable $a^3\Sigma^+_u$ State of the Helium Molecule,"
J. Chem. Phys. 61, 2200-12

(n.p. 95) M. L. Ginter

Hg$_2$

Hg$_2$

Methods of Production and Experimental Technique

Absorption in mercury vapor.

Emission from a Tesla discharge, hollow cathode discharge, argon and mercury discharge, electrodeless glow discharge.

Fluorescence (primarily from 2537Å mercury resonance line), resonance irradiation at low temperatures.

Band Systems

A. Continuous Spectra

I. $A^30_u^- \rightarrow X^1\Sigma_g^+$ System

Visible region — "4850Å continuum" (73.33, 73.32, 73.31, 52.23, 49.21, 44.20, 31.9, 31.8)

Emission: 6000-3800Å, maximum at ~ 5200Å
Fluorescence: 5300-4000Å, maximum at ~ 4850Å

II. $A^31_u \rightarrow X^1\Sigma_g^+$ System

Near ultraviolet region — "3350Å continuum" (73.34, 73.31, 52.23, 49.21, 44.20, 31.9)

Emission and fluorescence: 3700-3000Å region, maximum at 3350Å
Absorption (32.14)

III. ~ 2650Å

Emission at ~ 2650 ±30Å (27.1)

IV. Ultraviolet Region

Emission and Absorption: 2345-1808Å region (31.12, 31.10, 31.8, 30.4)

V. Far Ultraviolet Region

Absorption: Three continua at 1403, 1692, and 1808Å (31.12, 31.10)
Emission: 1692Å only

B. Band Spectra

I. $B^32_u \rightarrow A^30_u^-$ System

Triplet: 5461, 4358 and 4047Å (73.34, 73.31, 68.27, 33.15)

Emission (λ): 5492.7|5483.6|5473.5|4364.8|4061.1|4060.8|4050.4|
4023.8

II. Takeyama Bands

Emission

a. 5411-5369Å – group of 9 weak bands (52.23)

b. 5313-4988Å – group of 47 bands. Band heads of high intensity (λ)

v', v''	0	1	2	3	4	5	6	7
0	5096.8	5133.7	5170.2	5208.3	5246.4			
1	5066.0		5138.6	5175.3	5213.2	5250.5		
2	5036.2	5071.7	5108.5	5144.5	5180.5	5218.0	5254.5	5292.0
3	5006.5	5041.5	5076.5	5112.5	5149.5	5186	5223.2	5259.0
4		5012.3	5047	5082.0		5154.1		5227.2
5			5017.6		5087		5159	
6			4987.7			5092		5164

c. 4962-4926Å – 3 weak bands at 4962, 4938, and 4926Å

d. 4905-4768Å – 12 bands, (λ)
|4905|4890|4874|4855|4839|4834|4817|4811|4788|4782|4773|
4768|

e. 4751-4400Å – group of 61 bands. Band heads of high intensity (λ)

v', v''	0	1	2	3	4	5	6
0	4513.0	4541.9	4571.7	4601.3	4631.5	4662.0	
1	4488.4	4517.5	4546.5	4576.0	4605.6	4635.3	4666.0
2	4464.4	4493	4522.9	4551	4580.5	4610.0	4639.4
3	4441	4470.0	4498.2	4527	4555.7		4614.0
4	4417.8	4446	4475	4503		4560.3	
5		4423.0			4507.5		4564.5

f. 4321-3680Å – group of 117 bands (33.16). Bands of highest intensity (λ)
|4281.7|4277.6|4252.8|4251.0|4227.6|4225.5|4202.3|4200.2|
4181.6|4179.0|

g. 4310-4298Å – 2 large intense sets of bands; one at 4309 and 4305Å, the other at 4302 and 4298Å

h. 4133-4120Å – bands between 4132.3-4130.0 and 4122.4-4120.4Å

i. 4011-3685Å – diffuse bands (30.3). Bands of highest intensity at 3984 and 3946Å

j. 3663-3342Å – group of 26 bands partially superimposed on the continuum at 3350Å. Bands of highest intensity (λ)
|3497|3486|3474|3455|3443|

k. 3342-3014Å – heads of intense bands distinct at 3135.9, 3134.4, and 3112.3Å

l. 3014-2537Å – several structures discovered by (52.23) superimposed on continuous bands

III. Rayleigh Bands

a. 3014-2857Å "Core Bands" – the bands form a convergent series up to a point near the 2537Å resonance line

Fluorescence (λ) (32.13, 31.11)	Emission (λ) (31.11, 27.1)
3012.0	3014.3
2993.8	2994.4
2977.6	2976.1
2959.4	2960.2
2944.4	2945.4
2930.1	2929.7
2911.1	2913.2
--	2896.9
2883.3	2884.1
2871.6	2871.6
2857.4	2860.8

b. 2930-2782Å "Wing Bands" – the bands converge at ~ 2645Å

Fluorescence (λ) (32.13, 31.11)	Absorption (λ) (27.1)
2931.1	2930.7
2917.4	2918.2
2904.7	2907.0
2892.9	2894.8
2882.7	2883.8
2871.2	2873.6
2860.6	2863.6
2851.5	2853.1
2841.8	2843.2
2832.3	2833.2
2823.6	2825.3
2814.9	2815.3
2806.4	2807.4
2798.6	2799.2
2790.2	2790.8
2781.9	2784.3

IV.   <u>2540Å Bands</u>

Emission (49.22) at high dispersion
λ|2540.467|2540.453|2540.424|2540.378|2540.345|2540.290|2540.235|
bordered by 2540.524 and 2539.823Å

Absorption (49.22, 32.14).   Very weak bands apparently of the
same system
λ|2542.5|2541.5|2539.6|2538.5|

V.   <u>Wood Bands 2345-2311Å</u>

Emission (31.8)
Fluorescence (32.13)
Absorption (32.14, 29.2)

Band heads:

Absorption (λ)	Emission (λ)
	2345.5
2341.50	
2340.48	
2339.48	
2338.47	2338.6
2337.38	
2336.40	
2335.38	
2334.37	
2333.33	2333.8
2331.33	
2329.35	2329.0
2328.41	
2327.36	
2326.44	
2325.39	
2324.50	2324.4
2323.54	
2322.58	
2321.59	
2320.66	2320.8
2319.76	
2318.87	
2317.87	2318.0
2316.89	
2316.03	
2315.03	2315.1
2314.23	
2313.25	
2312.51	2312.2
2311.54	

VI.   <u>Steubing-Kremenevsky Bands 2167-2038Å</u>

Emission (31.8)
Absorption (31.12)
Very diffuse bands

Hg$_2$

Spectroscopic Constants
================

Dissociation energy = 0.14 ± 0.02 eV, 3.2 kcal/mole, 1120 cm^{-1}

Perturbations and General Information
================

The ground state has been shown to be a repulsive state.

Potential energy curves (73.34)

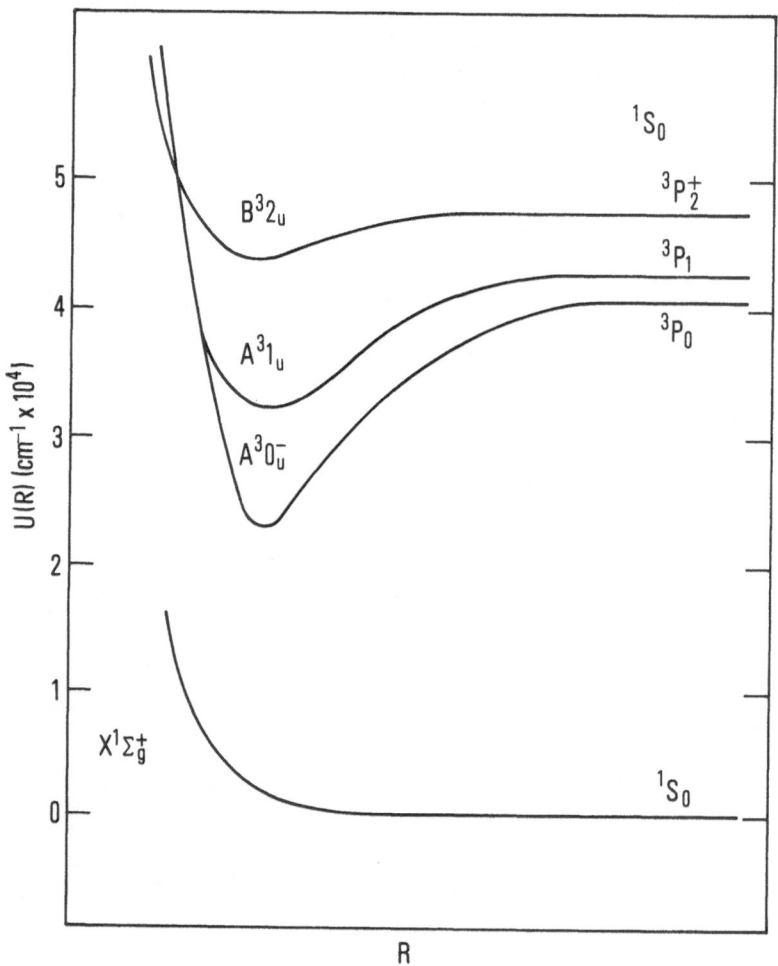

BIBLIOGRAPHY

(27. 1)   "Core Bands" in Emission and "Wing Bands" in Absorption,
          O. W. Richardson,
          Proc. Roy. Soc. A 115, 528-48

(29. 2)   Wood Bands in Absorption,
          J. M. Walter and S. Barratt,
          Proc. Roy. Soc. A 122, 201-10

(30. 3)   Emission,
          H. Volkringer,
          Ann. Physique 14, 15-81

(30. 4)   Ultraviolet Absorption,
          S. Mrozowski,
          Phys. Rev. 36, 1168-71

(31. 5)   Visible Continuum in Fluorescence,
          Lord Rayleigh,
          Nature 127, 10

(31. 6)   Arc Emission in the Ultraviolet,
          Lord Rayleigh,
          Nature 127, 125

(31. 7)   "Core Bands" in Fluorescence,
          Lord Rayleigh,
          Nature 127, 854

(31. 8)   Visible and Ultraviolet Spectra in Hollow Cathode,
          H. Hamada,
          Philos. Mag. 12, 50-67

(31. 9)   Visible Continuum in Fluorescence,
          P. D. Foote, A. E. Ruark, and R. L. Chenault,
          Phys. Rev. 37, 1685

(31. 10) Absorption in Ultraviolet Region,
          J. G. Winans,
          Phys. Rev. 37, 897-901

(31. 11) "Core" and "Wing Bands" in Fluorescence,
          Lord Rayleigh,
          Proc. Roy. Soc. A 132, 650-67

(31. 12)   Absorption in Ultraviolet Region,
N. Kremenevsky,
Z. Physik 71, 792-7

(32. 13)   "Core" and "Wing Bands" in Fluorescence,
Lord Rayleigh,
Proc. Roy. Soc. A 135, 617-27

(32. 14)   Ultraviolet Absorption and Potential Curves,
N. Kuhn and K. Freudenberg,
Z. Physik 76, 38-54

(33. 15)   Visible Emission,
E. Matuyama,
Nature 131, 58

(33. 16)   Visible Emission,
J. Okubo and E. Matuyama,
Science Rep. Tohoku Univ. 22, 383-92

(37. 17)   Potential Curves,
S. Mrozowski,
Z. Physik 104, 228-47

(37. 18)   Potential Curves,
S. Mrozowski,
Z. Physik 106, 458-62

(38. 19)   Ionization Potential,
F. L. Arnot and M. B. M'ewen,
Proc. Roy, Soc. A 165, 133-47

(44. 20)   Potential Curves,
S. Mrozowski,
Rev. Mod. Phys. 16, 153-74

(49. 21)   Visible and Ultraviolet Continuum in Fluorescence,
T. Holstein, D. Alpert, and A. O. MacCoubrey,
Phys. Rev. 76, 1259

(49. 22)   2540Å Bands at High Dispersion
S. Mrozowski,
Phys. Rev. 76, 1714-6

(52. 23)   Emission,
H. Takeyama,
J. Science Hiroshima Univ. A 15, 235-62

(52. 24)   Dissociation Energy,
           J. G. Winans,
           Z. Physik 133, 291-6

(59. 25)   I. Agirbiceanu and A. Ichimescu,
           "Observatii Asupra Spectrului de Absorbtie al Hg$_2$ in Ultraviolet,"
           Bul. Inst. Polit. Bucuresti 21, 41-48

(68. 26)   J. E. McAlduff, D. D. Drysdale, and D. J. LeRoy,
           "The Role of Hg 6^3P_0 Atoms in Mercury Photosensitization.   IV.
           Origin of the 4850Å and 3350Å Bands,"
           Can. J. Chem. 46, 199-206

(68. 27)   R. J. Carbone and M. M. Litvak,
           "Intense Mercury-Vapor Green-Band Emission,"
           J. Appl. Phys. 39, 2413-6

(69. 28)   R. B. Merrithew, G. V. Marusak, and C. E. Blount,
           "Absorption Spectra of Metal Atoms in Solid Xenon,"
           J. Molec. Spectrosc. 29, 54-65

(72. 29)   M. Stupavsky, G. W. F. Drake, and L. Krause,
           "Molecular Fluorescence of Mercury,"
           Phys. Letters 39A, 349-50

(73. 30)   R. A. Phaneuf, J. Skonieczny, and L. Krause,
           "Formation and Decay of Excited Hg$_2$ Molecules in Hg-N$_2$ Mixtures,"
           Phys. Rev. A 8, 2980-9

(73. 31)   A. C. Vikis and D. J. LeRoy,
           "The Hg$_2$ $\left(A^3 0_u^- \rightarrow X^1\Sigma_g^+\right)$ and $\left(A^3 1_u \rightarrow X^1\Sigma_g^+\right)$ Fluorescence,"
           Phys. Letters 44A, 325-6

(73. 32)   A. G. Ladd, C. G. Freeman, M. J. McEwan, R. F. C. Claridge, and
           L. F. Phillips,
           "Phase-Shift Studies of Hg$\left(^3P_0\right)$ Reactions, Part 4 — Observations of
           the Hg$_2$ Emission Bands at 335 and 485 nm,"
           J. Chem. Soc. Faraday Trans. II 69, 849-55

(73. 33)   R. M. Hill, D. J. Eckstrom, D. C. Lorents, and H. H. Nakano,
           "Measurements of Negative Gain for Hg$_2$ Continuum Radiation,"
           Appl. Phys. Letters 23, 373-4

(73. 34)   L. F. Phillips,
           "The Origin of the 550 nm Emission Band of Hg$_2$,"
           Chem. Phys. Letters 21, 28-9

Hg$_2$

(73. 35)   J. B. West,
           "Comment on 3884-Å Band Emission From Electrodeless Glow Dis-
           charges in Binary Mixtures Containing Mercury Vapor,"
           J. Chem. Phys. 58, 5844

(73. 36)   B. Chakraborti, M. Z. Hoffman, N. N. Lichten, and D. A. Sacks,
           "3884Å Band Emission From Electrodeless Glow Discharges in
           Binary Mixtures Containing Mercury Vapor,"
           J. Chem. Phys. 58, 405-10

Ho$_2$

Spectroscopic Constants

Dissociation energy = 1.98 ± 0.74 eV, 16 kcal/mole, 15970 cm^{-1} (72.2, 71.1).

Ho$_2$

BIBLIOGRAPHY

(71. 1)   S. Lin and A. Kant,
           "Dissociation Energies of Diatomic Rare Earth Molecules Dy$_2$, Ho$_2$,
           Er$_2$, Tm$_2$ and Yb$_2$,"
           TR No. AMMRC TR 71-34, Army Materials and Mechanics Research
           Center, Watertown, Mass.

(72. 2)   A. Kant and S. Lin,
           "Dissociation Energies of the Homonuclear Diatomic Molecules of
           the Rare Earths,"
           Monatshefte für Chemie 103, 757-63

I_2

Methods of Production and Experimental Technique

Absorption (between 0 and 1100°C).

Emission from discharge into I_2 in the presence of foreign gases.

Resonance and laser-excited fluorescence.

BAND SYSTEMS

	System	Transition	Sources	Wavelength Limits	Degrading	Band Head, $\nu_{0,0}$	Remarks	Bibliography
	I	$A^3\Pi_u \leftarrow X^1\Sigma_g^+$ (1_u) (0_g^+)	Absorption	9300-8300	R	11803		(70.73, 64.49, 31.11)
	II	$^1\Pi \rightleftharpoons X^1\Sigma_g^+$ (1_u)		6000-4500			Continuum	(73.118)
	III	$B^3\Pi_u \rightleftharpoons X^1\Sigma_g^+$ (0_u^+)	Absorption, emission, and fluorescence	8600-4300	R	15724.95		(73.119, 73.111, 73.107, 73.105, 72.87, 72.86, 72.85, 72.81, 70.74, 70.73, 70.67, 67.58, 65.53)
	IV	$C^3\Sigma_u^+ \rightarrow X^1\Sigma_g^+$ (1_u)	Emission, photofragment spectroscopy	2500-2400				(73.113)
	V	$D^3\Sigma_g^- \rightleftharpoons X^1\Sigma_g^+$ $(1_g, 0_g^+)$	Absorption, fluorescence	3460-3100	R			(72.87, 71.80)
	VI	$E^3\Sigma_u^- \rightarrow X^1\Sigma_g^+$ (1_u)	Discharge, emission	2400-2240	R	45170		(58.39)

I_2

System	Transition	Sources	Wavelength Limits	Degrading	Band Head, $\nu_{0,0}$	Remarks	Bibliography
VII	$F\,^1\Sigma_g^+ \leftarrow X\,^1\Sigma_g^+$ (0_g^+)	Absorption	2740-2530	R	47158.6		(72.87, 35.17, 34.15, 29.8, 29.7, 28.6)
VIII	$G\,^3\Sigma_u^- \rightleftarrows X\,^1\Sigma_g^+$ (0_u^+)	Emission	2730-2486	R	47148	Absorption superimposed on $D \leftarrow X$ and $F \leftarrow X$ systems	(58.40, 51.30, 47.27, 40.21, 35.17, 29.8, 29.7, 28.6)
IX	$H\,^1\Sigma_u^+ \rightleftarrows X\,^1\Sigma_g^+$ (0_u^+)	Discharge, absorption	2240-1950	R	45543(0, 12) 45030(1, 15)		(58.39, 46.23, 35.19, 29.8, 29.7)
X	$I\,^1\Sigma_u^+ \rightleftarrows X\,^1\Sigma_g^+$ (0_u^+)	Resonance fluorescence	2370-1770				(70.73, 56.37)
XI	$c\Pi \leftarrow X\,^1\Sigma_g^+$ (1_u)	Absorption	1770-1723	V			(70.73)
XII	$d\Delta \leftarrow X\,^1\Sigma_g^+$ (1_u)	Absorption	1736-1707	V			(70.73)
XIII	$e\Sigma^+ \leftarrow X\,^1\Sigma_g^+$ (0_u^+)	Absorption	1703-1688	V			(70.73)

System	Transition	Sources	Wavelength Limits	Degrading	Band Head, $\nu_{0,0}$	Remarks	Bibliography
XIV	$L\,^3\Pi \leftarrow X\,^1\Sigma_g^+$ (1_u)	Absorption	1625-1611	R			(70.73)
XV	$\alpha\Sigma_u^- \leftarrow X\,^1\Sigma_g^+$ $(1_u,\ 0_u^+)$	Absorption	1625-1593	R			(70.73)
XVI	$f\Pi \leftarrow X\,^1\Sigma_g^+$ (0_u^+)	Absorption	1600-1590	R			(70.73)
XVII	$M\,^3\Pi_u \leftarrow X\,^1\Sigma_g^+$ (0_u^+)	Absorption	1602-1574	R			(70.73)
XVIII	$g\Pi \leftarrow X\,^1\Sigma_g^+$ (1_u)	Absorption	1592-1564	R			(70.73)
XIX	$h\Pi \leftarrow X\,^1\Sigma_g^+$ (1_u)	Absorption	1579-1556	V			(70.73)
XX	$i\Sigma_u^+ \leftarrow X\,^1\Sigma_g^+$ (1_u)	Absorption	1578-1561	V			(70.73)

	System	Transition	Sources	Wavelength Limits	Degrading	Band Head, $\nu_{0,0}$	Remarks	Bibliography
	XXI	$j\Sigma_u^- \leftarrow X^1\Sigma_g^+$ $(1_u, 0_u^+)$	Absorption	~ 1555	V			(70,73)
	XXII	$k\Sigma_u^+ \leftarrow X^1\Sigma_g^+$ (0_u^+)	Absorption	1558-1539	R			(70,73)
	XXIII	$1\Pi \leftarrow X^1\Sigma_g^+$ (1_u)	Absorption	~ 1540	V			(70,73)
	XXIV	$N^1\Pi_u \leftarrow X^1\Sigma_g^+$ (1_u)	Absorption	1539-1533	R			(70,73)
	XXV	$m\Pi \leftarrow X^1\Sigma_g^+$ (1_u)	Absorption	1504-1495	V			(70,73)
	XXVI	$\beta\Sigma^- \leftarrow X^1\Sigma_g^+$ (1_u)	Absorption	1488-1464	R			(70,73)
	XXVII	$\beta'\Sigma^- \leftarrow X^1\Sigma_g^+$ (0_u^+)	Absorption	1495-1480	R			(70,73)

	System	Transition	Sources	Wavelength Limits	Degrading	Band Head, $\nu_{0,0}$	Remarks	Bibliography
	XXVIII	$o\Pi \leftarrow X^1\Sigma_g^+$ (1_u)	Absorption	1491-1477	R			(70,73)
	XXIX	$p\Sigma^+ \leftarrow X^1\Sigma_g^+$ (0_u^+)	Absorption	1480-1472	R			(70,73)
	XXX	$o'\Pi \leftarrow X^1\Sigma_g^+$ (0_u^+)	Absorption	1486-1467	V			(70,73)
	XXXI	$q\Sigma^+ \leftarrow X^1\Sigma_g^+$ (1_u)	Absorption	~ 1472	V			(70,73)
	XXXII	$\gamma\Sigma^+ \leftarrow X^1\Sigma_g^+$ (1_u)	Absorption	1467-1459	R			(70,73)
	XXXIII	$\gamma'\Sigma^+ \leftarrow X^1\Sigma_g^+$ (0_u^+)	Absorption	~ 1445	R			(70,73)
	XXXIV	$r\Sigma^- \leftarrow X^1\Sigma_g^+$ $(1_u, 0_u^+)$	Absorption	1445-1428	R			(70,73)

System	Transition	Sources	Wavelength Limits	Degrading	Band Head, $\nu_{0,0}$	Remarks	Bibliography
XXXV	$\delta\Sigma^- \leftarrow X^1\Sigma_g^+$ (1_u)	Absorption	~ 1436	R			(70.73)
XXXVI	$\delta'\Sigma^- \leftarrow X^1\Sigma_g^+$ (0_u^+)	Absorption	~ 1434				(70.73)
XXXVII	$\zeta\Pi \leftarrow X^1\Sigma_g^+$ (1_u)	Absorption	1442-1423	R			(70.73)
XXXVIII	$\zeta'\Pi \leftarrow X^1\Sigma_g^+$ (0_u^+)	Absorption	1439-1427				(70.73)
XXXIX	$s'\Pi \leftarrow X^1\Sigma_g^+$ (1_u)	Absorption	~ 1427	V			(70.73)
XL	$s\Pi \leftarrow X^1\Sigma_g^+$ (0_u^+)	Absorption	1425-1407	V			(70.73)
XLI	$t\Pi \leftarrow X^1\Sigma_g^+$ (1_u)	Absorption	1418-1397	R			(70.73)

System	Transition	Sources	Wavelength Limits	Degrading	Band Head, $\nu_{0,0}$	Remarks	Bibliography
XLII	$t'\Pi \leftarrow X^1\Sigma_g^+$ (0_u^+)	Absorption	1423-1397	R			(70.73)
XLIII	$\eta\Sigma^+ \leftarrow X^1\Sigma_g^+$ (1_u)	Absorption	1401-1388	R			(70.73)
XLIV	$\eta'\Sigma^+ \leftarrow X^1\Sigma_g^+$ (0_u^+)	Absorption	1398-1381	R			(70.73)
XLV	$\theta\Pi \leftarrow X^1\Sigma_g^+$ $(1_u, 0_u^+)$	Absorption	1390-1369	R			(70.73)
XLVI	$u\Pi \leftarrow X^1\Sigma_g^+$ (1_u)	Absorption	1382-1364	V			(70.73)
XLVII	$u'\Pi \leftarrow X^1\Sigma_g^+$ (0_u^+)	Absorption	1385-1367	V			(70.73)
XLVIII	$v \leftarrow X^1\Sigma_g^+$	Absorption	1374-1365			Very weak	(70.73)

System	Transition	Sources	Wavelength Limits	Degrading	Band Head, $\nu_{0,0}$	Remarks	Bibliography
XLIX	$\omega\Sigma^+ \leftarrow X^1\Sigma_g^+$ (1_u)	Absorption	1370-1349				(70.73)
L	Repulsive states $\leftarrow X^1\Sigma_g^+$	Absorption	1461-1447				(70.73)
LI	$D^3\Sigma_g^- \to B^3\Pi_u$ $\left(1_g,\,0_g^+\right)\left(0_u^+\right)$	Emission	4420-4000	R			(70.73, 51.30, 47.27)
LII	$E^3\Sigma_u^- \to B^3\Pi_u$ (1_u)	Emission	4360-4000	R			(72.87)
LIII	$F^1\Sigma_g^+ \to B^3\Pi_u$ $\left(0_g^+\right)$	Emission	3460-3015	R			(58.40, 33.14)
LIV	$J^3\Pi_g \to B^3\Pi_u$ $\left(1_g,\,0_g^+\right)$	Emission	2785-2731				(58.40)

System	Transition	Sources	Wavelength Limits	Degrading	Band Head, $\nu_{0,0}$	Remarks	Bibliography
LV	$a\Pi(2,\,1_g) \to$ Repulsive states	Emission	4520-2910			5 groups of bands	(70.73)
LVI	$b\Pi(1,\,0_g) \to$ Repulsive states	Emission	3775-2730			5 groups of bands	(70.73)
LVII	$H^1\Sigma_u^+\left(0_u^+\right) \to$ Repulsive states	Emission	2730-2486			5 groups of bands	(70.73, 46.24, 15.1)
LVIII	$I^1\Sigma_u^+\left(0_u^+\right) \to$ Repulsive states	Emission	4155-2687			5 groups of bands	(70.73, 46.24)
LIX	$K^1\Pi(1_g) \to$ Repulsive states	Emission	2672-2448			3 groups of bands	(70.73, 47.26, 47.25)

I_2

I. $\underline{A^3\Pi_u(1_u) \leftarrow X^1\Sigma_g^+(0_g^+)\ \text{System}}$

Band heads, λ (31.11):

v', v''	2	3	4	5
0		8950	9124	9299
1		8917	9084	9265
2		8887	9055	9230
3			9024	9197
4	8667	8828	8994	9168
5	8641	8801	8967	9137
...				
21	8384	8534		
22	8376			

III. $\underline{B^3\Pi_u(0_u^+) \rightleftarrows X^1\Sigma_g^+(0_g^+)\ \text{System}}$

Band heads, λ (n.p. 136, 74.125, 73.119, 73.118, 65.53, 65.51, 23.2):

v', v''	4	5	6	7	8	9	10	11	12
0	6721.0	6816.4	6914.0	7013.9	7115.9	7220.4	7327.2	7436.5	7548.4
1	6665.4	6759.2	6855.2	6953.3	7053.6	7156.2	7261.2	7368.5	7478.3
2	6611.3	6703.7	6798.1	6894.5	6993.1	7094.0	7197.1	7312.5	7410.4
3	6558.9	6649.7	6742.6	6837.5	6934.4	7033.6	7134.9	7238.5	7344.5
4	6505.3	6597.3	6688.7	6782.0	6877.5	6975.0	7074.6	7176.5	7280.6
5	6458.3	6546.4	6636.4	6728.3	6822.1	6918.1	7016.1	7116.3	7218.7
6	6410.2	6496.9	6585.5	6676.0	6768.4	6862.9	6959.3	7057.9	7158.6
7	6363.4	6448.9	6536.2	6625.3	6716.3	6809.3	6904.2	7001.2	7100.3
8	6318.0	6402.2	6488.3	6576.1	6665.7	6757.3	6850.8	6946.3	7043.8
9	6273.8	6356.9	6441.7	6528.3	6616.6	6706.8	6798.9	6893.0	6989.0
10	6231.0	6312.9	6396.6	6481.9	6569.0	6657.9	6748.6	6841.3	6935.8
11	6189.4	6270.2	6352.7	6436.9	6522.7	6610.4	6699.8	6791.1	6884.3
12	6149.0	6228.8	6310.1	6393.2	6477.9	6564.3	6652.5	6742.5	6834.3
13	6109.7	6188.5	6268.8	6350.8	6434.4	6519.6	6606.6	6695.4	6785.9
14	6071.6	6149.4	6228.6	6309.6	6392.1	6476.3	6562.1	6649.7	6739.0
15	6034.7	6111.5	6189.9	6269.7	6351.2	6434.3	6519.0	6605.4	6693.5

VI. $E^3\Sigma_u^-(1_u) \to X^1\Sigma_g^+\left(0_g^+\right)$ System

Band heads, λ (58.39):

v', v''	6	7	8	9	10	11	12	13	14
0		2287.9	2298.7		2320.2	2330.7	2341.4	2352.1	2362.6
1		2283.1	2293.9	2304.0	2315.2	2325.6	2336.2	2347.0	2357.5
2	2267.4		2289.0	2299.3	2310.2	2320.9		2342.0	2352.6
3	2262.7	2276.2	2284.3	2294.8	2305.4	2316.2	2326.7	2337.2	2347.6
4	2258.4	2269.1		2290.2	2300.6	2311.6			2342.9
5	2254.3	2264.9	2275.5	2285.9		2306.4			2338.3
6	2250.0	2760.7	2271.3	2281.6			2312.5	2323.2	2333.6
7	2245.9	2256.5	2267.0			2297.4	2308.1		2329.3
8	2241.9								
9	2237.7								

VII. $F^1\Sigma_g^+\left(0_g^+\right) \leftarrow X^1\Sigma_g^+\left(0_g^+\right)$ System

Band heads, λ (72.87, 51.30):

v', v''	43	44	45	46	47	48	49	50	51
0			2568.6	2577.9	2587.3	2597.3	2606.3	2615.6	2624.8
1				2571.6	2581.0	2590.4	2599.8	2609.0	2618.2
2		2546.3	2555.8	2565.3	2574.8	2584.1	2593.4	2602.6	2611.8
3	2530.6	2540.2	2549.7	2559.1	2568.5				
4	2524.7	2534.2	2543.7	2553.1	2562.4				2599.2
5	2518.8	2528.3							
6	2512.9				2550.6	2559.6			
7					2544.6				
8									
9			2514.7						

I_2

VIII. $\quad G^3\Sigma_u^- \left(0_u^+\right) \rightleftarrows X^1\Sigma_g^+ \left(0_g^+\right)$ System

Band heads, λ (Intensity) (58.40, 51.30, 40.21):

v', v''	λ	Intensity	v', v''	λ	Intensity
5, 69	2735.6	3	0, 58	2685.8	4
4, 66	2721.0	5	1, 58	2678.9	3
3, 65	2720.5	5	0, 57	2677.4	4
4, 65	2713.8	6	0, 56	2668.7	4
3, 64	2713.1	6	0, 55	2660.0	4
3, 63	2705.3	5	0, 54	2651.2	4
2, 62	2704.5	5	0, 53	2642.2	3
1, 61	2703.5	5	1, 53	2635.5	3
1, 60	2695.5	4	0, 52	2633.1	3
0, 59	2694.1	3	1, 52	2626.5	2
1, 59	2687.3	4			

IX. $\quad H^1\Sigma_u^+ \left(0_u^+\right) \rightleftarrows X^1\Sigma_g^+ \left(0_g^+\right)$ System

Band heads, λ (58.39):

v', v''	5	6	7	8	9	10	11	12	13
0								2195.0	
1						2172.2	2181.8	2191.3	
2						2168.8	2178.4	2187.7	2197.2
3						2165.2	2174.6	2184.2	2193.7
4						2161.6	2171.0	2180.3	2190.0
5						2158.3	2167.4	2176.7	
6					2146.3	2165.0	2164.1		
7					2142.6	2151.8	2160.9		
8					2193.3	2148.6	2167.5		
9					2136.1	2145.2			
10					2132.7				
11			2111.2	2120.4	2129.7	2139.3			
12					2126.6	2136.1			
13			2105.5	2114.3	2123.4	2132.7			
14			2102.4	2111.2	2120.4				
15	2081.7	2090.5	2099.3	2108.1	2117.1				
16	2078.8	2087.6	2096.3	2105.0					
17	2075.7	2084.6	2093.3						
18	2072.9	2081.7							
19	2069.9	2078.8							
20	2066.7	2075.7							

X. $I\,^1\Sigma_u^+\left(0_u^+\right) \rightleftarrows X\,^1\Sigma_g^+\left(0_g^+\right)$ System

Band heads, λ (70.73, 58.40, 46.23, 35.19, 29.8):

"Cordes Bands"

(3, 4)	1948.1	(6, 2)	1914.7	(10, 1)	1885.2
(2, 3)	1945.9	(10, 2)	1892.3	(21, 3)	1845.0
(1, 2)	1944.7	(9, 1)	1890.8	(23, 4) } (18, 1) }	1843.5
(5, 2)	1920.7	(8, 0) } (12, 3) }	1888.6	(17, 0)	1841.8
(4, 1)	1918.6	(11, 2)	1887.0	(22, 3)	1840.4
(3, 0)	1916.6			(28, 0)	1794.3

XI. $c\Pi(1_u) \leftarrow X\,^1\Sigma_g^+\left(0_g^+\right)$ System

Band heads (70.73)

Represents a Rydberg Series $\nu_{oo} = 75814 - \left[R/(n-3.588)^2\right]$ where $n = 6,\ 7,\ 8,\ \cdots 36$

λ (Intensity) for $n = 6$

v', v''	0	1	2
0	1756.1(10)	1762.7(6)	1769.3(4)
1	1749.4(8)		
2	1743.9(8)		
3	1738.1(6)	1744.6(6)	
4	1731.7(4)		
5	1725.4(4)		

XII. $d\Delta(1_u) \leftarrow X\,^1\Sigma_g^+\left(0_g^+\right)$ System

Band heads, λ (Intensity) (70.73):

v', v''	0	1
0	1730.0(2)	1736.3(4)
1	1722.5(2)	
2	1723.1(4)	1721.3(4)
3	1707.3(4)	1713.6(4)

I_2

XIII. $e\Sigma^+\left(0_u^+\right) \leftarrow X^1\Sigma_g^+\left(0_g^+\right)$ System

Band heads (70.73)

Represents a Rydberg Series $\nu_{oo} = 75814 - \left[R/(n-3.449)^2\right]$ where n = 6, 7, 8, $\cdots$ 34

λ (Intensity) for n = 6

v', v''	0	1
0	1696.3(10)	
1	1688.1(8)	1694.2(4)
2	1682.1(8)	
3	1676.1(6)	

XIV. $L^3\Pi(1_u) \leftarrow X^1\Sigma_g^+\left(0_g^+\right)$ System

Band heads, λ (Intensity) (70.73):

v', v''	0	1	2
0	1616.9(8)	1622.5(4)	
1	1614.0(8)	1619.6(6)	1625.2(2)
2	1611.3(8)		
3			
4	1605.8(4)		

XV. $\alpha\Sigma_u^-\left(1_u, 0_u^+\right) \leftarrow X^1\Sigma_g^+\left(0_g^+\right)$ System

Band heads, λ (Intensity) (70.73)

Probably represents a Rydberg state

v', v''	0	1	2	3	4
0	1613.6(6)	1619.2(4)	1624.9(2)		
1	1609.2(5)	1614.7(4)	1620.2(2)		
2	1604.9(4)	1610.5(2)	1616.0(4)	1621.4(2)	1626.9(2)
3				1614.5(4)	
4	1596.6(6)			1610.3(2)	
5	1592.6(4)	1597.9(4)			

XVI. $f\Pi\left(0_u^+\right) \leftarrow X^1\Sigma_g^+\left(0_g^+\right)$ System

Band heads (90.73)

Represents a Rydberg Series $\nu_{oo} = 80895 - \left[R/(n-3.544)^2\right]$ where n = 6, 7, 8

λ (Intensity) for n = 6

v', v''	0	1
0	1594. 9(8)	1600. 4(6)
1	1589. 5(4)	1595. 0(6)

XVII. $M^3\Pi_u\left(0_u^+\right) \leftarrow X^1\Sigma_g^+\left(0_g^+\right)$ System

Band heads, λ (Intensity) (70.73):

v', v''	0	1	2
0	1591. 2(8)	1596. 7(8)	1602. 1(2)
1		1593. 3(4)	1598. 7(2)
2	1584. 2(8)	1589. 6(6)	
3			
4	1577. 3(6)		
5	1574. 0(6)		

XVIII. $g\Pi(1_u) \leftarrow X^1\Sigma_g^+\left(0_g^+\right)$ System

Band heads (70.73)

Represents a Rydberg Series $\nu_{oo} = 75814 - \left[R/(n-1.048)^2\right]$ where n = 4, 5, 6, $\cdots$ 33

v', v''	0	1	2	3	4
0	1581. 9(8)	1587. 2(2)	1592. 6(4)		
1	1577. 2(8)	1582. 6(8)			
2		1577. 7(8)	1583. 0(4)		
3		1573. 1(6)			
4		1568. 5(6)	1573. 8(4)		
5		1564. 1(6)	1569. 3(4)	1574. 6(4)	
6					1575. 3(4)

$$I_2$$

XIX. $h\Pi(1_u) \leftarrow X^1\Sigma_g^+\left(0_g^+\right)$ System

Band heads, λ (Intensity) (70.73)

Represents a Rydberg Series although only n = 6 has been observed

v', v''	0	1	2	3
0	1573. 9(6)	1579. 2(6)	1584. 5(8)	
1	1567. 3(8)	1572. 6(6)		
2	1561. 4(6)	1566. 6(6)		
3		1560. 7(8)	1566. 0(6)	1571. 3(6)

XX. $i\Sigma_u^+(1_u) \leftarrow X^1\Sigma_g^+\left(0_g^+\right)$ System

Band heads (70.73)

Represents a Rydberg Series $\nu_{oo} = 80895 - \left[R/(n-3.484)^2\right]$ where n = 6, 7, 8

λ (Intensity) for n = 6

v', v''	0	1
0	1573. 0(8)	
1	1566. 9(8)	1572. 1(4)
2	1561. 1(8)	1566. 4(6)

XXI. $j\Sigma_u^-\left(1_u, 0_u^+\right) \leftarrow X^1\Sigma_g^+\left(0_g^+\right)$ System

Band head (70.73)

Represents a Rydberg Series $\nu_{oo} = 80895 - \left[R/(n-3.343)^2\right]$ where n = 6, 7, 8

The existence of this is uncertain

For n = 6, (0, 0) 1555. 6(2)

XXII. $\quad$ $\underline{k\Sigma_u^+\left(0_u^+\right) \leftarrow X^1\Sigma_g^+\left(0_g^+\right) \text{ System}}$

Band heads (70.73)

Represents a Rydberg Series $v_{oo} = 75814 - \left[R/(n-0.900)^2\right]$ where n = 4, 5, 6, $\cdots$ 16

λ (Intensity) for n = 4

v', v''	0	1	2
0	1552. 9(4)	1558. 0(4)	
1	1548. 3(6)	1553. 4(4)	
2	1543. 6(4)		1548. 7(4)
3	1539. 2(4)		
4			
5	1530. 8(4)		

XXIII. $\quad$ $\underline{1\Pi(1_u) \leftarrow X^1\Sigma_g^+\left(0_g^+\right) \text{ System}}$

Band heads, λ (Intensity) (70.73)

Represents a Rydberg Series $v_{oo} = 75814 - \left[R/(n-0.843)^2\right]$

Only two bands of the n = 4 system have been identified

(v', v'')	(0, 0)	(1, 0)
λ	1543. 1	1538. 1
Intensity	2	2

XXIV. $\quad$ $\underline{N^1\Pi_u(1_u) \leftarrow X^1\Sigma_g^+\left(0_g^+\right) \text{ System}}$

Band heads, λ (Intensity) (70.73):

v', v''	0	1
0	1539. 5(4)	1544. 6(2)
1	1536. 1(6)	1541. 0(2)
2	1532. 7(4)	

XXV. $\quad$ $m\Pi(1_u) \leftarrow X^1\Sigma_g^+\left(0_g^+\right)$ System

Band heads, λ (Intensity) (70.73)

Represents a Rydberg Series, although only n = 5 has been observed

(v', v'')	(0, 1)	(0, 0)	(1, 0)
λ	1505.0	1500.2	1495.3
Intensity	6	6	4

XXVI. $\quad$ $\beta\Sigma^-(1_u) \leftarrow X^1\Sigma_g^+\left(0_g^+\right)$ System

Band heads, λ (Intensity) (70.73)

Probably represents a Rydberg Series

v', v''	0	1	2
0	1484.3(2)	1489.0(2)	1493.7(2)
1	1480.7(4)		1490.0(6)

XXVII. $\quad$ $\beta'\Sigma^-\left(0_u^+\right) \leftarrow X^1\Sigma_g^+\left(0_g^+\right)$ System

Band heads, λ (Intensity) (70.73)

Probably represents a Rydberg Series

v', v''	0	1
0	1483.5(2)	
1	1479.9(2)	
2	1476.4(8)	
3	1472.8(2)	
4	1469.3(4)	1474.0(2)
5		1470.5(2)

XXVIII. $o\Pi(1_u) \leftarrow X^1\Sigma_g^+\left(0_g^+\right)$ System

Band heads (70.73)

Represents a Rydberg Series $\nu_{oo} = 80895 - \left[R/(n-1.139)^2\right]$ where n = 4, 5

λ (Intensity) for n = 4

v', v''	0	1	2
0	1481.9(6)	1486.5(6)	1491.2(6)
1	1477.2(6)	1481.9(6)	1486.6(4)

XXIX. $p\Sigma^+\left(0_u^+\right) \leftarrow X^1\Sigma_g^+\left(0_g^+\right)$ System

Band heads, λ (Intensity) (70.73)

Only four bands have been identified for n = 5

(v', v'')	(1, 1)	(0, 0)	(1, 0)	(2, 0)
λ	1480.7	1480.2	1476.1	1472.0
Intensity	4	6	6	10

XXX. $o'\Pi\left(0_u^+\right) \leftarrow X^1\Sigma_g^+\left(0_g^+\right)$ System

Band heads (70.73)

Represents a Rydberg Series $\nu_{oo} = 80895 - \left[R/(n-1.113)^2\right]$ where n = 4, 5

λ (Intensity) for n = 4

v', v''	0	1	2
0	1476.6(10)	1481.2(8)	1485.9(2)
1	1471.7(4)	1476.3(6)	1481.0(2)
2	1467.0(10)	1471.6(6)	

I_2

XXXI. $q\Sigma^+(1_u) \leftarrow X^1\Sigma_g^+\left(0_g^+\right)$ System

Band heads (70.73)

Represents a Rydberg Series $v_{oo} = 80895 - \left[R/(n-1.093)^2\right]$ where n = 4, 5

λ (Intensity) for n = 4

v', v''	0	1
0	1472. 1(8)	1476. 7(6)
1	1471. 8(8)	1467. 2(4)

XXXII. $\gamma\Sigma^+(1_u) \leftarrow X^1\Sigma_g^+\left(0_g^+\right)$ System

Band heads, λ (Intensity) (70.73)

Probably represents a Rydberg Series

(v', v'')	(1, 1)	(0, 0)	(1, 0)
λ	1463. 6	1462. 5	1459. 0
Intensity	4	8	8

XXXIII. $\gamma'\Sigma^+\left(0_u^+\right) \leftarrow X^1\Sigma_g^+\left(0_g^+\right)$ System

Band heads, λ (Intensity) (70.73)

Probably represents a Rydberg Series

(v', v'')	(1, 1)	(0, 0)	(1, 0)	(2, 0)
λ	1446. 3	1445. 4	1441. 8	1438. 4
Intensity	4	8	4	2

XXXIV. $r\Sigma^-\left(1_u,\ 0_u^+\right) \leftarrow X^1\Sigma_g^+\left(0_g^+\right)$ System

Band heads (70.73)

Represents a Rydberg Series $v_{oo} = 80895 - \left[R/(n-0.909)^2\right]$ where n = 4, 5

λ (Intensity) for n = 4

v', v''	0	1	2
0	1440. 7(8)	1445. 2(6)	
1		1440. 8(8)	1446. 4(6)
2	1432. 3(6)	1436. 7(4)	1441. 0(2)
3	1428. 0(10)		

XXXV. $\delta\Sigma^-(1_u) \leftarrow X^1\Sigma_g^+\left(0_g^+\right)$ System

Band heads, λ (Intensity) (70.73)

Probably represents a Rydberg Series

(v', v'')	(0, 1)	(0, 0)	(1, 0)
λ	1440. 2	1435. 8	1432. 3
Intensity	4	4	6

XXXVI. $\delta'\Sigma^-\left(0_u^+\right) \leftarrow X^1\Sigma_g^+\left(0_g^+\right)$ System

Band heads, λ (Intensity) (70.73)

Probably represents a Rydberg Series

(v', v'')	(0, 1)	(1, 1)	(0, 0)	(1, 0)
λ	1438. 7	1435. 3	1434. 4	1431. 0
Intensity	4	2	2	4

XXXVII. $\underline{\zeta\Pi(1_u) \leftarrow X^1\Sigma_g^+\left(0_g^+\right) \text{ System}}$

Band heads, λ (Intensity) (70.73)

Probably represents a Rydberg Series

v', v''	0	1	2
0	1433. 3(6)	1437. 6(4)	1442. 0(2)
1	1429. 7(2)	1434. 0(2)	
2		1430. 4(4)	1434. 9(2)
3	1422. 5(2)	1426. 9(4)	1431. 2(6)

XXXVIII. $\underline{\zeta'\Pi\left(0_u^+\right) \leftarrow X^1\Sigma_g^+\left(0_g^+\right) \text{ System}}$

Band heads, λ (Intensity) (70.73)

Probably represents a Rydberg Series

(v', v'')	(3, 3)	(0, 0)	(1, 0)
λ	1435. 5	1430. 5	1427. 2
Intensity	2	4	4

XXXIX. $\underline{s'\Pi(1_u) \leftarrow X^1\Sigma_g^+\left(0_g^+\right) \text{ System}}$

Band heads (70.73)

Represents a Rydberg Series $v_{oo} = 80895 - \left[R/(n-0.814)^2\right]$ where $n = 4,\ 5$

λ (Intensity) for $n = 4$

(v', v'')	(0, 1)	(0, 2)	(0, 0)
λ	1431. 6	1436. 0	1427. 3
Intensity	4	2	6

XL. $s\Pi\left(0_u^+\right) \leftarrow X\,{}^1\Sigma_g^+\left(0_g^+\right)$ System

Band heads (70.73)

Represents a Rydberg Series $\nu_{oo} = 80895 - \left[R/(n-0.768)^2\right]$ where n = 4, 5

λ (Intensity) for n = 4

v', v''	0	1	2	3
0	1420.7(6)	1425.0(6)	1429.3(2)	
1	1416.0(6)	1420.3(4)	1424.6(2)	1429.0(4)
2	1411.4(4)	1415.7(2)	1420.0(4)	1415.3(4)
3	1406.9(2)	1411.1(2)		

XLI. $t\Pi(1_u) \leftarrow X\,{}^1\Sigma_g^+\left(0_g^+\right)$ System

Band heads, λ (Intensity) (70.73)

Represents a Rydberg Series, although only n = 6 has been observed

v', v''	0	1
0	1413.8(6)	1418.1(6)
1	1409.6(8)	
2	1405.4(8)	1409.6(8)
3	1401.3(8)	1405.6(8)
4	1397.4(2)	

XLII. $t'\Pi\left(0_u^+\right) \leftarrow X\,{}^1\Sigma_g^+\left(0_g^+\right)$ System

Band heads, λ (Intensity) (70.73)

Represents a Rydberg Series, although only n = 6 has been observed

v', v''	0	1
0	1409.3(6)	1413.6(4)
1	1405.2(6)	
2	1401.1(6)	1405.3(6)
3	1397.0(2)	1401.2(6)

XLIII. $\underline{\eta\Sigma^+(1_u) \leftarrow X^1\Sigma_g^+\left(0_g^+\right) \text{ System}}$

Band heads, λ (Intensity) (70.73)

Probably represents a Rydberg Series

(v', v'')	(0, 0)	(1, 0)	(2, 0)	(3, 0)	(4, 0)
λ	1401.1	1397.8	1394.6	1391.3	1388.2
Intensity	8	6	6	4	4

XLIV. $\underline{\eta'\Sigma\left(0_u^+\right) \leftarrow X^1\Sigma_g^+\left(0_g^+\right) \text{ System}}$

Band heads, λ (Intensity) (70.73)

Probably represents a Rydberg Series

(v', v'')	(1, 0)	(2, 0)	(3, 0)	(4, 0)	(5, 0)
λ	1394.7	1391.2	1387.8	1384.5	1381.3
Intensity	6	4	4	8	2

XLV. $\underline{\theta\Pi\left(1_u,\, 0_u^+\right) \leftarrow X^1\Sigma_g^+\left(0_g^+\right) \text{ System}}$

Band heads, λ (Intensity) (70.73)

Probably represents a Rydberg Series

v', v''	0	1	2
0	1385.9(6)	1390.0(4)	
1	1382.5(4)		
2		1383.3(4)	1387.4(4)
3	1376.0(6)	1380.0(6)	

XLVI. $u\Pi(1_u) \leftarrow X^1\Sigma_g^+\left(0_g^+\right)$ System

Band heads, λ (Intensity) (70.73)

Represents a Rydberg Series, although only n = 5 has been observed

v', v''	0	1	2
0	1378. 1(6)	1382. 1(4)	
1	1373. 5(4)	1377. 5(4)	1381. 5(2)
2	1369. 1(2)	1373. 0(2)	
3	1364. 6(4)	1368. 6(2)	
4		1368. 4(2)	

XLVII. $u'\Pi\left(0_u^+\right) \leftarrow X^1\Sigma_g^+\left(0_g^+\right)$ System

Band heads, λ (Intensity) (70.73)

Represents a Rydberg Series, although only n = 5 has been observed

v', v''	0	1	2	3
0	1376. 5(10)	1380. 6(8)	1384. 6(6)	1388. 7(2)
1	1372. 2(10)	1376. 2(4)	1380. 2(2)	
2	1367. 9(6)	1371. 9(6)		

XLVIII. $v \leftarrow X^1\Sigma_g^+$ System

Band heads, λ (Intensity) (70.73)

Probably a forbidden transition

v', v''	0	1
0	1374. 4(2)	1378. 5(2)
1	1369. 9(2)	
2	1365. 4(2)	

I_2

XLIX. $\omega\Sigma^+(1_u) \leftarrow X^1\Sigma_g^+\left(0_g^+\right)$ System

Band heads, λ (Intensity) (70.73)

Represents a Rydberg Series, although only n = 5 has been observed

v', v''	0	1	2	3
0	1361.8(6)			
1	1357.8(2)		1365.7(2)	1369.7(2)
2	1353.8(2)			1365.6(4)
3	1349.9(2)	1353.7(2)		

L.   Repulsive State $\leftarrow X^1\Sigma_g^+\left(0_g^+\right)$ Systems (70.73)

The bands at 1447.1, 1451.9, 1456.6, and 1461.3Å are due to a transition from the v'' = 0, 1, 2, and 3 of the ground state to a repulsive state dissociating into $p^5\ ^2P_{3/2} + p^4s\ ^2P_{3/2}$ atoms at 68545 cm^{-1}. Also, the bands at 1446.9, 1450.9, 1449.5, and 1454.4Å could be explained as due to transitions from the ground state with v'' = 0 and v'' = 1 to two different repulsive states dissociating again into $p^5\ ^2P_{3/2} + p^4s\ ^2P_{3/2}$ atoms at 68545 cm^{-1}.

LI. $D^3\Sigma_g^-\left(1_g, 0_g^+\right) \rightarrow B^3\Pi_u\left(0_u^+\right)$ System

Band heads, λ (70.73, 51.30, 47.27):

v', v''	16	17	18	19	20	21	22	23	24
0	4177.3	4194.6	4211.4	4229.0	4245.9	4262.8	4279.0	4295.2	4310.2
1	4159.7	4177.3	4193.9			4242.9	4259.9	4275.9	4291.7
2	4143.5							4256.9	4273.5
3					4191.1	4207.3		4239.1	4254.8
4	4108.5					4189.7	4205.2		
5						4173.6		4204.2	4333.7
6							4170.8	4186.7	
7							4154.9		
8									4167.7
9	4110.3		4060.4						4151.6
10			4044.7				4105.4	4119.3	

LII. $E^3\Sigma_u^-(1_u) \rightarrow B^3\Pi_u(0_u^+)$ System

Band heads, λ (Intensity) for $^{127}I_2$ (72.87):

v', v''	16	17	18	19	20	21	22
0	4194.6(4)	4211.9(9)	4228.8(9)	4245.6(9)	4262.1(9)	4278.4(9)	4294.4(10)
1	4177.0(7)	4194.0(8)	4210.9(6)	4227.5(3)		4260.1(4)	4276.1(3)
2	4159.6(8)	4176.2(1)		4209.6(1)	4225.8(4)	4242.1(3)	4257.3(1)
3			4175.4(1)	4191.9(3)	4208.2(2)		4239.6(3)
4			4158.5(1)	4174.1(2)	4190.0(2)	4206.5(2)	
5					4173.1(2)		4203.6(2)

LIII. $F^1\Sigma_g^+(0_g^+) \rightarrow B^3\Pi_u(0_u^+)$ System

Band heads, λ (70.73, 58.40, 34.15, 33.14):

v', v''	0	1	2	3	4	5	6	7	8
0	3301.2	3315.3	3329.0	3342.4	3355.8	3369.4	3382.8	3396.0	3409.7
1	3289.7	3303.8	3317.5	3331.1	3344.5	3357.7	3371.3	3385.0	3397.8
2	3278.8	3292.3	3305.6	3319.9	3333.3	3346.2	3359.7	3373.0	3385.6
3	3268.2	3281.5	3294.9	3308.0	3321.7	3334.7	3347.9	3360.9	3374.3
4	3257.1	3270.0	3283.9	3297.0	3310.5	3323.5	3336.3	3349.6	3362.2
5	3246.0	3259.6	3272.9	3286.0	3299.0	3311.8	3325.0	3337.9	3350.4
6	3234.8	3248.5	3261.8	3274.6	3287.8	3301.2	3313.7	3326.7	3338.6
7	3224.3	3237.6	3250.7	3263.5	3276.8	3289.7	3302.9	3315.3	3328.0
8	3214.3	3227.9	3239.8	3252.5	3265.8	3278.6	3290.9	3303.8	3316.2
9	3203.6	3216.8	3229.5	3242.2	3254.9	3267.6	3280.0	3292.5	3304.6
10	3192.8	3206.0	3219.2	3231.3	3244.1	3257.5	3269.6	3281.9	3294.2

LIV. $J^3\Pi_g(1_g, 0_g^+) \rightarrow B^3\Pi_u(0_u^+)$ System

Band heads, λ (70.73, 58.40):

v', v''	0	1	2	3	4	5	6	7	8
0	2762.5	2772.0	2782.0						
1	2753.8	2763.8	2773.0						
2	2745.8	2755.1	2765.1	2774.4					
3	2737.5	2747.3	2756.7	2766.1	2775.4	2784.2			
4		2739.7	2748.9	2758.3	2767.6	2776.6			
5		2731.7	2740.9	2750.0	2759.4	2768.6		2785.7	
6			2733.1	2742.1		2760.1	2769.3	2778.2	
7				2734.0	2743.4	2752.4	2760.8	2770.2	2779.2
8					2736.4	2745.2		2762.5	2771.2
9						2737.5	2746.1		2763.2
10							2738.4	2747.3	2755.9

I_2

LV. $a\Pi(2, 1_g) \to$ Repulsive States

Five groups of diffuse semicontinuous emission bands arise in this level and have for their lower levels different repulsive states dissociating into $^2P + ^2P$ iodine atoms (70.73)

Group	λ	Transition
I	4519.0	
	4473.5	
	4431.8	$a\Pi(2, 1_g) \to {}^3\Sigma_u^+(1_u)$
	4391.4	
II	4357.4	
	4319.1	
	4283.4	$a\Pi(2, 1_g) \to {}^3\Sigma_u^+(0_u^-)$
	4243.8	
	4202.9	
III	3332.7	
	3307.9	
	3282.3	
	3258.3	
	3234.8	$a\Pi(2, 1_g) \to {}^3\Delta_u(1_u^-)$
	3211.4	
	3189.5	
	3169.7	
IV	3143.1	
	3123.4	
	3103.6	$a\Pi(2, 1_g) \to {}^3\Delta_u(2_u)$
	3085.5	
V	3062.8	
	3043.8	
	3025.8	
	3007.2	
	2987.9	$a\Pi(2, 1_g) \to {}^1\Pi_u(1_u)$
	2970.5	
	2953.6	
	2940.7	
	2919.5	

LVI. $b\Pi(1, 0_g) \rightarrow$ Repulsive States

Five groups of semicontinuous bands arise in this level (70.73)

Group	λ	Transition
I	3774.3	
	3721.5	
	3671.9	
	3624.4	$b\Pi(1, 0_g) \rightarrow {}^3\Sigma_u^+(1_u^-)$
	3579.6	
	3537.6	
	3500.4	
II	3741.0	
	3687.6	
	3641.8	
	3601.8	$b\Pi(1, 0_g) \rightarrow {}^3\Sigma_u^+(0_u^-)$
	3516.7	
	3480.0	
III	2904.3	
	2875.1	$b\Pi(1, 0_g) \rightarrow {}^3\Delta_u(1_u)$
	2847.0	
IV	2821.0	
	2793.5	
	2769.0	$b\Pi(1, 0_g) \rightarrow {}^3\Delta_u(2_u)$
	2745.9	
V	2754.8	
	2732.0	$b\Pi(1, 0_g) \rightarrow {}^1\Pi_u(1_u)$

$$I_2$$

LVII. $\underline{H^1\Sigma_u^+\left(0_u^+\right) \rightarrow \text{Repulsive States}}$

"MacLennan Bands" are observed due probably to transitions from the $H^1\Sigma_u^+\left(0_u^+\right)$ to lower unstable states (70.73, 46.25, 15.1)

λ	Intensity	λ	Intensity
4154.4	6	3919.4	3
4124.2	5	3879.9	6
4094.6	8	3855.8	6
4067.6	7	3821.8	3
4044.6	8	3798.0	4
4018.0	8	2712.0	4
3991.1	3	2699.6	3
3945.6	8	2687.5	4

LVIII. $\underline{I^1\Sigma_u^+\left(0_u^+\right) \rightarrow \text{Repulsive States}}$

Five groups of diffuse emission bands arise in this level and have for their lower levels different repulsive states dissociating into $^2P + ^2P$ iodine atoms

Bands seen in emission, λ (Intensity) (70.73, 46.25)

Group		Transition
I	4154.4(6)	
	4124.2(5)	
	4094.6(8)	
	4067.6(7)	$I^1\Sigma_u^+\left(0_u^+\right) \rightarrow {}^1\Sigma_g^+\left(0_g^+\right)$
	4044.6(8)	
	4018.0(8)	
	3991.1(3)	
II	3945.6(8)	$I^1\Sigma_u^+\left(0_u^+\right) \rightarrow {}^3\Sigma_g^-\left(0_g^+\right)$
	3919.4(3)	
III	3879.9(6)	$I^1\Sigma_u^+\left(0_u^+\right) \rightarrow {}^3\Sigma_g^-(1_g)$
	3855.8(6)	

Bands seen in emission, λ (Intensity) (70.73, 46.25)

Group		Transition
IV	3821.8(3)	$I\,^1\Sigma_u^+\left(0_u^+\right) \rightarrow {}^1\Pi_g(1_g)$
	3798.0(4)	
V	2712.0(4)	$I\,^1\Sigma_u^+\left(0_u^+\right) \rightarrow {}^3\Pi_g\left(0_g^+\right)$
	2699.6(3)	
	2687.5(4)	

LIX. $K\,^1\Pi(1_g) \rightarrow$ Repulsive States

Three groups of diffuse emission bands arise in this level and have for their lower level three different repulsive states dissociating into $^2P_{3/2} + {}^2P_{1/2}$ iodine atoms (70.73)

Group	λ	Transition
I	2671.1	
	2661.1	
	2652.5	
	2642.6	
	2633.3	$K\,^1\Pi(1_g) \rightarrow {}^3\Delta_u(1_u)$
	2623.6	
	2616.6	
	2607.5	
II	2593.3	
	2584.4	
	2575.9	
	2569.2	
	2562.2	
	2555.1	$K\,^1\Pi(1_g) \rightarrow {}^3\Delta_u(2_u)$
	2548.9	
	2541.6	
	2534.1	
	2526.7	

I_2

Group	λ	Transition
III	2522.7	
	2516.0	
	2508.6	
	2501.4	
	2494.1	
	2487.2	$K\,^1\Pi(1_g) \rightarrow {}^1\Pi_u(1_u)$
	2481.0	
	2474.1	
	2467.9	
	2461.8	
	2454.9	
	2448.8	

SPECTROSCOPIC CONSTANTS

State	T_o	ω_o	$x_o\omega_o$	B_e	$\alpha_e \times 10^3$	$D_e \times 10^6$	r_e	Remarks	Bibliography
$\omega\Sigma^+$ (1_u)	73432	218						Rydberg	(73.101, 70.73)
v	72757	241						Rydberg	(73.101, 70.73)
$u'\Pi$ (0_u^+)	72647	230						Rydberg	(73.101, 70.73)
$u\Pi(1_u)$	72565	240						Rydberg	(73.101, 70.73)
$\theta\Pi$ $(1_u$ or $0_u^+)$	72157	174						Possibly Rydberg	(73.101, 70.73)
$\eta'\Sigma^+$ (0_u^+)	71522	179						Possibly Rydberg	(73.101, 70.73)
$\eta\Sigma^+$ (1_u)	71372	168						Possibly Rydberg	(73.101, 70.73)
$t'\Pi$ (0_u^+)	70955	211						Rydberg	(73.101, 70.73)
$t\Pi$ (1_u)	70730	212						Rydberg	(73.101, 70.73)

I_2

SPECTROSCOPIC CONSTANTS

State	T_o	ω_o	$\omega_o x_o$	B_e	$\alpha_e \times 10^3$	$D_e \times 10^6$	r_e	Remarks	Bibliography
$s\Pi(0_u^+)$	70388	233						Rydberg	(73.101, 70.73)
$s'\Pi(1_u)$	70064							Rydberg	(73.101, 70.73)
$\zeta'\Pi(0_u^+)$	69904	164						Possibly Rydberg	(73.101, 70.73)
$\zeta\Pi(1_u)$	69771	175						Possibly Rydberg	(73.101, 70.73)
$\delta'\Sigma^-(0_u^+)$	69717	166						Possibly Rydberg	(73.101, 70.73)
$\delta\Sigma^-(1_u)$	69649	171						Possibly Rydberg	(73.101, 70.73)
$r\Sigma^-(1_u, 0_u^+)$	69410	209						Rydberg	(73.101, 70.73)
$\gamma'\Sigma^+(0_u^+)$	69184	172						Possibly Rydberg	(73.101, 70.73)
$\gamma\Sigma^+(1_u)$	68375	164						Possibly Rydberg	(73.101, 70.73)
$q\Sigma^+(1_u)$	67930	229						Rydberg	(73.101, 70.73)

294

SPECTROSCOPIC CONSTANTS

State	T_o	ω_o	$x_o\omega_o$	B_e	$\alpha_e \times 10^3$	$D_e \times 10^6$	r_e	Remarks	Bibliography
$o'\Pi(0_u^+)$	67723	226						Rydberg	(73.101, 70.73)
$p\Sigma^+(0_u^+)$	67559	188						Rydberg	(73.101, 70.73)
$o\Pi(1_u)$	67483	212						Rydberg	(73.101, 70.73)
$\beta'\Sigma^-(0_u^+)$	67410	164						Possibly Rydberg	(73.101, 70.73)
$\beta\Sigma^-(1_u)$	67373	164						Possibly Rydberg	(73.101, 70.73)
$m\Pi(1_u)$	66660	214						Rydberg	(73.101, 70.73)
$N^1\Pi_u(1_u)$	64956	145							(73.101, 70.73)
$1\Pi(1_u)$	64803	214						Rydberg	(73.101, 70.73)
$k\Sigma_u^+(0_u^+)$	64395	195						Rydberg	(73.101, 70.73)
$j\Sigma_u^-(1_u,0_u^+)$	64282							Rydberg	(73.101, 70.73)

SPECTROSCOPIC CONSTANTS

State	T_o	ω_o	$\omega_o x_o$	B_e	$\alpha_e \times 10^3$	$D_e \times 10^6$	r_e	Remarks	Bibliography
$i\Sigma_u^+(1_u)$	63574	248						Rydberg	(73.101, 70.73)
$h\Pi(1_u)$	63538	265						Rydberg	(73.101, 70.73)
$g\Pi(1_u)$	63216	190						Rydberg	(73.101, 70.73)
$M^3\Pi_u \left(0_u^+\right)$	62844	144							(73.101, 70.73)
$f\Pi\left(0_u^+\right)$	62700	211						Rydberg	(73.101, 70.73)
$\alpha\Sigma_u^- \left(1_u, 0_u^+\right)$	61972	173	1.5					Possibly Rydberg	(73.101, 70.73)
$L^3\Pi \left(1_u\right)$	61847	110	1.9						(73.101, 70.73)
$e\Sigma^+ \left(0_u^+\right)$	58953	284						Rydberg	(73.101, 70.73)
$K^1\Pi \left(1_g\right)$	58578	120							(73.101, 70.73)

SPECTROSCOPIC CONSTANTS

I_2

State	T_o	ω_o	$x_o\omega_o$	B_e	$\alpha_e \times 10^3$	$D_e \times 10^6$	r_e	Remarks	Bibliography
$d\Delta(1_u)$	57805	250							(73.101, 70.73)
$c\Pi(1_u)$	56944	219						Rydberg	(73.101, 70.73)
$b\Pi(1,0_g)$	56000	360							(73.101, 70.73)
$J^3\Pi_g$ $(1_g, 0^+_g)$	51923	111.7	0.705						(73.101, 70.73)
$a\Pi(2, 1_g)$	51528	215						Rydberg	(73.101, 70.73)
$I^1\Sigma^+_u(0^+_u)$	51405	168.47	0.938						(73.101, 70.73)
$H^1\Sigma^+_u(0^+_u)$	48072	79							(73.101, 70.73)
$G^3\Sigma^-_u(0^+_u)$	47148	96.0	0.505						(73.101, 70.73)

297

I_2

SPECTROSCOPIC CONSTANTS

State	T_o	ω_o	$x_o\omega_o$	B_e	$\alpha_e \times 10^3$	$D_e \times 10^6$	r_e	Remarks	Bibliography
$F\,^1\Sigma_g^+$ $\left(0_g^+\right)$	46009	103.6	0.095				3.6		(73.101, 72.87, 70.73)
$E\,^3\Sigma_u^-$ $\left(1_u\right)$	45230	93.4	0.6				3.7		(73.101, 72.87, 70.73)
$D\,^3\Sigma_g^-$ $\left(1_g,\,0_g^+\right)$	41483.1	101.88	0.3				4.1		(73.101, 70.73, 70.70)
$C\,^3\Sigma_u^+$ $\left(1_u\right)$	40473.5	203.49	0.42						(73.101, 70.73)
$B\,^3\Pi_u$ $\left(0_{u.}^+\right)$	(a) 15770.45	(b) 124.97	0.693		289.944	0.004381	3.0309		(74.125, 73.118, 73.101, 70.73)
$A\,^3\Pi_u$ $\left(1_u\right)$	11803	43	1.0						(73.118, 70.73)
$X\,^1\Sigma_g^+$ $\left(0_g^+\right)$	0	(b) 214.50	(c) 0.615	0.037389	11.298	0.00425	2.6663	$y_e\omega_e = -1.30 \times 10^{-3}$	(73.105, 73.101, 70.73, 52.33)

SPECTROSCOPIC CONSTANTS

State	T_o	ω_o	$x_o\omega_o$	B_e	$\alpha_e \times 10^3$	$D_e \times 10^6$	r_e	Remarks	Bibliography

(a) T_e, (b) ω_e, (c) $x_e\omega_e$

Dissociation energy = 1.54 eV, 35.57 kcal/mole, 12440.1 cm^{-1} (73.105).

Perturbations and General Information

Potential energy curves – RKR potential (73.105):

	State	v	$E(v) cm^{-1}$	$r_{min}(\text{Å})$	$r_{max}(\text{Å})$
$T_e = 15770.45$	$B^3\Pi\left(0_u^+\right)$	0	62.66	2.963	3.093
		1	186.82	2.921	3.148
		2	309.44	2.895	3.189
		3	430.49	2.874	3.224
		4	549.97	2.856	3.256
		5	667.88	2.842	3.286
		6	784.18	2.828	3.314
		7	898.83	2.816	3.341
		8	1011.82	2.806	3.367
		9	1123.17	2.796	3.393
		10	1232.80	2.786	3.418
		11	1340.72	2.778	3.444
		12	1446.90	2.770	3.469
		13	1551.33	2.762	3.493
		14	1653.97	2.755	3.518
		15	1754.80	2.749	3.543

Potential energy curves for low-lying systems (73.118):

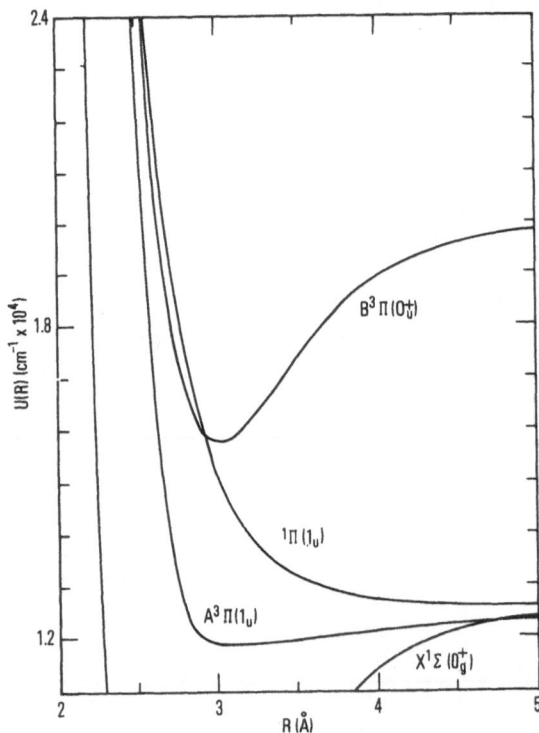

Predissociation of the $B^3\Pi\left(0_u^+\right)$ state is observed from the crossing by the $^1\Pi(1_u)$ repulsive state at low v' values (v' < 6) (74.135, 73.106, 72.88, 72.83).

Vibrational deactivation and quenching rates of the $B^3\Pi\left(0_u^+\right)$ state by various gases (73.110).

Z_o^k - rate constant for vibrational deactivation between first and zeroth vibrational levels

Z_o^T - rate of electronic quenching by collision with gas reservoir (independent of vibrational level)

System	Z_o^k, cm^3/mole sec	σ_k^2, (Å)^2	Z_o^T, cm^3/mole sec	σ_T^2, (Å)^2
I_2 - He	1.058×10^{-12}	0.0839	0.745×10^{-12}	0.0593
I_2 - Ar	1.493×10^{-12}	0.3480	4.870×10^{-12}	1.1310
I_2 - N$_2$	2.260×10^{-12}	0.4480	6.360×10^{-12}	1.0730

Radiative lifetimes of the $B^3\Pi\left(0_u^+\right)$ state have been studied and are found to be strongly dependent on v', varying from less than 0.4 μsec for v' = 6 to greater than 7 μsec for v' = 56 (73.103).

Self quenching cross sections are also given in (73.103) varying from 47 to 90×10^{-16} cm^2.

Foreign gas quenching cross sections have been measured for Ar, Br$_2$, CO$_2$, H$_2$, He, Kr, N$_2$, NO, Ne, O$_2$, SF$_6$ and Xe showing a dependence on v' and the particular gas used (73.103).

A study of the photo dissociation quantum yields for the $^1\Pi(1u) \rightleftharpoons X^1\Sigma_g^+$ and the $B^3\Pi_u\left(0_u^+\right) \rightleftharpoons X^1\Sigma_g^+$ systems is given in (72.88a).

I_2

Franck-Condon factors for $B^3\Pi\left(0_u^+\right) \rightleftarrows X^1\Sigma_g^+\left(0_g^+\right)$ system (n.p. 136):

v', v''	4	5	6	7	8	9	10
0	.2831-04	.1308-03	.4903-03	.1532-02	.4087-02	.9419-02	.1900-01
1	.2585-03	.1043-02	.3364-02	.8899-02	.1961-01	.3639-01	.5704-01
2	.1201-02	.4196-02	.1150-01	.2523-01	.4465-01	.6353-01	.7131-01
3	.3778-02	.1131-01	.2590-01	.4582-01	.6205-01	.6187-01	.4086-01
4	.9864-02	.2284-01	.4270-01	.5841-01	.5551-01	.3282-01	.5151-02
5	.1740-01	.3656-01	.5395-01	.5310-01	.2919-01	.3617-02	.4642-02
6	.2807-01	.4779-01	.5282-01	.3242-01	.5198-02	.3638-02	.2664-01
7	.3872-01	.5163-01	.3901-01	.1015-01	.1142.02	.2192-01	.3393-01
8	.4628-01	.4584-01	.1958-01	.9323-04	.1438-01	.3240-01	.1797-01
9	.4814-01	.3240-01	.4448-02	.5548-02	.2799-01	.2372-01	.1669-02
10	.4338-01	.1666-01	.1021-03	.1849-01	.2864-01	.7278-02	.3315-02
11	.3326-01	.4600-02	.6177-02	.2740-01	.1713-01	.1380-05	.1651-01
12	.2076-01	.1539-04	.1674-01	.2601-01	.4356-02	.6391-02	.2365-01
13	.9433-02	.3049-02	.2461-01	.1638-01	.6198-04	.1740-01	.1748-01
14	.2106-02	.1071-01	.2552-01	.5625-02	.5504-02	.2189-01	.5783-04
15	.2616-04	.1868-01	.1971-01	.1862-03	.1457-01	.1667-01	.2146-04

BIBLIOGRAPHY

(15. 1)   Ultraviolet Emission,
J. C. MacLennan,
Proc. Roy. Soc. A 91, 23-9

(23. 2)   B ⇄ X System in Absorption and Fluorescence,
R. Mecke,
Ann. Physik 71, 104-34

(25. 3)   B ← X System, Convergence Interpretation,
J. Franck,
Trans. Faraday Soc. 21, 536-42

(27. 4)   B ← X System, Vibrational and Rotational Analysis,
F. W. Loomis,
Phys. Rev. 29, 112-34

(28. 5)   B → X System, Resonance Series,
R. W. Wood and F. W. Loomis,
Philos. Mag. 6, 231-8

(28. 6)   Ultraviolet Absorption at High Temperature,
P. Pringsheim and B. Rosen,
Z. Physik 50, 1-14

(29. 7)   Far Ultraviolet Absorption,
M. Kimura and M. Miyanishi,
Sci. Papers Inst. Phys. Chem. Res. Japan 10, 32-42

(29. 8)   Far Ultraviolet Absorption,
H. Sponer and W. W. Watson,
Z. Physik 56, 184-96

(30. 9)   B → X System, Discharge and Fluorescence,
L. A. Turner,
Z. Physik 65, 480-8

(31. 10)   B ← X System, Convergence,
W. G. Brown,
Phys. Rev. 38, 709-11

(31. 11)   A ← X System,
W. G. Brown,
Phys. Rev. 38, 1187-9

I_2

(32. 12)   Fluorescence by Ultraviolet Excitation at High Temperature,
           E. Hirschlaff,
           Z. Physik 75, 315-24

(33. 13)   General Ultraviolet Emission,
           W. E. Curtis and S. F. Evans,
           Proc. Roy. Soc. A 141, 603-25

(33. 14)   E. Skorko,
           Nature 131, 366-7

(34. 15)   Ultraviolet Absorption at 1000°C,
           E. Skorko,
           Acta Phys. Polon. 3, 191-6

(34. 16)   Theoretical Discussion,
           R. S. Mulliken,
           Phys. Rev. 46, 549-71

(35. 17)   Ultraviolet Absorption Between 20-1150°C,
           D. T. Warren,
           Phys. Rev. 47, 1-6

(35. 18)   Resonance Series in the Ultraviolet at Elevated Temperatures,
           F. Duschinsky, E. Hirschlaff, and P. Pringsheim,
           Physica 2, 439-48

(35. 19)   Far Ultraviolet Absorption,
           H. Cordes,
           Z. Physik 97, 603-24

(36. 20)   B → X System, Discharge,
           Y. Uchida,
           Sci. Papers Inst. Phys. Chem. Res. 30, 71-82

(40. 21)   Ultraviolet Fluorescence in the Presence of Foreign Gases,
           A. Elliott,
           Proc. Roy. Soc. A 174, 273-85

(46. 22)   B → X System in Fluorescence,
           D. H. Rank,
           J. Opt. Soc. Am. 36, 299-301

(46. 23)   "Cordes Bands," Vibrational Analysis,
           P. Venkateswarlu,
           Proc. Indian Acad. Sci. A 24, 473-9

(46.24)   MacLennan Bands, Interpretation,
          P. Venkateswarlu,
          Proc. Indian Acad. Sci. A 24, 480-6

(47.25)   P. Venkateswarlu,
          Proc. Indian Acad. Sci. A 25, 119-27

(47.26)   P. Venkateswarlu,
          Proc. Indian Acad. Sci. A 25, 133-7

(47.27)   Ultraviolet Emission in the Presence of Foreign Gases,
          J. Waser and K. Wieland,
          Nature 160, 643-4

(50.28)   B ← X System, Extinction as a Function of Temperature,
          P. Sulzer,
          Helv. Phys. Acta 23, 530-5

(51.29)   Constants of the Ground State,
          D. H. Rank and W. M. Baldwin,
          J. Chem. Phys. 19, 1210-1

(51.30)   Ultraviolet Emission in the Presence of Foreign Gases, D → X and
          F → X Systems,
          P. Venkateswarlu,
          Phys. Rev. 81, 821-9

(51.31)   W. Luck,
          Z. Naturforsch. A 6, 313-9

(52.32)   Absorption Intensity,
          P. Sulzer and K. Wieland,
          Helv. Phys. Acta 25, 653-76

(52.33)   Emission in the Presence of Foreign Gases,
          K. Wieland and J. Waser,
          Phys. Rev. 85, 385-6

(54.34)   Ultraviolet Fluorescence,
          C. V. N. Rao and V. R. Rao,
          Indian J. Phys. 28, 403-22

(54.35)   Vacuum Ultraviolet Absorption,
          N. S. Bayliss and J. V. Sullivan,
          J. Chem. Phys. 22, 1615-6

(55.36)   Absorption in Inert Solvents,
          D. F. Evans,
          J. Chem. Phys. 23, 1424-6

I_2

(56. 37)   Electronic States and Potential Energy Diagrams,
            L. Mathieson and A. L. G. Rees,
            J. Chem. Phys. 25, 753-61

(57. 38)   Ionization Potential,
            K. Watanabe,
            J. Chem. Phys. 26, 542-7

(58. 39)   Emission Spectra,
            P. B. V. Haranath and P. T. Rao,
            J. Molec. Spectrosc. 2, 428-63

(58. 40)   Emission in the Presence of Argon,
            R. D. Verma,
            Proc. Indian Acad. Sci. A 48, 197-226

(60. 41)   Emission Spectra,
            P. B. V. Haranath and T. A. P. Rao,
            Indian J. Phys. 34, 123-30

(60. 42)   Resonance Series in the Ultraviolet,
            R. D. Verma,
            J. Chem. Phys. 32, 738-49

(63. 43)   Lifetimes in Fluorescence,
            L. Brewer, R. A. Berg, and G. M. Rosenblatt,
            J. Chem. Phys. 38, 1381-8

(63. 44)   B - X System, Franck-Condon Factors,
            R. W. Nicholls,
            J. Chem. Phys. 38, 1029-30

(63. 45)   Potential Energy Curve of the Ground State,
            S. Weissman, J. T. Vanderslice, and R. Battino,
            J. Chem. Phys. 39, 2226-8

(64. 46)   Energy Transfer to the B State,
            L. Brown and W. Klemperer,
            J. Chem. Phys. 41, 3072-89

(64. 47)   Intensity of the Resonance Series,
            R. N. Zare,
            J. Chem. Phys. 40, 1934-44

(64. 48)   Visible Absorption, Effects of Pressure,
            C. A. Goy and H. O. Pritchard,
            J. Molec. Spectrosc. 12, 38-44

(64. 49)   Constants of the Ground State,
           D. H. Rank and B. S. Rao,
           J. Molec. Spectrosc. 13, 34-42

(64. 50)   Potential Energy Curves,
           W. G. Richards and R. F. Barrow,
           Trans. Faraday Soc. 60, 797-800

(65. 51)   Isotope Effect in the B - X System,
           R. L. Brown and T. C. James,
           J. Chem. Phys. 42, 33-5

(65. 52)   Energy Transfer Into the B State,
           J. I. Steinfeld and W. Klemperer,
           J. Chem. Phys. 42, 3475-97

(65. 53)   B - X System,
           J. I. Steinfeld, R. N. Zare, L. Jones, M. Lesk, and W. Klemperer,
           J. Chem. Phys. 42, 25-33

(65. 54)   Pressure Effects on Spectra,
           E. A. Ogryzlo and G. G. Thomas,
           J. Molec. Spectrosc. 17, 198-202

(65. 55)   F. Jenc,
           "The Reduced Potential Curves of Heavy Diatomic Molecules. The
           Reduced Potential Curves of Halogens and Interhalogens,"
           Coll. Czech. Chem. Comm. 30, 3772-84

(66. 56)   M. Halmann, I. Laulicht, and J. I. Steinfeld,
           "Franck-Condon Calculations for Iodine Using a Morse Potential,"
           J. Molec. Spectrosc. 21, 328-332

(66. 57)   A. Nobs and K. Wieland,
           "The Ultraviolet Absorption Spectrum of Iodine (I_2) Vapour — A
           Forgotten Problem of Old Time Spectroscopy,"
           Helv. Phys. Acta 39, 564-6

(67. 58)   A. W. Richardson and R. A. Powell,
           "The 5350Å Band of Iodine: A Simple Resolution Test,"
           J. Molec. Spectrosc. 24, 379-81

(68. 59)   E. Wasserman, W. E. Falconer, and W. A. Yager,
           "Direct Predissociation of I_2 B ($^3\Pi_{0_u}+$),"
           J. Chem. Phys. 49, 1971-2

I_2

(68. 60)   S. Ezekiel and R. Weiss,
"Laser-Induced Fluorescence in a Molecular Beam of Iodine, "
Phys. Rev. Letters 20, 91-3

(68. 61)   E. W. Abrahamson, D. Husain, and J. R. Wiesenfeld,
"Time-Resolved Studies of Emission From $I_2(B^3\Pi_{0u}+)$, "
Trans. Faraday Soc. 64, 833-9

(69. 62)   Dissociation Limits,
J. I. Steinfeld, J. D. Campbell, and N. A. Weiss,
J. Molec. Spectrosc. 29, 204-15

(69. 63)   A. Chutjian,
"Calculation of Predissociation Rates of the B $^3\Pi_0{}^+{}_u$ State of I_2, "
J. Chem. Phys. 51, 5414-9

(69. 64)   A. Chutjian and T. C. James,
"Intensity Measurements in the $B^3\Pi_0{}^+{}_u - X^1\Sigma_0{}^+{}_g$ System of I_2, "
J. Chem. Phys. 51, 1242-9

(69. 65)   K. Sakurai and H. P. Broida,
"Observation and Identification of I_2 Fluorescence Excited by a
5682-Å Krypton-Ion Laser, "
J. Chem. Phys. 50, 557-8

(70. 66)   J. A. Myer and J. A. R. Samson,
"Absorption Cross Section and Photoionization Yield of I_2 Between
1050 and 2200 Å, "
J. Chem. Phys. 52, 716-8

(70. 67)   M. Kroll,
"A Calculation of the Absorption Continuum Associated With the
Visible Iodine Spectrum, "
J. Molec. Spectrosc. 36, 44-52

(70. 68)   E. Menke,
"On the Laser-Induced Resonance Fluorescence in the $B^3\Pi_{0u}{}^+$ -
$X^1\Sigma_g^+$ System of Iodine Molecules, "
Z. Naturforsch. 25a, 442

(70. 69)   Th. Halldorsson and E. Menke,
"Interferometric Measurements of the Laser-Induced Resonance
Fluorescence in Iodine Molecule, "
Z. Naturforsch. 25a, 1356-8

(70. 70)   R. S. Mulliken,
"The Role of Kinetic Energy in the Franck-Condon Principle; With
Applications to the Iodine Molecule Emission Spectrum, "
Chem. Phys. Letters 7, 11-4

(70.71)   R. J. LeRoy and R. B. Bernstein,
"Dissociation Energies of Diatomic Molecules From Vibrational
Spacings of Higher Levels: Application to the Halogens,"
Chem. Phys. Letters 5, 42-4

(70.72)   I. R. Beattie, G. A. Ozin, and R. O. Berry,
"The Gas-Phase Raman Spectra of P_4, P_2, As_4, and As_2. The
Resonance Fluorescence Spectrum of $^{80}Se_2$. Resonance Fluorescence-
Raman Effects in the Gas-Phase Spectra of Sulphur and I_2. The Effect
of Pressure on the Depolarization Ratios for I_2.
J. Chem. Soc. A 12, 2071-4

(70.73)   P. Venkateswarlu,
"Vacuum Ultraviolet Spectrum of the Iodine Molecule,"
Can. J. Phys. 48, 1055-80

(70.74)   M. Kroll and K. K. Innes,
"Molecular Electronic Spectroscopy by Fabry-Perot Interferometry.
Effect of Nuclear Quadrupole Interactions on the Line Widths of the
$B^3\Pi_0+ - X^1\Sigma_g^+$ Transition of the I_2 Molecule,"
J. Molec. Spectrosc. 36, 295-309

(70.75)   R. J. LeRoy,
"Molecular Constants and Internuclear Potential of Ground-State
Molecular Iodine,"
J. Chem. Phys. 52, 2683-9

(71.76)   M. Kroll and D. Swanson,
"The Resonance Fluorescence of I_2 Excited by a Single Mode Laser,"
Chem. Phys. Letters 9, 115-8

(71.77)   A. B. Cornford, D. C. Frost, C. A. McDowell, J. L. Ragle, and
I. A. Stenhouse,
"Photoelectron Spectra of the Halogens,"
J. Chem. Phys. 54, 2651-7

(71.78)   G. R. Hanes, J. Lapierre, P. R. Bunker, and K. C. Shotton,
"Nuclear Hyperfine Structure in the Electronic Spectrum of $^{127}I_2$
by Saturated Absorption Spectroscopy, and Comparison With Theory,"
J. Molec. Spectrosc. 39, 506-15

(71.79)   J. D. Knox and Y. H. Pao,
"High-Resolution Saturation Spectra of the Iodine Isotope $^{129}I_2$ in
the 633-nm Wavelength Region,"
Appl. Phys. Letters 18, 360-2

(71.80)   R. S. Mulliken,
"Iodine Revisited,"
J. Chem. Phys. 55, 288-309

I_2

(72.81)   J. Vigue and J. C. Lehmann,
"Assignment of the Transitions Induced in Molecular Iodine Vapour by a Krypton Ion Laser,"
Chem. Phys. Letters <u>16</u>, 385-7

(72.82)   W. Kiefer and H. J. Bernstein,
"Vibrational-Rotational Structure in the Resonance Raman Effect of Iodine Vapor,"
J. Molec. Spectrosc. <u>43</u>, 366-81

(72.83)   G. D. Chapman and P. R. Bunker,
"Magnetic Quenching of Iodine Fluorescence Excited by a 6328 Å He/Ne Laser,"
J. Chem. Phys. <u>57</u>, 2951-9

(72.84)   R. Solarz and D. H. Levy,
"Non-Linear Level Crossing in the $X^1\Sigma_g^+$ State of Iodine,"
Chem. Phys. Letters <u>17</u>, 35-8

(72.85)   M. Levenson,
"Hyperfine Interactions in Molecular Iodine,"
Thesis, Stanford University

(72.86)   M. S. Sorem,
"Spectroscopy by Saturated Fluorescence and Absorption in Molecular Iodine,"
Thesis, Stanford University

(72.87)   K. Weiland, J. B. Tellinghuisen, and A. Nobs,
"The Band Systems E→B (4000-4360 Å) and F→X (2530-2740 Å) of $^{127}I_2$ and $^{129}I_2$, and the Corresponding System E⇌B of Br_2 and Cl_2,"
J. Molec. Spectrosc. <u>41</u>, 69-83

(72.88)   J. Tellinghuisen,
"Spontaneous Predissociation in I_2,"
J. Chem. Phys. <u>57</u>, 2397-2402

(72.88a)   L. Brewer, J. Tellinghuisen,
"Quantum Yield for Unimolecular Dissociation of I_2 in Visible Absorption,"
J. Chem. Phys. <u>56</u>, 3929-38

(72.89)   M. S. Sorem, T. W. Hansch, and A. L. Schawlow,
"Nuclear Quadrupole Coupling in the $^1\Sigma_g^+$ and $^3\Pi_{0u}^+$ States of Molecular Iodine,"
Chem. Phys. Letters <u>17</u>, 300-2

(72.90)   G. A. Capelle and H. P. Broida,
"Magnetic Field Dependence of I_2 B-State Lifetimes,"
J. Chem. Phys. <u>57</u>, 5027-9

(72.91)   G. A. Capelle,
          "Radiative Lifetime and Quenching Cross Section Measurements of
          I_2 and Other Small Molecules, Using Tunable Dye Laser Excitation,"
          Thesis, University of California at Santa Barbara

(73.92)   M. S. Child and R. B. Bernstein,
          "Diatomic Interhalogens: Systematics and Implications of Spectro-
          scopic Interatomic Potentials and Curve Crossings,"
          J. Chem. Phys. <u>59</u>, 5916-25

(73.93)   M. E. Akopyan, F. I. Vilesov, and Yu. L. Sergeev,
          "Photoionization Processes in Molecular Iodine,"
          Opt. Spectrosc. <u>35</u>, 472-5

(73.94)   J. C. Keller, M. Broyer, and J. C. Lehmann,
          "Direct Measurement of the Lifetime of the Landé Factors of the
          $^3\Pi_{0_u^+}$, v' = 62, J' = 27 Level of the I_2 Molecule,"
          C. R. Acad. Sci. B <u>277</u>, 369-72

(73.95)   R. F. Barrow, D. F. Broyd, L. B. Pederson, and K. K. Yee,
          "The Dissociation Energies of Gaseous Br_2 and I_2,"
          Chem. Phys. Letters <u>18</u>, 357-8

(73.96)   R. L. St. Peters, S. D. Silverstein, M. Lapp, and C. M. Penney,
          "Resonant Raman Scattering or Resonance Fluorescence in I_2
          Vapor?"
          Phys. Rev. Letters <u>30</u>, 191-2

(73.97)   K. K. Yee,
          "Analysis of RKR Long-Range Potential of the B $^3\Pi_{0^+{}_u}$ State of I_2,"
          Chem. Phys. Letters <u>21</u>, 334-7

(73.98)   B. R. Higginson, D. R. Lloyd, and P. J. Roberts,
          "Variable Temperature Photoelectron Spectroscopy. The Adiabatic
          Ionization Potential of the Iodine Molecule,"
          Chem. Phys. Letters <u>19</u>, 480-2

(73.99)   L. S. Wall, K. G. Bartlett, and D. F. Edwards,
          "Selective Excitation of Molecular Iodine,"
          Chem. Phys. Letters <u>19</u>, 274-8

(73.100) M. Broyer, J. Vigue, and J. C. Lehmann,
          "Two Unexpected Effects Concerning Perturbations of the $B^3\Pi_{0_u^+}$
          State of Molecular Iodine,"
          Chem. Phys. Letters <u>22</u>, 313-6

(73.101) F. J. Comes, U. Nielsen, and W. H. E. Schwarz,
          "Inner Electron Excitation of Iodine in the Gaseous and Solid Phase,"
          J. Chem. Phys. <u>58</u>, 2230-7

I_2

(73. 102) M. Broyer, F. W. Dalby, J. Vigue, and J. C. Lehmann,
"Observation of a New Effect on the Fluorescence of Molecular
Iodine Excited by Ionized Argon or Krypton Lasers,"
Can. J. Phys. 51, 226-8

(73. 103) G. A. Capelle and H. P. Broida,
"Lifetimes and Quenching Cross Sections of $I_2(B^3\Pi_{0u}^+)$,"
J. Chem. Phys. 58, 4212-22

(73. 104) A. Gelb, R. Kapral, and G. Burns,
"Dissociation of Vibrationally-Rotationally Excited $I_2(B^3\Pi_{0u}^+)$,"
J. Chem. Phys. 59, 2980-5

(73. 105) R. F. Barrow and K. K. Yee,
$B^3\Pi_{0u}^+ - X^1\Sigma_g^+$ System of $^{127}I_2$: Rotational Analysis and Long-
Range Potential of the $B^3\Pi_{0u}^+$ State,"
Faraday Trans. II 69, 684-700

(73. 106) M. S. Child,
"Direct Inversion of Magnetic Fluorescence Quenching Data for the
$B^3\Pi(0_u^+)$ State of Iodine,"
J. Molec. Spectrosc. 45, 293-7

(73. 107) J. Tellinghuisen,
"Continuous Absorption Below the Band Convergence Limit in the
I_2 B ← X Transition,"
J. Chem. Phys. 59, 849-52

(73. 108) R. Hansen, V. Calder, and D. K. Hoffman,
"A Diatomic Potential Function Based on Observed Regularities in
Spectroscopic Data,"
Spectrochim. Acta 29A, 531-41

(73. 109) F. Spieweck,
$^{127}I_2$ Absorption by 3 Krypton II-Laserlines,"
Metrologia 9, 24-5

(73. 110) E. D. Bugrim, S. N. Makrenko, and I. L. Tsikora,
"Spectroscopic Study of Vibrational-Energy Scattering in I_2 Mole-
cules Excited by a He-Ne Laser,"
Opt. Spectrosc. 34, 35-7

(73. 111) J. D. Brown, G. Burns, and R. J. LeRoy,
"Improved Spectroscopic Data Synthesis for $I_2(B^3\Pi_{0u}^+)$ and Pre-
dictions of J Dependence for $B(^3\Pi_{0u}^+) - X(^1\Sigma_g^+)$ Transition Intensities,"
Can. J. Phys. 51, 1664-7

(73. 112)   D. G. Youmans, L. A. Hackel, and S. Ezekiel,
            "High-Resolution Spectroscopy of I$_2$ Using Laser-Molecular-Beam
            Techniques,"
            J. Appl. Phys. <u>44</u>, 2319-21

(73. 113)   R. D. Clear and K. R. Wilson,
            "Assignment of Continuous Spectra by Photofragment Spectroscopy:
            C State of I$_2$,"
            J. Molec. Spectrosc. <u>47</u>, 39-44

(73. 114)   M. K. Matzen, G. V. Calder, and D. K. Hoffman,
            "An Accurate Semi-Empirical Potential Function for Diatomic
            Molecules,"
            Spectrochim. Acta <u>29A</u>, 2005-16

(73. 115)   S. A. Bazhutin, V. S. Letokhov, A. A. Makarov, and V. A. Semchishen,
            "Selective Predissociation of Ortho-I$_2$ Molecules by Laser Radiation,"
            JETP Letters <u>18</u>, 515-9

(73. 116)   E. P. Gordeev, S. Ya. Umansky, and A. I. Voronin,
            "Calculations of the Iodine Molecule Potential Curves,"
            Chem. Phys. Letters <u>23</u>, 524-8

(73. 117)   D. J. Ruben, S. G. Kukolich, L. A. Hackel, D. G. Youmans, and
            S. Ezekiel,
            "Laser-Molecular Beam Measurement of Hyperfine Structure in
            the I$_2$ Spectrum,"
            Chem. Phys. Letters <u>22</u>, 326-30

(73. 118)   J. Tellinghuisen,
            "Resolution of the Visible-Infrared Absorption Spectrum of I$_2$ Into
            Three Contributing Transitions,"
            J. Chem. Phys. <u>58</u>, 2821-34

(73. 119)   S. M. Singh and J. Tellinghuisen,
            "The Visible Emission Spectrum of Iodine,"
            J. Molec. Spectrosc. <u>47</u>, 409-19

(73. 120)   W. R. Harshbarger and M. B. Robin,
            "The Quenching of Excited Iodine Atoms by Oxygen Molecules as
            Studied by Opto-Acoustic Spectroscopy,"
            Chem. Phys. Letters <u>21</u>, 462-5

(74. 121)   B. Garetz, M. Rubinson, and J. I. Steinfeld,
            "Potential Surface Crossing in Classical Trajectory Calculations:
            Application to Collision-Induced Predissociation in Electronically
            Excited I$_2$,"
            Chem. Phys. Letters <u>28</u>, 120-4

I_2

(74. 122) R. D. Verma and R. J. LeRoy,
"Comment on the uv Resonance Spectrum and Ground-State
Dissociation Energy of I_2,"
J. Chem. Phys. <u>61</u>, 438

(74. 123) R. J. LeRoy,
"Long-Range Potential Coefficients From RKR Turning Points:
C_6 and C_8 for $B(^3\Pi_{0u}{}^+)$-State Cl_2, Br_2, and I_2,"
Can. J. Phys. <u>52</u>, 246-56

(74. 124) P. F. Williams, D. L. Rousseau, and S. H. Dworetsky,
"Resonance Fluorescence and Resonance Raman Scattering: Life-
times in Molecular Iodine,"
Phys. Rev. Letters <u>32</u>, 196-9

(74. 125) J. Wei and J. Tellinghuisen,
"Parameterizing Diatomic Spectra: "Best" Spectroscopic Constants
for the I_2 B $\rightleftharpoons$ X Transition,"
J. Molec. Spectrosc. <u>50</u>, 317-32

(74. 126) J. Vigue, M. Broyer, and J. C. Lehmann,
"Quantum Interference Effect Between the Magnetic and Natural
Predissociations in the $B^3\Pi_0{}^+{}_u$ State of I_2. A New Experimental
Proof,"
J. Phys. B: Atom. Molec. Phys. <u>7</u>, L158-L161

(74. 127) R. Clark, S. R. Jeyes, A. J. McCaffery, and R. A. Shatwell,
"Circular Emission From Molecular Iodine,"
Chem. Phys. Letters <u>25</u>, 74-7

(74. 128) R. C. Oldenborg, J. L. Gole, and R. N. Zare,
"Chemiluminescent Spectra of Alkali-Halogen Reactions,"
J. Chem. Phys. <u>60</u>, 4032-42

(74. 129) J. A. Piper,
"Increased Efficiency and New CW Transitions in the Helium-
Iodine Laser System,"
J. Phys. D: Appl. Phys. <u>7</u>, 323-8

(74. 130) R. D'alterio, R. Mattson, and R. Harris,
"Potential Curves for the I_2 Molecule,"
J. Chem. Ed. <u>51</u>, 282-4

(74. 131) R. Wallenstein, J. A. Paisner, and A. L. Schawlow,
"Observation of Zeeman Quantum Beats in Molecular Iodine,"
Phys. Rev. Letters <u>32</u>, 1333-6

(74. 132)   M. Brith, O. Schnepp, and P. Stephens,
"Magnetic Circular Dichroism Spectra of the Halogen Molecules,
I_2, Br_2 and Cl_2. Resolution of Overlapping $0_u^+(^3\Pi)$ and $^1\Pi$ Bands, "
Chem. Phys. Letters 26, 549-52

(74. 133)   D. L. Rousseau and P. F. Williams,
"Discrete and Diffuse Emission Following Two-Photon Excitation
of the E State in Molecular Iodine, "
Phys. Rev. Letters 33, 1368-72

(74. 134)   J. D. Brown and G. Burns,
"The Thermal Emission of Iodine, "
Can. J. Phys. 52, 1862-71

(74. 135)   J. A. Paisner and R. Wallenstein,
"Rotational Lifetimes and Self-Quenching Cross Sections in the B^3
$\Pi_{0_u}^+$ State of $^{127}I_2$, "
J. Chem. Phys. 61, 4317-20

(n.p. 136)   J. I. Steinfeld,
Private Communication

In$_2$

Methods of Production and Experimental Technique

Absorption.

Emission from a King furnace, Tesla discharge in (Hg + In).

BAND SYSTEMS

System	Transition	Sources	Wavelength Limits	Degrading	Band Head, $v_{0,0}$	Remarks	Bibliography
I		Emission	> 5500	R		Feature-less	(65.5)
II		Emission	5460-5500	V		Line-like bands	(65.5)
III		Emission	5370-5480	R, V			(65.5)
IV		Emission	5320-5350	V		Q-like bands	(65.5)
V		Emission	5100-5310	(V)		Line-like bands	(65.5)
VI		Emission	4950-5100	R, V		Diffuse	(65.5)
VII		Absorption	3680-3818				(37.2, 37.1)
VIII		Absorption	2336-2339				(37.2, 37.1)

316

I.   5500Å System

No apparent Q branches.   Bands exhibit much overlapped rotational structure. λ (65.5).

v' | v''	m	m+1	m+2	m+3
n	5552.1	5596.5	5641.6	5687.1
n+1			5605.2	5650.2
⋮				
n+3	5447.3			
n+4	5412.9			
n+5	5379.6	5421.3	5463.6	

VII., VIII.   Absorption Bands (37.2, 37.1)

Continuum bands: $\lambda \sim 3818 | 3734 | 3680$

Line-like bands: $\lambda_{max} \sim 3548 | 3544 | 3541 | 3536 | 3534 | 3526 | 3523 |$
$3290 | 3280 | 3262 | 3259$

Bands: $\lambda \sim 2339 | 2337.5 | 2336$

SPECTROSCOPIC CONSTANTS

State	T_e	ω_e	$x_e\omega_e$	B_e	$\alpha_e \times 10^3$	$D_e \times 10^6$	r_e	Remarks	Bibliography
$X(^3\Sigma_g^-)$	18025.1	115.0[a]							(65.5)
	0.0	143.1	0.1						(65.5)

[a] ω_o

Dissociation energy = 0.88 ± 0.15 eV, 20.3 kcal/mole, 7100 cm^{-1} if the ground state is $^3\Sigma^-$, or:

1.00 ± 0.11 eV, 23.2 kcal/mole, 8100 cm^{-1} if the ground state is $\Omega = 0$ (59.4).

BIBLIOGRAPHY

(37. 1)   Absorption,
          R. Wajnkranc,
          Z. Physik 104, 122-31

(37. 2)   Absorption,
          R. Wajnkranc,
          Z. Physik 105, 516

(40. 3)   Fluorescence,
          J. G. Winans, F. J. Davis, and V. A. Leitzke,
          Phys. Rev. 57, 70-1

(59. 4)   Dissociation Energy,
          G. DeMaria, J. Drowart, and M. G. Inghram,
          J. Chem. Phys. 31, 1076-81

(65. 5)   Emission Systems,
          D. E. S. Ginter,
          "Electronic Spectra of the Homonuclear Molecules of the Group III
          Metals,"
          Thesis, Vanderbuilt University

$$K_2$$

Methods of Production and Experimental Technique

Absorption.

Emission from a heat pipe, laser fluorescence.

BAND SYSTEMS

System	Transition	Sources	Wavelength Limits	Degrading	Characteristic Bands, λ	Remarks	Bibliography
I	$A\,^1\Sigma_u^+ \rightleftarrows X\,^1\Sigma_g^+$	Heat Pipe	8850-7700	R			(71.47, 30.10)
II	$B\,^1\Pi_u \rightleftarrows X\,^1\Sigma_g^+$	Absorption, laser fluorescence	6950-6250	R	6583.2(0,2) 6544.0(0,1) 6473.6(1,0)		(68.39, 32.15, 31.12)
III	$C\,^1\Pi_u \rightleftarrows X\,^1\Sigma_g^+$	Absorption	4510-4220	R	4343.5(1,0)		(61.32, 48.29)
IV	$D\left(^1\Pi_u\right) \leftarrow X\,^1\Sigma_g^+$	Absorption	4160-3940	R	4082.7(1,2)		(48.29)
V	$E\left(^1\Pi_u\right) \leftarrow X\,^1\Sigma_g^+$	Absorption	3925-3700	R	3797.6(2,3) 3793.7(1,2)		(50.31)
VI	$F \leftarrow X\,^1\Sigma_g^+$	Absorption	3700-3600	R			(37.20, 37.19)
VII	$G \leftarrow X\,^1\Sigma_g^+$	Absorption	3600-3480	R			(37.19)

I. $\underline{A\,^1\Sigma_u^+ \rightleftharpoons X\,^1\Sigma_g^+ \text{ System}}$

Most characteristic bands, λ (30.10):

(v', v'')	(0, 3)	(0, 2)	(1, 2)	(0, 1)	(0, 0)	(1, 0)	(2, 0)
λ	8773.15	8702.00	8651.79	8634.43	8566.30	8515.70	8468.23

II. $\underline{B\,^1\Pi_u \rightleftharpoons X\,^1\Sigma_g^+ \text{ System}}$

Most intense band heads, λ (Intensity) (32.15, 31.12):

v', v''	(0, 2)	(0, 1)	(1, 1)	(1, 0)	(2, 0)
λ	6583.19	6544.00	6512.19	6473.58	6443.00
(Intensity)	9	8	5	10	8

III. $\underline{C\,^1\Pi_u \rightleftharpoons X\,^1\Sigma_g^+ \text{ System}}$

Most intense band heads, λ (Intensity) (61.32, 48.29):

v', v''	(0, 0)	(1, 0)	(2, 0)	(3, 0)	(4, 0)
λ	4355.1	4343.5	4332.3	4320.9	4310.0
(Intensity)	8	10	7	7	7

IV. $\underline{D\left(^1\Pi_u\right) \leftarrow X\,^1\Sigma_g^+ \text{ System}}$

Possibly two independent systems, λ (Intensity) (27.20, 37.19):

v', v''	0	1	2	3	4	5
0			4092.3(8)	4107.3(7)	4122.7(6)	
1		4067.0(8)	4082.7(10)	4097.4(7)	4112.8(8)	
2				4087.5(6)	4103.0(6)	
3	4033.5(6)			4078.2(6)		4108.6(6)
4	4024.9(6)					

V.   <u>$E\left(^1\Pi_u\right) \leftarrow X\,^1\Sigma_g^+$ System</u>

Most intense band heads, λ (Intensity) (50.31):

v', v''	0	1	2.	3
0				
1			3793. 7(10)	3806. 2(7)
2		3771. 5(8)	3784. 4(7)	3797. 6(10)
3		3762. 8(7)	3776. 0(7)	3789. 2(7)
4				
5	3733. 8(7)	3746. 6(7)		
6		3738. 1(7)		

VI.   <u>$F \leftarrow X\,^1\Sigma_g^+$ System</u>

Most intense band heads, analysis uncertain, λ (37.20, 37.19):

v', v''	0	1	2	3
0		3639. 5	3651. 7	
1		3631. 6	3643. 4	
2	3611. 2	3623. 5	3635. 3	3647. 3
3	3603. 2			

VII.   <u>$G \leftarrow X\,^1\Sigma_g^+$ System</u>

Most intense band heads, λ (Intensity) (37.19):

(v', v'')	(0, 2)	(1, 2)	(2, 2)	(3, 2)	(4, 2)	(3, 1)	(4, 1)
λ	3583. 7	3575. 6	3567. 6	3559. 9	3553. 4	3548. 6	3541. 1
(Intensity)	4	4	4	3	4	3	3

SPECTROSCOPIC CONSTANTS

State	T_e	ω_e	$x_e\omega_e$	$B_e \times 10^2$	$\alpha_e \times 10^4$	$D_e \times 10^8$	r_e	Remarks	Bibliography
G	28091	64.9	0.05						(37.19)
F	27571	62.2	0.24						(37.19)
E($^1\Pi_u$)	26493.0	60.6	0.15						(50.31)
D($^1\Pi_u$)	24627.7	61.6	0.90					(a)	(48.29)
C$^1\Pi_u$	22969.7	61.48	0.14	4.404	1.10		4.43		(61.32, 50.31)
B$^1\Pi_u$	15376.4	92.021	0.2829	5.6743	1.65	8.63	4.23	(b)	(68.39, 32.15, 31.13)
A$^1\Sigma_u^+$	11682.6	69.09	0.153						(30.10)
X$^1\Sigma_g^+$	0	92.64		5.622	2.19	8.28	3.92	(c)	(61.32, 48.29)

(a) $y_e\omega_e = 0.001$, $z_e\omega_e = -0.0003$; (b) $y_e\omega_e = -0.002055$, $\gamma_e = -7.2 \times 10^{-6}$, $\delta_e = 1.5 \times 10^{-7}$, $\beta = -7.4 \times 10^{-10}$;
(c) $\beta = -8.3 \times 10^{-11}$

Dissociation energy = 0.51 ± 0.05 eV, 11.8 kcal/mole, 4114 cm^{-1}.

Perturbations and General Information

Radiative lifetime of $B^1\Pi_u$ state (70.44, 70.41):
$$\tau\left(B^1\Pi_u\right) = 9.65 \pm 0.3 \text{ nsec.}$$

Absolute absorption cross sections (68.37, 66.35).

Potential energy curves, RKR potentials (69.40):

	State	v	U(cm^{-1})	r_{min}(Å)	r_{max}(Å)
$T_e = 0.0$	$X^1\Sigma_g^+$	0	46.2	3.7906	4.0643
		1	138.2	3.6996	4.1752
		2	229.4	3.6394	4.2554
		3	319.9	3.5918	4.3230
		4	409.7	3.5516	4.3835
		5	498.8	3.5164	4.4391
		6	587.2	3.4848	4.4912
		7	674.9	3.4560	4.5407
		8	761.9	3.4294	4.5880
		9	848.1	3.4047	4.6337
		10	933.7	3.3815	4.6780
		11	1018.5	3.3597	4.7212
		12	1102.7	3.3389	4.7633
		13	1186.1	3.3192	4.8047
		14	1268.9	3.3003	4.8453
		15	1350.9	3.2822	4.8852
$T_e = 15376.4$ cm^{-1}	$B^1\Pi_u$	0	37.4	4.0886	4.3929
		1	111.6	3.9885	4.5179
		2	185.1	3.9225	4.6089
		3	257.9	3.8706	4.6861
		4	330.0	3.8269	4.7553
		5	401.3	3.7886	4.8192
		6	472.0	3.7544	4.8794
		7	542.0	3.7233	4.9367
		8	611.2	3.6945	4.9917
		9	679.8	3.6678	5.0451
		10	747.6	3.6427	5.0970
		11	814.7	3.6190	5.1478
		12	881.0	3.5965	5.1978
		13	946.5	3.5749	5.2471
		14	1011.2	3.5542	5.2959
		15	1075.1	3.5341	5.3445

BIBLIOGRAPHY

(08. 1)   B ← X System,
R. W. Wood and T. S. Carter,
Phys. Rev. 27, 107-16

(23. 2)   A, B ← X Systems,
J. C. MacLennan and D. S. Ainslie,
Proc. Roy. Soc. A 103, 304-14

(24. 3)   B ← X System, Incorrect Analysis,
H. G. Smith,
Proc. Roy. Soc. A 106, 400-15

(27. 4)   B ← X System, Incorrect Analysis,
W. R. Fredrickson and W. W. Watson,
Phys. Rev. 30, 429-38

(27. 5)   B ← X System, Incorrect Analysis,
P. Pringsheim and B. Rosen,
Z. Physik 43, 519-23

(28. 6)   A, B ← X Systems,
R. Ritschl and D. Villars,
Naturwissenschaften 16, 219-20

(28. 7)   C ⇄ X System,
J. M. Walter and S. Barratt,
Proc. Roy. Soc. A 119, 257-75

(29. 8)   C ⇄ X System, Incorrect Constants,
H. Yamamoto,
Japan J. Phys. 5, 153-6

(30. 9)   C ⇄ X System,
W. Weizel and M. Kulp,
Ann. Phys. 4, 971-84

(30. 10)   A ← X System,
W. O. Crane and A. Christy,
Phys. Rev. 36, 421-9

(30. 11)   Theory,
E. Hutchinson,
Phys. Rev. 36, 410-20

(31. 12) B ← X System, Rotational Analysis,
F. W. Loomis,
Phys. Rev. 38, 2153-61

(31. 13) B ← X System,
F. W. Loomis and R. W. Wood,
Phys. Rev. 38, 854-6

(32. 14) B ← X System,
F. W. Loomis,
Phys. Rev. 39, 189

(32. 15) B ← X System, Vibrational Analysis, Dissociation Energy,
F. W. Loomis and R. E. Nusbaum,
Phys. Rev. 39, 89-98

(32. 16) Band Polarization,
H. Kuhn,
Z. Physik 76, 782-92

(36. 17) Band Polarization,
B. K. Chakraborti,
Indian J. Phys. 10, 155-62

(37. 18) A ← X System,
T. Carroll,
Phys. Rev. 52, 822-35

(37. 19) Ultraviolet Systems, Uncertain Analysis,
H. Yoshinaga,
Proc. Phys. -Math. Soc. Japan 19, 847-59

(37. 20) Ultraviolet System, Uncertain Analysis,
H. Yoshinaga,
Proc. Phys. -Math. Soc. Japan 19, 1073-83

(37. 21) Theory,
C. H. D. Clark,
Trans. Faraday Soc. 33, 1398-401

(37. 22) Theory,
C. H. D. Clark and C. W. Scaife,
Trans. Faraday Soc. 33, 1394-8

(39. 23) B ← X System,
W. W. Watson and W. F. Meggers,
J. Res. Nat. Bur. Stand. 20, 125-8

(40. 24)   Theory,
           G. B. B. M. Sutherland,
           J. Chem. Phys. 8, 161-4

(40. 25)   Theory,
           R. F. Barrow,
           Trans. Faraday Soc. 36, 624-5

(40. 26)   Theory,
           C. H. D. Clark,
           Trans. Faraday Soc. 36, 370-6

(41. 27)   Theory,
           H. M. Hulburt and J. O. Hirschfelder,
           J. Chem. Phys. 9, 61-9

(47. 28)   Preliminary Note to (48. 29),
           R. W. B. Pearse and S. P. Sinha,
           Nature 160, 159

(48. 29)   C, D ⇌ X Systems,
           S. P. Sinha,
           Proc. Phys. Soc. 60, 436-43

(49. 30)   S. P. Sinha,
           Indian J. Phys. 23, 229-36

(50. 31)   C ← X System, Vibrational Analysis,
           S. P. Sinha,
           Proc. Phys Soc. A 63, 952-6

(61. 32)   C ← X System, Rotational Analysis,
           E. W. Robertson and R. F. Barrow,
           Proc. Chem. Soc. 329-30

(64. 33)   RKR Potential Energy Curves for B and X States,
           D. K. Rai and A. N. Tripathi,
           Can. J. Chem. 42, 452-7

(64. 34)   Nuclear Magnetic Moment,
           R. A. Brooks, C. H. Anderson, and N. F. Ramsey,
           Phys. Rev. A 132, 62-8

(66. 35)   Absorption Cross-Sections,
           M. Lapp and L. P. Harris,
           "Absorption Cross Sections of Alkali-Vapor Molecules: I.   Cs$_2$ in
           the Visible, II.   K$_2$ in the Red,"
           J. Quant. Spectrosc. Radiative Transfer 6, 169-79

(67. 36) Nonlinear Absorption of Ruby Laser Light,
V. G. Abramov, O. V. Konstantinov, N. N. Kostin, and V. A. Khadovoi,
"Nonlinear Absorption of Ruby Laser Light by Molecular Potassium
Vapor,"
Zh. Exsp. Teor. Fiz. <u>53</u>, 822-30

(68. 37) Absorption Cross-Sections,
D. M. Creek and G. V. Marr,
"Some Ultraviolet Cross-Section Measurements on Molecular Alkali-
Metal Vapours,"
J. Quant. Spectrosc. Radiative Transfer <u>8</u>, 1431-6

(68. 38) Theory,
L. Szasz and G. McGinn,
"Atomic and Molecular Calculations With the Pseudopotential Method.
III.   The Theory of Li$_2$, Na$_2$, K$_2$, LiH, NaH, and KH,"
J. Chem. Phys. <u>48</u>, 2997-3008

(68. 39) B ← X System, Spectroscopic Analysis,
W. J. Tango, J. K. Link, and R. N. Zare,
"Spectroscopy of K$_2$ Using Laser-Induced Fluorescence,"
J. Chem. Phys. <u>49</u>, 4264-8

(69. 40) Potential Curves B, X States, RKR Potential,
D. C. Jain and R. C. Sahni,
"Reduced Potential Energy Curves of Some Electronic States of
Alkali Molecules,"
Trans. Faraday Soc. <u>65</u>, 897-903

(70. 41) Radiative Lifetime B State,
W. J. Tango and R. N. Zare,
"Radiative Lifetime of the B $^1\Pi_u$ State of K$_2$,"
J. Chem. Phys. <u>53</u>, 3094-100

(70. 42) Reaction With Cl,
W. S. Struve, T. Kitagawa, and D. R. Herschbach,
"Chemiluminescence in Molecular Beams: Electronic Excitation in
Reactions of Cl Atoms With Na$_2$ and K$_2$ Molecules,"
J. Chem. Phys. <u>54</u>, 2759-61

(70. 43) Potential Curves,
A. C. Roach and P. Baybutt,
"Potential Curves of Alkali Diatomic Molecules and the Origins of
Bonding Anomalies,"
Chem. Phys. Letters <u>7</u>, 7-10

K_2

(70.44)   Radiative Lifetime B State,
G. Baumgartner, W. Demtroder, and M. Stock,
"Lifetime-Measurements of Alkali-Molecules Excited by Different Laserlines,"
Z. Physik 232, 462-72

(71.45)   Production in Supersonic Expansion,
R. J. Gordon, Y. T. Lee, and D. R. Herschbach,
"Supersonic Molecular Beams of Alkali Dimers,"
J. Chem. Phys. 54, 2393-409

(71.46)   Reactions With H and D Atoms,
Y. T. Lee, R. J. Gordon, and D. R. Herschbach,
"Molecular Beam Kinetics: Reactions of H and D Atoms with Diatomic Alkali Molecules,"
J. Chem. Phys. 54, 2410-23

(71.47)   Spectra From Heat Pipes,
P. P. Sorokin and J. R. Lankard,
"Emission Spectra of Alkali-Metal Molecules Observed With a Heat-Pipe Discharge Tube,"
J. Chem. Phys. 55, 3810-3

(71.48)   Optical Pumping,
G. Alzetta, A. Gozzini, and L. Moi,
"Effect of Atomic Orientation by Optical Pumping on the Formation of the K_2 Molecule,"
C. R. Acad. Sci. B 274, 39-42

(72.49)   Theory,
A. C. Roach,
"Theoretical Ground State and Excited State Potential Energy Curves for Alkali Diatomic Molecules,"
J. Molec. Spectrosc. 42, 27-37

(74.50)   Reactions With Halogens,
R. C. Oldenborg, J. L. Gole, and R. N. Zare,
"Chemiluminescent Spectra of Alkali-Halogen Reactions,"
J. Chem. Phys. 60, 4032-42

(74.51)   R. W. Molof, T. M. Miller, H. L. Schwartz, B. Pederson, and J. T. Park
"Measurements of the Average Electric Dipole Polarizabilities of the Alkali Dimers,"
J. Chem. Phys. 61, 1816-22

Kr_2

Methods of Production and Experimental Technique

Absorption.

Emission: positive columns, condensed discharge, microwave discharge, electron beam discharge, α-particle irradiation.

BAND SYSTEMS

System	Transition	Sources	Wavelength Limits	Degrading	Characteristic Bands, λ	Remarks	Bibliography
I	$^{1,3}\Sigma_u \rightarrow {}^1\Sigma_g^+$	α irradiation	1250-1850		Max. ~1480Å, 1280Å	Continuum	(73.9, 65.4, 55.3, 55.2)
II	$B(1_u) \leftarrow X^1\Sigma_g^+\left(0_g^+\right)$	Absorption	1252-1257	V	1254.8(3,4)		(73.10)
III	$C\left(0_u^+\right) \leftarrow X^1\Sigma_g^+\left(0_g^+\right)$	Absorption	1239-1245		1241.3(3,4) 1242.3(4,4)		(73.10)
IV	$D\left(0_u^+\right) \leftarrow X^1\Sigma_g^+\left(0_g^+\right)$	Absorption	1167-1169		1168.1(2,0) 1167.6(4,1)		(73.10)
V	$E \leftarrow X^1\Sigma_g^+$	Absorption	1161-1170				(73.10)
VI		Emission	2000-8000			Continuum	(67.7, 42.1)
VII			1064-1080			4 fragmented systems	(73.10)

Systems II - V correlate to separated atom limits in which one atom is excited to various levels of configuration $4p^55s$.

System VII systems are energetically close to various atom levels of configuration $4p^55p$.

II. $\underline{B(1_u) \leftarrow X^1\Sigma_g^+\left(0_g^+\right) \text{System}}$

Band heads, λ (Intensity) (73.10):

v', v''	0	1	2	3	4
0					1252. 3(2)
1				1252. 8(2)	1253. 1(6)
2				1253. 7(6)	1253. 9(8)
3		1254. 0(1)		1254. 6(8)	1254. 8(10)
4		1255. 0(0)		1255. 6(1)	1255. 8(3)

III. $\underline{C\left(0_u^+\right) \leftarrow X^1\Sigma_g^+\left(0_g^+\right) \text{System}}$

Band heads, λ (Intensity) (73.10):

v', v''	0	1	2	3	4
0				1239. 2(9)	1239. 5(9)
1		1239. 2(9)	1239. 5(7)	1239. 8(8)	1240. 0(10)
2	1239. 6(6)	1239. 9(7)	1240. 2(8)	1240. 4(8)	1240. 7(9)
3	1240. 2(4)	1240. 6(4)	1240. 9(5)	1241. 1(6)	1241. 3(10)
4	1241. 0(1)	1241. 4(3)	1241. 6(5)	1241. 9(8)	1242. 1(10)

IV. $\underline{D\left(0_u^+\right) \leftarrow X^1\Sigma_g^+\left(0_g^+\right) \text{System}}$

Band heads, λ (Intensity) (73.10):

v', v''	0	1	2	3	4
0	1169. 2(5)	1169. 5(4)	1169. 7(3)	1170. 0(2)	1170. 1(1)
1	1168. 6(8)	1168. 9(5)	1169. 2(5)	1169. 4(2)	1169. 6(2)
2	1168. 1(8)	1168. 4(2)	1168. 7(3)	1168. 9(5)	1169. 1(3)
3	1167. 7(6)	1168. 0(7)	1168. 2(2)	1168. 4(2)	1168. 7(3)
4	1167. 3(7)	1167. 6(8)	1167. 8(5)		1168. 2(2)

V. $\underline{E \leftarrow X^1\Sigma_g^+ \text{System}}$

Band heads in absorption, λ (Intensity) (73.10):

λ	1161.4	1162.3	1163.1	1163.7	1164.1	1164.4
(Intensity)	10	9	8	7	6	6

SPECTROSCOPIC CONSTANTS .

State	T_e	ω_e	$x_e\omega_e$	B_e	$\alpha_e \times 10^3$	$D_e \times 10^6$	r_e	Remarks	Bibliography
$D\left(0_u^+\right)$	85531.5 (a)	39.66 (b)							(73.10)
$C\left(0_u^+\right)$	80763.9 (a)	35.75 (b)							(73.10)
$B\left(1_u\right)$	79932.8	22.3 (b)							(73.10)
$X^1\Sigma_g^+$ $\left(0_g^+\right)$	0	23.99	1.3	0.024	1.0			$y_e\omega_e = 0.021$	(73.11, 73.10)

(a) T_o; (b) $\Delta G_{1/2}$

Dissociation energy = 0.02 eV, 0.38 kcal/mole, 138.4 cm^{-1} (73.10).

Kr$_2$

Perturbations and General Information

Laser action has been observed on the $^{1,3}\Sigma_u^+ \to X^1\Sigma_g^+$ transition at $1457 \pm 8\text{Å}$ (73.13).

BIBLIOGRAPHY

(42. 1)   Observation,
          B. Vogel,
          Ann. Physik <u>41</u>, 196-210

(55. 2)   Continuum in Condensed Discharge,
          Y. Tanaka,
          J. Opt. Soc. Am. <u>45</u>, 710-3

(55. 3)   Observation in Microwave Discharge,
          P. G. Wilkinson,
          J. Opt. Soc. Am. <u>45</u>, 1044-6

(65. 4)   Observation in Microwave Discharge,
          P. G. Wilkinson and E. T. Byram,
          Appl. Opt. <u>4</u>, 581-8

(65. 5)   Observation of Continuum,
          R. E. Huffman, J. C. Larrabee, and Y. Tanaka,
          Appl. Opt. <u>4</u>, 1581-8

(66. 6)   Ionization Potential,
          J. A. R. Samson and R. B. Cairns,
          J. Opt. Soc. Am. <u>56</u>, 769-75

(67. 7)   Formation of Molecule,
          J. F. Prince and W. W. Robertson,
          J. Chem. Phys. <u>46</u>, 3309-13

(68. 8)   R. Turner and H. D. Riccius,
          "Visible Afterglow Emission of Krypton,"
          J. Chem. Phys. <u>48</u>, 4351-6

(73. 9) $^{1,3}\Sigma_g^+ \rightarrow X^1\Sigma_g^+$ Emission,
          A. Gedanken, B. Raz, and J. Jortner,
          "Emission Spectra of Homonuclear Diatomic Rare Gas Molecules in
          Solid Neon,"
          J. Chem. Phys. <u>59</u>, 1630-3

(73. 10) Absorption,
          Y. Tanaka, K. Yoshino, and D. E. Freeman,
          "Vacuum Ultraviolet Absorption Spectra of the Van Der Waals
          Molecules Kr_2 and ArKr,"
          J. Chem. Phys. <u>59</u>, 5160-83

Kr$_2$

(73. 11) Ground State Potential,
          K. K. Docken and T. P. Schafer,
          "Spectroscopic Information in Ground-State Ar$_2$, Kr$_2$, and Xe$_2$ From
          Interatomic Potentials,"
          J. Molec. Spectrosc. 46, 454-9

(73. 12) Interatomic Potential,
          U. Buck, M. G. Dondi, U. Valbusa, M. L. Klein, and G. Scoles,
          "Determination of the Interatomic Potential of Krypton,"
          Phys. Rev. A 8, 2409-16

(73. 13) P. W. Hoff, J. C. Swingle, and C. K. Rhodes,
          "Observations of Stimulated Emission From High Pressure Krypton
          and Argon/Xenon Mixtures,"
          Appl. Phys. Letters 23, 245-8

(74. 14) D. W. Gough, E. B. Smith, and G. C. Maitland,
          "The Pair Potential Energy Function for Krypton,"
          Molec. Phys. 27, 867-72

La$_2$

Methods of Production and Experimental Technique

Thermal emission from a King furnace (T > 2000°C).

Band Systems

Bands in the region 6100-6040Å have been attributed to La$_2$. The bands are degraded principally to the violet, but the series convergence is degraded red (69.2).

Characteristic bands:

λ | 6075. 3 | 6074. 9 | 6074. 7 | 6074. 6 | 6069. 4 | 6068. 8 | 6049. 6 | 6049. 1

A vibrational analysis yields $\omega_o' = 82.6$ cm^{-1} and $\omega_o'' = 76.9$ cm^{-1}, but these values are in doubt.

Spectroscopic Constants

Dissociation energy = 2.50 ± 0.22 eV, 57.6 kcal/mole, 20200 cm^{-1} (64.1).

BIBLIOGRAPHY

(64. 1)   Dissociation Energy,
          G. Verhaegen, S. Smoes, and J. Drowart,
          J. Chem. Phys. 40, 239-41

(69. 2)   Emission, Vibrational Analysis,
          P. Carette and J. M. Blondeau,
          C. R. Acad. Sci. 269, 16-18

Li_2

Methods of Production and Experimental Technique

Absorption, magnetic rotation.

BAND SYSTEMS

	System	Transition	Sources	Wavelength Limits	Degrading	Characteristic Bands, λ	Remarks	Bibliography
	I	$A\,^1\Sigma_u^+ \leftarrow X\,^1\Sigma_g^+$	Absorption	7700-6550	R	6883.9(2,0)		(72.64, 29.16, 28.1)
	II	$B\,^1\Pi_u \leftarrow X\,^1\Sigma_g^+$	Absorption	5590-4500	R	4800.6(3,1) 4778.8(2,0)		(33.13, 31.9)
	III	$C\,^1\Pi_u \leftarrow X\,^1\Sigma_g^+$	Absorption	3500-3100	R	3358.6(0,2) 3315.6(0,1)		(60.36, 38.31)
	IV	$D\,^1\Pi_u \leftarrow X\,^1\Sigma_g^+$	Absorption	3100-2500	R			(60.36)

Several bands of the isotopic species $^7Li\,^6Li$ have been observed for Systems II and III.

$$\text{Li}_2$$

I. $\underline{A\,{}^1\Sigma_u^+ \leftarrow X\,{}^1\Sigma_g^+ \text{ System}}$

Most intense band heads, λ (Intensity) (36.16, 28.1):

(v', v'')	(0, 2)	(0, 1)	(1, 1)	(1, 0)	(2, 0)	(3, 0)
λ	7690.3	7309.2	7177.4	7003.7	6883.9	6768.7
(Intensity)	8	8	8	8	10	8

II. $\underline{B\,{}^1\Pi_u \leftarrow X\,{}^1\Sigma_g^+ \text{ System}}$

Most intense band heads of ${}^7\text{Li}_2$, λ (Intensity) (31.9):

(v', v'')	(2, 1)	(1, 0)	(3, 1)	(2, 0)	(4, 1)	(3, 0)
λ	4859.7	4838.2	4800.6	4778.8	4744.9	4722.0
(Intensity)	1.5	4	10	10	4	1.5

Most intense band heads of ${}^7\text{Li}{}^6\text{Li}$, λ (Intensity) (31.9):

(v', v'')	(0, 0)	(1, 0)	(4, 1)
λ	4901.8	4836.5	4739.7
(Intensity)	5	4	2

III. $\underline{C\,{}^1\Pi_u \leftarrow X\,{}^1\Sigma_g^+ \text{ System}}$

Most intense band heads, λ (Intensity) (60.36, 48.31):

(v', v'')	(0, 4)	(1, 4)	(0, 3)	(2, 4)	(0, 2)	(0, 1)	(0, 0)	(1, 0)
λ	3431.2	3404.4	3392.1	3378.5	3358.6	3315.6	3277.6	3253.1
(Intensity)	4	4	6	4	10	9	6	10

IV. $\underline{D\,{}^1\Pi_u \leftarrow X\,{}^1\Sigma_g^+ \text{ System}}$

Several systems are superimposed in the region 3100-2500Å. Simple Q branches here have been attributed to a $D\,{}^1\Pi_u \leftarrow X\,{}^1\Sigma_g^+$ system. The D state appears perturbed (60.36).

SPECTROSCOPIC CONSTANTS

State	T_e	ω_e	$x_e\omega_e$	B_e	$\alpha_e \times 10^3$	$D_e \times 10^6$	r_e	Remarks	Bibliography
$D\,^1\Pi_u$	≤34140	~205		0.465			3.18		(60.36)
$C\,^1\Pi_u$	30549	237.9	3.33	0.5068	9.39	9.9	3.08	$y_e\omega_e = 0.060$	(60.36)
$B\,^1\Pi_u$	20439.40	270.94	3.13	0.5577	8.88	9.45	2.93	$y_e\omega_e = -0.0637$	(33.13, 3.9)
$A\,^1\Sigma_u^+$	14069.9	255.50	1.59	0.4975	5.22		3.11	$y_e\omega_e = 0.0039$ (a)	(36.16, 28.1)
$X\,^1\Sigma_g^+$	0	351.43	2.55	0.672	6.8	9.87	2.67	(b)	(69.51, 36.16, 28.1)

(a) Spectroscopic constants for ^{6}Li$_2$ (72.64); (b) spectroscopic constants for ^{6}Li$_2$, ^{7}Li^{6}Li (69.31).

Dissociation energy = 1.026 ± 0.006 eV, 23.66 kcal/mole, 8275 cm^{-1} (69.51).

Li$_2$

Perturbations and General Information

Gyromagnetic ratio (g_j) = 0.10797 nuclear magnetons (64.39).

Transition probabilities (70.53):

Transition	υ	f
$A\,^1\Sigma_u^+ - X\,^1\Sigma_u^+$	14068	0.8688
$C\,^1\Pi_u - X\,^1\Sigma_u^+$	30558	0.0158

Average polarizability (990°K) = 34 × 10^{-24} cm^3 (74.68).

Potential energy curves — RKR potentials (69.50):

	State	v	U(cm^{-1})	r_{min}(Å)	r_{max}(Å)
$T_e = 0.0$	$X\,^1\Sigma_g^+$	0	175.1	2.5163	2.8480
		1	521.3	2.4131	2.9911
		2	862.3	2.3470	3.0980
		3	1198.0	2.2961	3.1906
		4	1528.4	2.2542	3.2752
		5	1853.5	2.2183	3.3548
		6	2173.2	2.1868	3.4309
		7	2487.5	2.1588	3.5046
		8	2796.4	2.1336	3.5766
		9	3099.7	2.1107	3.6475
		10	3397.6	2.0897	3.7175
		11	3689.9	2.0704	3.7872
		12	3976.6	2.0526	3.8566
		13	4257.7	2.0361	3.9260
		14	4533.2	2.0208	3.9956
		15	4802.9	2.0066	4.0656
		16	5067.0	1.9935	4.1361
$T_e = 14069.9$ cm^{-1}	$A\,^1\Sigma_u^+$	0	127.3	2.9237	3.3125
		1	379.7	2.8043	3.4812
		2	628.8	2.7281	3.6066
		3	874.9	2.6693	3.7142
		4	1117.9	2.6205	3.8116
		5	1357.7	2.5782	3.9021

	State	v	U(cm^{-1})	r$_{min}$(Å)	r$_{max}$(Å)
T$_e$ = 20439.40 cm^{-1}	B$^1\Pi_u$	0	134.2	2.7598	3.1389
		1	398.2	2.6448	3.3074
		2	656.1	2.5714	3.4354
		3	907.6	2.5148	3.5480
		4	1152.2	2.4675	3.6526
		5	1389.7	2.4263	3.7528
		6	1619.6	2.3893	3.8506
		7	1841.5	2.3552	3.9476
		8	2055.0	2.3232	4.0449
		9	2259.8	2.2927	4.1434
		10	2455.5	2.2631	4.2441
		11	2641.7	2.2339	4.3479
		12	2814.6	2.2016	4.4635
		13	2976.8	2.1704	4.5812
		14	3127.9	2.1384	4.7059

Li$_2$

BIBLIOGRAPHY

(28. 1)   A ← X System, Vibrational Analysis,
          K. Wurm,
          Naturwissenschaften 16, 1028

(29. 2)   B ← X System, Rotational Analysis,
          A. Harvey and F. A. Jenkins,
          Phys. Rev. 34, 1286

(29. 3)   B ← X System, Rotational Analysis,
          K. Wurm,
          Z. Physik 58, 562-9

(29. 4)   A ← X System, Rotational Analysis,
          K. Wurm,
          Z. Physik 59, 35-41

(30. 5)   B ← X System, Rotational Analysis,
          A. Harvey and F. A. Jenkins,
          Phys. Rev. 35, 132

(30. 6)   B ← X System, Rotational Analysis,
          A. Harvey and F. A. Jenkins,
          Phys. Rev. 35, 789-801

(31. 7)   Comparison With Theory,
          J. H. Bartlett and W. H. Furry,
          Phys. Rev. 37, 1712

(31. 8)   B ← X System, Rotational Analysis,
          F. W. Loomis and R. E. Nusbaum,
          Phys. Rev. 37, 1712

(31. 9)   B ← X System, Vibrational Analysis, Dissociation Energy,
          F. W. Loomis and R. E. Nusbaum,
          Phys. Rev. 38, 1447-57

(31. 10) Theory,
          W. R. Van Wijk and A. J. Van Koeveringe,
          Proc. Roy. Soc. A 132, 98-107

(33. 11) Theory,
          W. H. Furry,
          Phys. Rev. 43, 361-2

(33. 12) Comparison Between ^{6}Li$_2$ and ^{7}Li$_2$,
F. A. Jenkins and A. MacKellar,
Phys. Rev. **44**, 325-6

(33. 13) B ← X System, Rotational Analysis,
A. MacKellar,
Phys. Rev. **44**, 155-64

(35. 14) Preliminary Note to (36. 16),
G. M. Almy and G. R. Irwin,
Phys. Rev. **48**, 104-5

(35. 15) C ← X System, Ultraviolet Systems,
J. E. Vance and J. R. Huffman,
Phys. Rev. **47**, 215-6

(36. 16) A ← X System, Rotational Analysis,
G. M. Almy and G. R. Irwin,
Phys. Rev. **49**, 72-7

(37. 17) Theory,
H. Yoshinaga,
Proc. Phys. -Math. Soc. Japan **19**, 847-59

(37. 18) Theory,
H. Yoshinaga,
Proc. Phys. -Math. Soc. Japan **19**, 1073-83

(39. 19) Theory,
C. H. D. Clark,
Nature **144**, 285-6

(40. 20) Theory,
M. F. Mamotenko,
Acta Physicochim. **12**, 946-7

(40. 21) Theory,
G. B. B. M. Sutherland,
J. Chem. Phys. **8**, 161-4

(40. 22) Theory,
R. F. Barrow,
Trans. Faraday Soc. **36**, 624-5

(40. 23) Theory,
C. H. D. Clark,
Trans. Faraday Soc. **36**, 370-6

Li$_2$

(40.24)   Theory,
          J. W. Linnett,
          Trans. Faraday Soc. 36, 1123-34

(41.25)   Theory,
          H. M. Hulburt and J. O. Hirschfelder,
          J. Chem. Phys. 9, 61-9

(41.26)   Theory,
          C. H. D. Clark,
          Trans. Faraday Soc. 37, 299-302

(41.27)   Theory,
          C. H. D. Clark and K. R. Welb,
          Trans. Faraday Soc. 37, 293-8

(42.28)   Theory,
          H. Fajans,
          J. Chem. Phys. 10, 759-60

(42.29)   Theory,
          K. Fajans,
          J. Chem. Phys. 10, 761

(47.30)   Preliminary Note to (48.31),
          R. W. B. Pearse and S. P. Sinha,
          Nature 160, 159

(48.31)   C ← X System,
          S. P. Sinha,
          Proc. Phys. Soc. 60, 443-7

(53.32)   Calculation of Magnetic Properties,
          R. M. Sternheimer and H. M. Foley,
          Phys. Rev. 92, 1460-8

(57.33)   Theory,
          R. Fieschi,
          Ist. Naz. Fiz. Nucl. Milano 112, 195

(57.34)   Theory,
          E. Ishiguro, K. Kayama, M. Kotani, and Y. Mizuno,
          J. Phys. Soc. Japan 12, 1355-85

(58.35)   Theory,
          T. Arai and M. Sakamoto,
          J. Chem. Phys. 28, 32-48

(60.36) C, D - X Systems,
R. F. Barrow, N. Travis, and C. V. Wright,
Nature 187, 141-2

(60.37) LCAO-MO-SCF Calculations,
B. J. Ransil,
Rev. Mod. Phys. 32, 400-11

(63.38) VB Calculations,
C. Manneback,
Physica 29, 769-83

(64.39) Rotational Magnetic Moment,
R. A. Brooks, C. H. Anderson, and N. F. Ramsey,
Phys. Rev. 136, 62-8

(65.40) Magnetic Properties,
R. M. Stevens and W. N. Lipscomb,
J. Chem. Phys. 42, 4302-4

(66.41) LCAO-MO-SCF Calculations,
G. Das and A. C. Wahl,
J. Chem. Phys. 44, 87-96

(67.42) Extended Hartree-Fock Calculations,
G. Das,
J. Chem. Phys. 46, 1568-78

(67.43) Calculated Molecular Properties,
W. D. Lyon and J. O. Hirschfelder,
J. Chem. Phys. 46, 1788-96

(67.44) AMO Calculations,
D. K. Rai and J. L. Calais,
J. Chem. Phys. 47, 906-11

(67.45) J. R. De La Vega and H. F. Hameka,
"Calculation of Magnetic Susceptibilities of Diatomic Molecules. V.
Homonuclear Molecules, "
Physica 35, 313-22

(67.46) J. R. De La Vega and H. F. Hameka,
"Calculation of Magnetic Susceptibilities of Diatomic Molecules. VIII.
Anisotropies and Rotational Magnetic Moments, "
J. Chem. Phys. 47, 1834-6

(68.47)   M. A. Marchetti and S. R. LaPaglia,
"Theoretical $^1\Sigma_g^+ - ^1\Sigma_u^+$ Dipole Strengths of Some Homonuclear Diatomic
Molecules: Configuration Interaction,"
J. Chem. Phys. <u>48</u>, 434-9

(68.48)   L. Szasz and G. McGinn,
"Atomic and Molecular Calculations With the Pseudopotential Method.
III. The Theory of Li_2, Na_2, K_2, LiH, NaH, and KH,"
J. Chem. Phys. <u>48</u>, 2997-3008

(69.49)   P. Colmant and Ch. Manneback,
"Electronic Structure and Energy of the Homonuclear Molecule Li_2
in its $^1\Sigma_g^+$ Ground State,"
Bull. Cl.^{S} Sci. Acad. Rol. Belg. <u>55</u>, 55-84

(69.50)   D. C. Jain and R. C. Sahni,
"Reduced Potential Energy Curves of Some Electronic States of
Alkali Molecules,"
Trans. Faraday Soc. <u>65</u>, 897-903

(69.51)   R. Velasco, Ch. Ottinger, and R. N. Zare,
"Dissociation Energy of Li_2 From Laser-Excited Fluorescence,"
J. Chem. Phys. <u>51</u>, 5522-32

(70.52)   W. Kutzelnigg and M. Gelus,
"Potential Curve of the Li_2 Ground State for Large Internuclear
Distances. A Contribution to the Understanding of Interatomic Forces,"
Chem. Phys. Letters <u>7</u>, 296-302

(70.53)   A. N. Singh and D. K. Rai,
"Transition Probabilities for Electronic Spectra of Li_2 and C_2,"
Proc. Nat. Acad. Sci. A <u>40</u>, 101-2

(70.54)   J. N. Bardsley,
"Psuedopotential Calculations of Alkali Interactions,"
Chem. Phys. Letters <u>7</u>, 517

(71.55)   A. K. Chandra and R. Sundar,
"A View of Bond Formation in the Li_2 Molecule,"
Molec. Phys. <u>22</u>, 369-74

(71.56)   Ch. Ottinger and D. Poppe,
"Collision-Induced Rotational and Vibrational Quantum Jumps in
Electronically Excited Li_2,"
Chem. Phys. Letters <u>8</u>, 513-8

(72.57) W. von Niessen,
"Density Localization of Atomic and Molecular Orbitals. II. Homo-
nuclear Diatomic Molecules,"
Theoret. Chim. Acta 27, 9-23

(72.58) G. L. Bendazzoli, F. Bernardi, and A. Geremia,
"A. M. O. Calculations for Some First Row Diatomic Molecules,
Theoret. Chim. Acta 27, 63-8

(72.59) W. Kutzelnigg, V. Staemmler, and M. Gelus,
"Potential Curve of the Lowest Triplet State of Li$_2$,"
Chem. Phys. Letters 13, 496-500

(72.60) S. Kohda, K. Ohno, and H. Taketa,
"On the First Singlet and Triplet Excited States of the Lithium
Molecule,"
Bull. Chem. Soc. Japan 45, 2737-8

(72.61) G. C. Shukla,
"An Improved AMO Function for the Li$_2$ Molecule,"
Acta Phys. Polon. A41, 683-7

(72.62) A. C. Roach,
"Theoretical Ground State and Excited State Potential Energy Curves
for Alkali Diatomic Molecules,"
J. Molec. Spectrosc. 42, 27-37

(72.63) A. Hernandez and E. V. Ludena,
"Loge Localization Analysis of Diatomic Molecules: Li$_2$ and F$_2$,"
J. Chem. Phys. 57, 5350-3

(72.64) R. Velasco and F. Rivero,
"The A - X Bands System of the ^{6}Li$_2$ Molecule,"
Opt. Pura Appl. 5, 76-9

(73.65) T. Caves,
"Van Der Waals Interactions Between Excited Li Atoms,"
J. Chem. Phys. 59, 6177-82

(73.66) R. Velasco and V. Morales,
"The B - X Band System of the ^{6}Li$_2$ Molecule,"
Opt. Pura Appl. 6, 52-6

(74.67) G. C. Lie and E. Clementi,
"Study of the Electronic Structure of Molecules. XXII. Correlation
Energy Corrections as a Function of the Hartree-Fock Type Density
and its Application to the Homonuclear Diatomic Molecules of the
Second Row Atoms,"
J. Chem. Phys. 60, 1288-96

Li$_2$

(74. 68) R. W. Molof, T. M. Miller, H. L. Schwartz, B. Bederson, and J. T. Park, "Measurements of the Average Electric Dipole Polarizabilities of the Alkali Dimers," J. Chem. Phys. $\underline{61}$, 1816-22

(74. 69) G. Ennen and Ch. Ottinger, "Rotation – Vibration – Translation Energy Transfer in Laser Excited Li$_2\left(B^1\Pi_u\right)$," Chem. Phys. $\underline{3}$, 404-30

Mg_2

Methods of Production and Experimental Technique

Absorption (T ~ 800°C).

BAND SYSTEMS

	System	Transition	Sources	Wavelength Limits	Degrading	Characteristic Bands, λ	Remarks	Bibliography
	I	$A^1\Sigma_u^+ \leftarrow X^1\Sigma_g^+$	Absorption	3853-3140	V	3790.9(0, 2) 3764.7(0, 3)		(70.7)
	II	$(^1\Pi_u) \leftarrow X^1\Sigma_g^+$	Absorption	2852-2660	-			(70.7)

Mg_2

I. $\underline{A\,^1\Sigma_u^+ \leftarrow X\,^1\Sigma_g^+\ \text{System}}$

Band heads, λ (70.7):

v', v''	3	4	5	6	7
2	3790.9	3796.5	3801.6	3806.3	3810.7
3	3764.6	3770.2	3775.3	3779.9	3784.2
4	3739.2	3744.5	3749.5	3754.2	3758.3
5	3714.2	3719.5	3724.5	3729.0	

SPECTROSCOPIC CONSTANTS

State	T_e	ω_e	$x_e\omega_e$	B_e	$\alpha_e \times 10^3$	$D_e \times 10^6$	r_e	Remarks	Bibliography
$A\ ^1\Sigma_u^+$	26068.76	190.615	1.14562	0.147999	1.31642	0.334286	3.082		(70.7)
$X\ ^1\Sigma_g^+$	0	51.12	1.6448	0.0929	3.7758	1.2166	3.890		(70.7)

Dissociation energy = 0.05 eV, 1.15 kcal/mole, 404.1 cm^{-1} (73.10).

Mg_2

Perturbations and General Information

Potential energy curves — RKR potentials (72.8):

	State	v	$E(v)cm^{-1}$	$r_{min}(Å)$	$r_{max}(Å)$
$T_e = 0.0$	$X^1\Sigma_g^+$	0	25.156	3.6872	4.1626
		1	73.037	3.5698	4.4165
		2	117.757	3.5010	4.6260
		3	159.384	3.4509	4.8226
		4	197.971	3.4112	5.0166
		5	233.558	3.3786	5.2140
		6	266.168	3.3513	5.4195
		7	295.811	3.3285	5.6380
		8	322.482	4.4097	5.8750
		9	346.162	3.2948	6.1378
		10	366.806	3.2835	6.4364
		11	384.393	3.2762	6.7852
		12	398.831	3.2717	7.2110
$T_e = 26068.76\ cm^{-1}$	$A^1\Sigma_u^+$	0	95.021	2.9676	3.2111
		1	283.350	2.8915	3.3154
		2	469.404	2.8426	3.3927
		3	653.193	2.8048	3.4591
		4	834.728	2.7736	3.5193
		5	1014.020	2.7467	3.5754
		6	1191.078	2.7231	3.6286
		7	1365.915	2.7018	3.6796
		8	1538.541	2.6826	3.7290
		9	1708.965	2.6649	3.7771
		10	1877.199	2.6486	3.8242
		11	2043.254	2.6335	3.8704
		12	2207.139	2.6193	3.9159
		13	2368.867	2.6060	3.9609
		14	2528.446	2.5935	4.0055
		15	2685.889	2.5818	4.0497

Franck-Condon factors — RKR potentials (72.8):

$A\,^1\Sigma_u^+ - X\,^1\Sigma_g^+$

v', v''	0	1	2	3	4	5	6	7	8
0	0.0000	0.0000	0.0001	0.0003	0.0006	0.0010	0.0014	0.0019	0.0022
1	0.0001	0.0004	0.0012	0.0027	0.0047	0.0070	0.0092	0.0110	0.0121
2	0.0004	0.0020	0.0053	0.0102	0.0159	0.0211	0.0249	0.0266	0.0264
3	0.0016	0.0065	0.0148	0.0245	0.0326	0.0370	0.0371	0.0337	0.0283
4	0.0044	0.0157	0.0301	0.0412	0.0448	0.0406	0.0316	0.0216	0.0130
5	0.0103	0.0302	0.0471	0.0508	0.0416	0.0264	0.0126	0.0039	0.0004
6	0.0204	0.0480	0.0578	0.0452	0.0235	0.0067	0.0002	0.0014	0.0055
7	0.0350	0.0636	0.0550	0.0259	0.0045	0.0004	0.0067	0.0138	0.0169
8	0.0530	0.0707	0.0381	0.0061	0.0011	0.0121	0.0210	0.0212	0.0156
9	0.0722	0.0653	0.0160	0.0004	0.0147	0.0261	0.0228	0.0124	0.0039
10	0.0895	0.0485	0.0015	0.0120	0.0297	0.0248	0.0099	0.0009	0.0008

BIBLIOGRAPHY

(31. 1)   H. Hamada,
Nature 127, 555

(31. 2)   H. Hamada,
Philos. Mag. 12, 50-67

(55. 3)   Dissociation Energy,
J. R. Soulen, P. Sthapitanonda, and J. L. Margrave,
J. Phys. Chem. 59, 132-6

(64. 4)   S. Weniger,
J. Phys. 25, 946-9

(70. 5)   R. H. Ewing and A. M. Mellor,
"Further Calculations of Equilibrium Dimer Properties,"
J. Chem. Phys. 53, 2983-4

(70. 6)   W. C. Stwalley,
"Long-Range Analysis of the Internuclear Potential of Mg$_2$,"
Chem. Phys. Letters 7, 600-2

(70. 7)   W. J. Balfour and A. E. Douglas,
"Absorption Spectrum of the Mg$_2$ Molecule,"
Can. J. Phys. 48, 901-14

(72. 8)   W. J. Balfour and R. F. Whitlock,
"Rotational Dependence of Franck-Condon Factors in the $A\,^1\Sigma_u^+ \leftarrow X\,^1\Sigma_g^+$
System of ^{24}Mg$_2$,"
Can. J. Phys. 50, 1648-51

(72. 9)   A. C. Brett and C. Chan,
"The Mg$_2$ Potential: A Curve Fitting Study,"
Can. J. Phys. 50, 1587-9

(73. 10)   K. C. Li and W. C. Stwalley,
"Vibrational Levels Near Dissociation in Mg$_2$ and Long-Range Forces,"
J. Chem. Phys. 59, 4423-7

(73. 11)   M. K. Matzen, G. V. Calder, and D. K. Hoffman,
"An Accurate Semi-Empirical Potential Function for Diatomic
Molecules,"
Spectrochim. Acta 29A, 2005-16

Mn$_2$

Spectroscopic Constants

Dissociation energy = 0.22 $\pm$ 0.17 eV, 5 kcal/mole, 1750 cm^{-1} (68.2).

Mn$_2$

BIBLIOGRAPHY

(64. 1)   A. Kant and B. Strauss,
"Dissociation Energies of Diatomic Molecules of the Transition
Elements. II. Titanium, Chromium, Manganese, and Cobalt,"
J. Chem. Phys. 4, 3806-8

(68. 2)   A. Kant, S. Lin, and B. Strauss,
"Dissociation Energy of Mn$_2$,"
J. Chem. Phys. 49, 1983-5

N$_2$

Methods of Production and Experimental Technique

Absorption (in the vacuum ultraviolet).

Emission from discharge into air, pure N$_2$, or N$_2$ in rare gases, hollow
cathode discharge, high voltage arc, afterglow, aurora, laser emission,
electron beam emission.

BAND SYSTEMS

	System	Transition	Sources	Wavelength Limits	Degrading	Characteristic Bands, λ	Remarks	Bibliography
Vegard-Kaplan	I	$A\,^3\Sigma_u^+ \rightleftharpoons X\,^1\Sigma_g^+$	Luminescence	5060-2100	R	2760.8(0,6)		(71.105, 68.80, 68.75, 65.57, 62.40, 61.38, 59.31, 34.6, 34.5, 32.3)
Wilkinson	II	$B\,^3\Pi_g \leftarrow X\,^1\Sigma_g^+$	Absorption	1690-1630	R	1635(0,0) 1638(1,0)		(62.42)
Saum-Benesch	III	$W\,^3\Delta_u \leftarrow X\,^1\Sigma_g^+$	Absorption	4400-2400				(71.101)
Ogawa-Tanaka-Wilkinson	IV	$B'\,^3\Sigma_u^- \leftarrow X\,^1\Sigma_g^+$	Absorption	2240-1120	R			(65.51, 64.46)

	System	Transition	Sources	Wavelength Limits	Degrading	Characteristic Bands, λ	Remarks	Bibliography
Ogawa-Tanaka-Wilkinson-Mulliken	V	$a'^1\Sigma_u^- \rightleftharpoons X^1\Sigma_g^+$	Absorption: N$_2$ + Ar	2000-1080	R			(66.62, 65.54, 64.46, 60.35, 59.32, 59.30)
Lyman-Birge-Hopfield	VI	$a^1\Pi_g \rightleftharpoons X^1\Sigma_g^+$	Absorption and discharge	2600-1090	R	2125.0(5, 14) 2041.2(5, 13)		(66.63, 65.59, 65.55, 54.46, 56.25)
Tanaka	VII	$w^1\Delta_u \leftarrow X^1\Sigma_g^+$	Absorption	1400-1140	R			(64.46)
Tanaka	VIII	$C^3\Pi_u \leftarrow X^1\Sigma_g^+$	Absorption	1130-1070	R		5 heads	(65.53, 64.46)
	IX	$E^3\Sigma_g^+ \leftarrow X^1\Sigma_g^+$	Energy loss spectra	~1050		1043.9(0, 0)		(73.166)

	System	Transition	Sources	Wavelength Limits	Degrading	Characteristic Bands, λ	Remarks	Bibliography
Dressler-Lutz	X	$a''^1\Sigma_g^+ \leftarrow X^1\Sigma_g^+$	Absorption	~1010		1011.5(0, 0)		(73.166, 67.67)
	XI	$b^1\Pi_u \rightleftharpoons X^1\Sigma_g^+$	Absorption and discharge	995-855	R	979.5(2, 0)		(73.166, 69.83, 69.82, 69.81, 64.47)
	XII	$F^3\Pi_u \leftarrow X^1\Sigma_g^+$	Energy loss spectra	980-930		972.2(0, 0)		(73.166)
	XIII	$G^3\Pi_u \leftarrow X^1\Sigma_g^+$	Energy loss spectra	970-940		967.7(0, 0)		(73.166)
	XIV	$D^3\Sigma_u^+ \leftarrow X^1\Sigma_g^+$	Energy loss spectra	~960		965.4(0, 0)		(73.166)
	XV	$b'^1\Sigma_u^+ \rightleftharpoons X^1\Sigma_g^+$	Absorption and discharge	965-830	R			(69.83, 69.82, 69.81, 64.47)
	XVI	$c^1\Pi_u \rightleftharpoons X^1\Sigma_g^+$	Absorption and discharge	960-865	R			(69.83, 69.82, 69.81, 64.47)
	XVII	$c'^1\Sigma_u^+ \rightleftharpoons X^1\Sigma_g^+$	Absorption and discharge	960-840	R			(69.83, 69.82, 69.81, 64.47)

	System	Transition	Sources	Wavelength Limits	Degrading	Characteristic Bands, λ	Remarks	Bibliography
	XVIII	$o\,^1\Pi_u \rightleftharpoons X\,^1\Sigma_g^+$	Absorption and discharge	950-880	R			(69.83, 69.82, 69.81, 64.47)
	XIX	$e\,^1\Pi_u \leftarrow X\,^1\Sigma_g^+$	Energy loss spectra	~ 860		865.1(0, 0)		(69.90)
	XX	$e'\,^1\Sigma_u^+ \leftarrow X\,^1\Sigma_g^+$	Energy loss spectra	~ 860		863.8(0, 0)		(69.90)
First Positive	XXI	$B\,^3\Pi_g \rightarrow A\,^3\Sigma_u^+$	Positive column	Infrared - 4700	V	10510.1(0, 0) 8912.4(1, 0)		(61.38, 59.33)
Herman- Kaplan	XXII	$E\,^3\Sigma_g^+ \rightarrow A\,^3\Sigma_u^+$	Lumines- cence	2740-2130	V	2471.4(0, 4) 2391.6(0, 3)	Bands not resolved	(45.16, 35.9)
Wu- Benesch	XXIII	$W\,^3\Delta_u \rightleftharpoons B\,^3\Pi_g$	Discharge	69000-7000			Bands not resolved	(71.101, 68.73)
"Y" Bands	XXIV	$B'\,^3\Sigma_u^- \rightarrow B\,^3\Pi_g$	Lumines- cence from discharge	8920-6060	R		Complex structure	(64.45, 60.36, 60.34, 58.29)

	System	Transition	Sources	Wavelength Limits	Degrading	Characteristic Bands, λ	Remarks	Bibliography
Second Positive	XXV	$C\,^3\Pi_u \rightarrow B\,^3\Pi_g$	Positive column	5450-2680	V	3371.3(0, 0) 3576.9(0, 1)		(65.56, 64.49, 60.37, 59.33)
Goldstein- Kaplan	XXVI	$C'\,^3\Pi_u \rightarrow B\,^3\Pi_g$	Lumines- cence	5060-2860	R	4728.0(0, 11)		(64.45, 63.44, 61.38)
Fourth Positive	XXVII	$D\,^3\Sigma_u^+ \rightarrow B\,^3\Pi_g$	Lumines- cence from discharge	2910-2250	V	2448.0(0, 2)	5 heads	(40.11)
	XXVIII	$E\,^3\Sigma_g^+ \rightarrow B\,^3\Pi_g$	Electron impact	3180-2740	V	2740(0, 0)		(69.88)
MacFarlane Infrared	XXIX	$a\,^1\Pi_g \rightarrow a'\,^1\Sigma_u^-$	Laser emission	82000-33000				(65.58)
Fifth Positive	XXX	$x\,^1\Sigma_g^- \rightarrow a'\,^1\Sigma_u^-$	Discharge	2850-2030	V	2411.7(1, 4)		(56.26)

N$_2$

BAND SYSTEMS

	System	Transition	Sources	Wavelength Limits	Degrading	Characteristic Bands, λ	Remarks	Bibliography
First Kaplan	XXXI	$y\,^1\Pi_g \to a'\,^1\Sigma_u^-$	Discharge	2470-2070	V	2225.9(0, 1)		(57.28)
MacFarlane Infrared	XXXII	$w\,^1\Delta_u \to a\,^1\Pi_g$	Laser emission	36500				(66.64)
Gaydon-Herman	XXXIII	$b\,^1\Pi_u \to a\,^1\Pi_g$	Discharge	3420-2740	R			(69.82, 69.81, 57.27)
Gaydon-Herman	XXXIV	$b'\,^1\Sigma_u^+ \to a\,^1\Pi_g$	Discharge	2500	R			(69.82, 69.81, 57.27)
Gaydon-Herman	XXXV	$c\,^1\Pi_u \to a\,^1\Pi_g$	Discharge	3010-2220	R,V			(69.82, 69.81, 57.27)
Gaydon-Herman	XXXVI	$c'\,^1\Sigma_u^+ \to a\,^1\Pi_g$	Discharge	3660-2280	R,V			(69.82, 69.81, 57.27)

	System	Transition	Sources	Wavelength Limits	Degrading	Characteristic Bands, λ	Remarks	Bibliography
Gaydon-Herman	XXXVII	$d'\,^1?_u \to a\,^1\Pi_g$	Discharge	2550-2350				(69.82, 69.81, 57.27)
Gaydon-Herman	XXXVIII	$b\,^1\Pi_u \to a\,^1\Pi_g$	Discharge	2860-2720	R			(69.82, 69.81, 57.27)
Second Kaplan	XXXIX	$y\,^1\Pi_g \to w\,^1\Delta_u$	Discharge	2860-2260	V	2536.6(0, 2)		(57.28)
	XL	$z\,^1\Delta_g \to w\,^1\Delta_u$	Discharge	2480-2360	V			(57.27)
	XLI	$E\,^3\Sigma_g^+ \to C\,^3\Pi_u$	Electron impact	12850	V	12843.6(0, 0)	One band observed	(69.88)
Gaydon Green	XLII	?	Discharge	6340-5040	V	5815(0, 1)	Bands not resolved	(54.23, 53.22, 44.15)
Herman Infrared	XLIII	?	Discharge	8550-7000	V	8057(0, 0)		(53.22, 51.18)

362

	System	Transition	Sources	Wavelength Limits	Degrading	Characteristic Bands, λ	Remarks	Bibliography
Worley-Jenkins	XLIV	$X^2\Sigma_g^+ \leftarrow X^1\Sigma_g^+$ (N_2^+)	Absorption	< 960			Rydberg series	(69.82, 69.81, 67.66, 62.39, 53.21, 53.20, 43.14, 42.13)
Carroll-Yoshino	XLV	$X^2\Sigma_g^+ \leftarrow X^1\Sigma_g^+$ (N_2^+)	Absorption	< 960			Rydberg series	(69.82, 69.81, 67.66)
Worley	XLVI	$A^2\Pi_u \leftarrow X^1\Sigma_g^+$ (N_2^+)	Absorption	< 960			Rydberg series	(62.39, 53.21, 53.20)
Hopfield	XLVII	$B^3\Sigma_u^+ \leftarrow X^1\Sigma_g^+$ (N_2^+)	Absorption	< 960			Rydberg series	(62.39, 43.14, 42.13, 38.10, 34.9, 30.1)
	XLVIII	$C^2\Sigma_u^+ \leftarrow X^1\Sigma_g^+$ (N_2^+)	Absorption	570-470			Rydberg series	(66.60, 52.19)
	XLIX	Continuum	Absorption	1000-610				(73.151)

I. $\underline{A^3\Sigma_u^+ \rightleftarrows X^1\Sigma_g^+}$ (Vegard-Kaplan) System

Band heads, λ (61.38, 50.17):

v', v''	2	3	4	5	6	7	8	9
0	2215.1	2332.8	2461.6	2603.6	2760.8	2935.7		3351.5
1	2146.6	2257.2	2377.5	2509.8	2655.5	2817.1	2997.0	3197.5
2		2187.8	2300.7	2424.2	2560.1	2710.1		
3		2123.5	2229.9	2346.0	2472.5	2612.8	2766.9	
4			2164.5	2274.0		2523.4	2666.6	
5				2207.2	2319.7	2441.8	2576.0	2722.5

II. $\underline{B^3\Pi_g \leftarrow X^1\Sigma_g^+}$ (Wilkinson) System

Band heads: (v', v'')   (0,0)   (1,0)
                  λ 1685   1638

III. $\underline{W^3\Delta_u \leftarrow X^1\Sigma_g^+}$ (Saum-Benesch) System

Band heads, λ (70.101, 70.94):

v', v''	0	1	2	3	4	5	6	7	8
0	1683.6	1752.4	1826.0	1905.1	1990.2	2082.0	2181.4	2289.1	2406.4
1	1642.7	1708.1	1778.0	1852.9	1933.3	2019.9	2113.2	2214.2	2323.8
2	1604.4	1666.7	1733.2	1804.3	1880.4	1962.2	2050.2	2145.1	2247.7
3	1568.3	1627.8	1691.2	1758.8	1831.1	1908.5	1991.7	2081.1	2177.6
4	1534.4	1591.3	1651.8	1716.3	1785.0	1858.6	1937.3	2021.9	2112.8
5	1502.9	1557.0	1614.9	1676.4	1742.0	1811.9	1886.7	1966.8	2052.7
6	1472.8	1524.6	1580.1	1639.0	1701.6	1768.3	1839.4	1915.4	1996.9
7	1444.2	1494.1	1547.4	1603.8	1663.7	1727.4	1795.2	1867.5	1944.9
8	1416.9	1465.3	1516.5	1570.6	1628.0	1689.0	1753.7	1822.7	1896.3
9	1391.5	1438.1	1487.3	1539.4	1594.5	1652.9	1714.9	1780.8	1851.0
10	1367.3	1412.3	1459.8	1509.9	1562.8	1618.9	1678.3	1741.4	1808.5

IV. $\quad B'\,^3\Sigma_u^- \leftarrow X\,^1\Sigma_g^+$ (Ogawa-Tanaka-Wilkinson) System

Band heads, λ (66.61, 60.35, 59.30):

v', v''	0	1	2	3	4	5	6	7	8
0	1518.1			1695.6	1762.6	1834.2		1993.0	2081.2
1	1484.4		1593.9	1653.8	1717.5			1935.4	2018.6
2	1452.8						1808.6	1881.9	
3	1422.9								
4	1394.7								
5	1368.1								
6	1342.8								
7	1318.9								
8	1296.2								
9	1274.7								
10	1254.2								

V. $\quad a'\,^1\Sigma_u^- \rightleftarrows X\,^1\Sigma_g^+$ (Ogawa-Tanaka-Wilkinson-Mulliken) System

Band heads in absorption, λ (Intensity) (66.61):

(v', v'')	(0, 0)	(1, 0)	(2, 0)	(3, 0)	(4, 0)	(5, 0)	(6, 0)
λ	1477.1	1446.5	1414.7	1387.6	1360.5	1335.0	1310.7
(Intensity)	(2)	(4)	(8)	(16)	(22)	(30)	(30)

(v', v'')	(7, 0)	(8, 0)	(9, 0)	(10, 0)	(11, 0)	(12, 0)	(13, 0)
λ	1287.7	1265.8	1245.0	1225.3	1206.4	1188.5	1171.3
(Intensity)	(52)	(60)	(48)	(42)	(38)	(34)	(28)

(v', v'')	(14, 0)	(15, 0)	(16, 0)	(17, 0)	(18, 0)	(19, 0)
λ	1155.0	1139.3	1124.6	1110.0	1096.3	1083.2
(Intensity)	(24)	(20)	(16)	(10)	(6)	(4)

Band heads in emission, λ (Intensity) (60.33, 59.33):

(v', v'')	(0, 8)	(0, 7)	(0, 6)	(0, 5)	(0, 4)	(0, 3)
λ	2004.2	1922.2	1845.6	1774.0	1707.0	1643.8
(Intensity)	(1)	(2)	(3)	(4)	(4)	(3)

VI. $a^1\Pi_g \rightleftharpoons X^1\Sigma_g^+$ (Lyman-Birge-Hopfield) System

Band heads in emission, λ (66.61):

v', v''	9	10	11	12	13	14	15	16	17
0									
1	1972.6								
2		1988.9	2073.0						
3			2006.0	2089.7	2181.1	2278.3			
4			1944.3	2023.5	2108.1	2198.7	2296.1		
5				1961.8	2041.2	2125.9	2216.6	2314.0	2418.4
6					1979.5	2059.0	2144.0	2234.8	2332.2
7									2253.4

VII. $w^1\Delta_u \leftarrow X^1\Sigma_g^+$ (Tanaka) System

Band heads, λ (Intensity) (64.46):

(v', v'')	(0, 0)	(1, 0)	(2, 0)	(3, 0)	(4, 0)	(5, 0)
λ	1393.9	1364.7	1337.1	1311.0	1286.3	1262.9
(Intensity)		(1)	(2)	(3)	(3)	(4)

(v', v'')	(6, 0)	(7, 0)	(8, 0)	(9, 0)	(10, 0)	(11, 0)
λ	1240.6	1219.4	1199.3	1180.3	1162.1	1144.7
(Intensity)	(5)	(7)	(6)	(6)	(5)	(4)

VIII. $C^3\Pi_u \leftarrow X^1\Sigma_g^+$ (Tanaka) System

Band heads, λ (Intensity) (66.61):

(v', v'')	(0, 0)	(1, 0)	(2, 0)
λ	1124.2	1099.6	1076.3
(Intensity)	(45)	(60)	(30)

IX. $E^3\Sigma_g^+ \leftarrow X^1\Sigma_g^+$ System

Represents a part of a Rydberg series corresponding to a $N_2^+ \, X^2\Sigma_g^+$ core.

Band heads, λ (74.188, 73.166):

(v', v'')	(0, 0)	(1, 0)	(2, 0)
λ	1043.9	1020.7	998.9

X. $a''^1\Sigma_g^+ \leftarrow X^1\Sigma_g^+$ (Dressler-Lutz) System

Represents part of a Rydberg series corresponding to a N_2^+ $X^2\Sigma_g^+$ core.

Band heads, λ (67.67):

(v', v'')	(0, 0)	(1, 0)
λ	1011.5	990.9

XI. $b^1\Pi_u \rightleftharpoons X^1\Sigma_g^+$ System

Band heads, λ (73.166, 69.83, 69.82, 69.81):

(v', v'')	λ	(v', v'')	λ
(0, 0)	991.9	(8, 0)	935.1
(1, 0)	985.6	(9, 0)	929.0
(2, 0)	978.9	(10, 0)	922.7
(3, 0)	972.1	(11, 0)	916.4
(4, 0)	965.7	(12, 0)	910.5
(5, 0)	955.1	(13, 0)	904.7
(6, 0)	949.2	(14, 0)	899.2
(7, 0)	942.4	(15, 0)	895.9

XII. $F^3\Pi_u \leftarrow X^1\Sigma_g^+$ System

Represents a part of a Rydberg series corresponding to a N_2^+ $A^2\Pi_u$ core.

Band heads, λ (73.166):

(v', v'')	(0, 0)	(1, 0)	(2, 0)
λ	972.2	955.0	938.4

XIII. $G^3\Pi_u \leftarrow X^1\Sigma_g^+$ System

Represents part of a Rydberg series corresponding to a $N_2^+ \ X^2\Sigma_g^+$ core.

Band heads, λ (73.166):

(v', v'')	$(0, 0)$	$(1, 0)$
λ	967.7	949.2

XIV. $D^3\Sigma_u^+ \leftarrow X^1\Sigma_g^+$ System

Represents part of a Rydberg series corresponding to a $N_2^+ \ X^2\Sigma_g^+$ core.

Band head, λ (73.166):

(v', v'')	$(0, 0)$
λ	965.4

XV. $b'^1\Sigma_u^+ \rightleftarrows X^1\Sigma_g^+$ System

Band heads, λ (69.83, 69.82, 69.81, 64.47):

(v', v'')	λ	(v', v'')	λ
$(0, 0)$	964.6	$(8, 0)$	-
$(1, 0)$	957.7	$(9, 0)$	907.5
$(2, 0)$	951.0	$(10, 0)$	901.4
$(3, 0)$	944.6	$(11, 0)$	896.2
$(4, 0)$	937.9	$(12, 0)$	891.0
$(5, 0)$	931.9	$(13, 0)$	885.7
$(6, 0)$	926.1	$(14, 0)$	880.7
$(7, 0)$	917.8	$(15, 0)$	875.9

XVI. $c^1\Pi_u \rightleftarrows X^1\Sigma_g^+$ System

c_3 represents the first member of a Rydberg series corresponding to a $N_2^+ \ X^2\Sigma_g^+$ core.

Band heads, λ (69.83, 69.82, 69.81, 64.47):

(v', v'')	λ
$(0, 0)$	960.3
$(1, 0)$	920.0

XVII. $c'^1\Sigma_u^+ \rightleftarrows X^1\Sigma_g^+$ System

c_4' represents the first member of a Rydberg series corresponding to a $N_2^+\,X^2\Sigma_g^+$ core.

Band heads, λ (69.83, 69.82, 69.81, 64.47):

(v', v'')	λ	(v', v'')	λ
(0, 0)	958. 6	(4, 0)	886. 8
(1, 0)	940. 1	(5, 0)	870. 8
(2, 0)	921. 2	(6, 0)	856. 0
(3, 0)	903. 7	(7, 0)	841. 9

XVIII. $o^1\Pi_u \rightleftarrows X^1\Sigma_g^+$ System

Represents the first member of the Worley Rydberg series corresponding to a $N_2^+\,A^2\Pi_u$ core.

Band heads, λ (69.83, 69.82, 69.81, 64.47):

(v', v'')	λ
(0, 0)	946. 1
(1, 0)	928. 9
(2, 0)	912. 6
(3, 0)	897. 2
(4, 0)	882. 5

XIX. $e^1\Pi_u \leftarrow X^1\Sigma_g^+$ System

e_4 represents a member of the Worley-Jenkins Rydberg series corresponding to a $N_2^+\,X^2\Sigma_g^+$ core.

Band heads, λ (69.90):

(v', v'')	(0, 0)	(1, 0)	(2, 0)
λ	865. 1	849. 9	834. 2

XX. $e'^1\Sigma_u^+ \leftarrow X^1\Sigma_g^+$ System

e_4' represents a member of the Worley-Jenkins Rydberg series corresponding to a $N_2^+\,X^2\Sigma_g^+$ core.

Band head, λ (69.90):

(v', v'')	(0, 0)
λ	863. 8

XXI. $B^3\Pi_g \rightarrow A^3\Sigma_u^+$ (First Positive) System

Band heads, λ (Intensity) (50.17):

v', v''	0	1	2	3	4	5	6
0	10510.0(10)						
1	8912.4(10)						
2	7753.2(6)	8722.3(8)	9942.0(2)				
3	6875.0(2)	7626.2(7)	8541.8(6)	9682.1(3)			
4	6186.8(3)	6788.6(6)	7503.9(7)	8369.2(2)	9436.4(3)		
5	5632.7(1)	6127.4(3)	6704.8(8)	7386.6(5)	8204.8(3)	9203.9(2)	
6		5592.9(1)	6069.7(7)	6623.6(9)	7273.3(3)	8047.4(2)	
7		5553.7(1)	6013.6(7)	6544.8(10)	7164.8(2)	7896.4(2)	
8			5515.6(2)	5959.0(8)	6468.5(10)	7059.0(2)	
9				5478.5(2)	5906.0(8)	6394.7(9)	
10					5442.3(3)	5854.4(8)	
11					5053.6	5407.1(3)	
12						5030.8	

XXII. $E^3\Sigma_g^+ \rightarrow A^3\Sigma_u^+$ (Herman-Kaplan) System

Band heads, λ (74.188, 45.16, 35.9):

v', v''	0	1	2	3	4	5	6	7
0		2242.3	2315.3	2391.6	2471.4	2554.9	2642.1	2733.2
1		2137.6	2203.8	2272.9		2419.8	2497.8	

XXIII. $W^3\Delta_u \rightleftarrows B^3\Pi_g$ (Wu-Benesch) System

Band heads, λ (n.p. 218, 71.101, 70.92, 68.73):

v', v''	0	1	2	3	4	5	6
0	629373.3	-65875.5	-31578.9	-20889.5	-15675.8	-12589.4	-10549.8
1	61962.2	586939.5	-58422.4	-30011.2	-20307.6	-15412.6	-12462.8
2	32833.3	73057.5	-357305.2	-52623.6	-28633.1	-19777.0	-15169.8
3	22450.3	36005.1	88595.4	-203381.4	-47987.9	-27413.9	-19292.1
4	17124.2	24002.4	39775.2	111892.7	-143172.7	-44201.2	-26329.2
5	13885.3	18099.6	25797.6	44328.0	150660.7	-111088.6	-41053.2
6	11708.4	14568.8	19174.3	27817.1	49931.5	227885.3	-91169.0
7	10145.4	12225.3	15311.3	20363.6	30133.7	56992.5	456755.3
8	8969.2	10557.0	12781.6	16120.4	21686.6	32816.8	66157.9
9	8052.4	9309.4	10997.3	13382.0	17005.2	23166.3	35959.1
10	7318.0	8341.7	9671.8	11469.3	14031.7	17976.5	24831.6

XXIV. $B'^3\Sigma_u^- \rightarrow B^3\Pi_g$ ("Y" Bands) System

Band heads, λ (Intensity) (50.17):

v', v''	0	1	2	3	4
4	8058(2)				
5	7243(2)	8262(5)			
6	6587(1)	7420(6)	8473(8)		
7	6062(1)	6744(6)	7602(10)	8691(10)	
8		6203(3)	6905(10)	7791(10)	8917(2)

XXV. $C^3\Pi_u \rightarrow B^3\Pi_g$ (Second Positive) System

Band heads, λ (Intensity) (50.17):

v', v''	0	1	2	3	4	5
0	3371.3(10)	3576.9(10)	3804.9(10)	4059.4(8)	4343.6(4)	4667.3(0)
1	3159.3(9)	3338.9(2)	3536.7(8)	3755.4(10)	3998.4(9)	4269.7(5)
2	2976.8(6)	3136.0(8)	3309 (2)	3500.5(4)	3710.5(8)	3943.0(8)
3	2819.8(1)	2962.0(6)	3116.7(6)	3285.3(3)	3469 (0)	3671.9(6)
4	2687	2814.3(1)	2953.2(6)	3104.0(3)	3268.1(4)	3446 (0)

XXVI. $C'^3\Pi_u \rightarrow B^3\Pi_g$ (Goldstein-Kaplan) System

Band heads, λ (50.17):

v', v''	2	3	4	5	6	7	8	9	10
0	2863.5	3005.4	3159.2	3326.1	3504.0	3707.1	3925.4	4166.0	4432.2
1			3025.8	3178.4					

XXVII. $D^3\Sigma_u^+ \rightarrow B^3\Pi_g$ (Fourth Positive) System

Band heads, λ (Intensity) (50.17):

(v', v'')	(0, 6)	(0, 5)	(0, 4)	(0, 3)	(0, 2)	(0, 1)	(0, 0)
λ	2903.9	2777.9	2660.5	2550.7	2448.0	2351.4	2260.8
(Intensity)	(1)	(2)	(5)	(8)	(10)	(6)	(2)

XXVIII. $E^3\Sigma_g^+ \to B^3\Pi_g$ System

Band heads, λ (69.88):

(v', v'')	(0,0)	(0,1)	(0,2)	(0,3)
λ	2740	2880	3020	3180

XXIX. $a^1\Pi_g \to a'^1\Sigma_u^-$ (MacFarlane Infrared) System

Band heads, λ (65.58):

(v', v'')	(0, 0)	(1, 0)	(2, 1)
λ	82489.2	34739.4	33214.5

XXX. $x^1\Sigma_g^- \to a'^1\Sigma_u^-$ (Fifth Positive) System

Band heads, λ (Intensity) (50.17):

v', v''	0	1	2	3	4	5	6
0	2198.9(4)	2274.3(6)	2353.6(4)		2525.6(2)	2619.3(4)	
1	2112.1(5)	2181.5(4)		2331.3(2)	2411.8(7)	2496.7(3)	2586.6(7)
2	2033.6(5)	2097.9(2)	2165.2(5)	2235.9(3)		2387.9	2469.9(4)

XXXI. $y^1\Pi_g \to a'^1\Sigma_u^-$ (First Kaplan) System

Band heads, λ (Intensity) (50.17):

v', v''	0	1	2	3	4
0	2153.6(4)	2225.9(5)	2301.9(4)	2381.7(3)	2466.0(2)
1	2077.3			2288.6(1)	2366.4(2)

XXXII. $w^1\Delta_u \to a^1\Pi_g$ (MacFarlane Infrared) System

Band head, λ (66.64):

(v', v'')	(0, 0)
λ	36399.5

XXXIII. $b^1\Pi_u \to a^1\Pi_g$ (Gaydon-Herman) System

Band heads, λ (69.82, 69.81, 57.27):

v', v''	0	1	2	3	4
0					
1	3075. 1	3241. 3			
...					
5	2795. 4	2932. 0	3079. 9	3240. 8	3416. 5
6	2746. 2	2877. 9	3020. 3	3175. 0	

XXXIV. $b'^1\Sigma_g^+ \to a^1\Pi_g$ (Gaydon-Herman) System

Band head, λ (69.82, 69.81, 57.27):   (v', v'')   (0, 7)
                                                 λ 2498. 6

XXXV. $c^1\Pi_u \to a^1\Pi_g$ (Gaydon-Herman) System

c_3 and c_4 are the two first members of a $c^1\Pi_u$ Rydberg series that converges at $N_2^+ X^2\Sigma_g^+$.

Band heads, λ (69.82, 69.81, 57.27):

$$c_3{}^1\Pi_u \to a^1\Pi_g$$

v', v''	0	1	2	3	4
0	2839. 4	2980. 1			
1					
2	2516. 0	2626. 2	2744. 3	2871. 1	3008. 1

$$c_4{}^1\Pi_u \to a^1\Pi_g$$

v', v''	0	1	2	3	4	5	6
0		2224. 4	2308. 6	2397. 8	2492. 4	2592. 8	2699. 9

XXXVI. $c'^1\Sigma_u^+ \rightarrow a^1\Pi_g$ (Gaydon-Herman) System

c_4' is the first member of a $c'^1\Sigma_u^+$ Rydberg series that converges at $N_2^+ \, X^2\Sigma_g^+$.

Band heads, λ (69.82, 69.81, 57.27):

v', v''	0	1	2	3	4	5
0	2827.1	2967.0	3118.6	3283.3	3463.3	3661.1
1	2671.7	2796.0				
2	2524.9		2753.8			
3	2397.1	2496.8	2603.3			
4	2281.5	2371.6	2467.7	2569.6	2678.5	2795.6

XXXVII. $d'(^1\Sigma_u^-$ or $^1\Delta_u$?$) \rightarrow a^1\Pi_g$ (Gaydon-Herman) System

Band heads, λ (69.82, 69.81, 57.27):

v', v''	0	1	2
0	2358.8	2455.1	2558

XXXVIII. $o^1\Pi_u \rightarrow a^1\Pi_g$ (Gaydon-Herman) System

o is the first member of the Worley Rydberg series that converges at $N_2^+ \, A^2\Pi_u$.

Band heads, λ (69.82, 69.81, 57.27):

(v', v'')	(0, 0)	(0, 1)
λ	2723.6	2853.3

XXXIX. $y^1\Pi_g \rightarrow w^1\Delta_u$ (Second Kaplan) System

Band heads, λ (Intensity) (50.17):

v', v''	0	1	2	3	4	5	6
0	2354.5(4)		2536.6(5)	2636.2(5)	2741.9(3)	2854.9	
1	2263.4(4)		2431.0	2522.3(3)	2619.3(5)	2722.0(3)	2831.7

XL. $z\,^1\Delta_g \rightarrow w\,^1\Delta_u$ System

Band heads, λ (57.27):

(v', v'')	(n, 2)	(n+3, 4)	for n = 2 ?
λ	2477.3	2368.8	

XLI. $E\,^3\Sigma_g^+ \rightarrow C\,^3\Pi_u$ System

Band head, λ (69.88):

(v', v'')	(0, 0)
λ	12843.6

XLII.   Gaydon Green System

Band heads, λ (54.23, 53.22, 44.15):

v', v''	0	1	2	3	4	5
0	5574.4(9)	5815.1(10)	6068.6(8)	6336.3(5)		
1	5308.6(8)	5527.1(2)	5755.1(3)	5994.5(6)	6246.3(5)	
2	5073.4(4)	5272.0(5)	5479.6(6)		5924 (1)	6160.5(3)
3		5047.0(2)		5435.0(3)	5640 (1)	

XLIII.   Herman Infrared System

Band heads, λ (53.22, 51.18):

v', v''	0	1	2
0	8057.6(10)	8549 (2)	
1	7521.0(0)		8397 (1)
2	7061.7(6)	7435.0(5)	7828.5(8)
3		7001.2(4)	

XLIV. $X\,^2\Sigma_g^+\left(N_2^+\right) \leftarrow X\,^1\Sigma_g^+$ (Worley-Jenkins) System

Represents a $^1\Pi_u$ Rydberg series, the first state of which is $c\,^1\Pi_u$
(69.82, 69.81, 67.66, 62.39)

$$\upsilon = 125665.8 - R\left[m + 0.3450 - (0.1000/m) - (0.100/m^2)\right]^{-2}$$
    where m = 2, 3, $\cdots$ 26

XLV. $X^2\Sigma_g^+\left(N_2^+\right) \leftarrow X^1\Sigma_g^+$ (Carroll-Yoshino) System

Represents a $^1\Sigma_u^+$ Rydberg series, the first member of which is
$c'\,^1\Sigma_u^+$ (69.82, 69.81, 67.66)

m	2	3	4	5	6	7
λ	958.559	(863.6)	833.746	820.592	(813.2)	808.672
n*	2.2675		4.3776	5.3713		7.394

XLVI. $A^2\Pi_u\left(N_2^+\right) \leftarrow X^1\Sigma_g^+$ (Worley) System

Represents a $^1\Pi_u$ Rydberg series, the first of which is $o\,^1\Pi_u$ (62.39, 53.21, 53.20)

$\upsilon = 136607 - R\left[m - 0.0441 - (0.018/m^2)\right]^{-2}$ where m = 2, 3, $\cdots$ 6

XLVII. $B^2\Sigma_u^+\left(N_2^+\right) \leftarrow X^1\Sigma_g^+$ (Hopfield) System

Represents a Rydberg series (62.39, 43.14, 42.13, 38.10, 24.8, 30.1)

$\upsilon = 151240 - R(m - 0.092)^{-2}$ where m = 3, 4, $\cdots$ 7

XLVIII. $C^2\Sigma_u^+\left(N_2^+\right) \leftarrow X^1\Sigma_g^+$ System

Represents a Rydberg series

Band heads, λ (66.60, 52.19):

n* = 3.040	(v', v'')	λ	(v', v'')	λ
	(0, 0)	560.48	(7, 0)	520.46
	(1, 0)	554.10	(8, 0)	515.61
	(2, 0)	548.00	(9, 0)	510.93
	(3, 0)	542.11	(10, 0)	506.35
	(4, 0)	536.41	(11, 0)	502.02
	(5, 0)	530.86	(12, 0)	497.77
	(6, 0)	525.54	(13, 0)	493.71

n* = 4.059   (v', v'') λ

	(v', v'')	λ
	(3, 0)	527.33
	(4, 0)	521.89
	(5, 0)	516.71
	(6, 0)	
	(7, 0)	506.71
	(8, 0)	502.02

n* = 5.05   (v', v'') λ

	(v', v'')	λ
	(8, 0)	496.15

XLIX.   <u>Continuum</u>

There are two weak continua between 825 and 1000 Å with maximums of approximately 5 cm^{-1} at 970 Å and 15 cm^{-1} at 910 Å. At approximately 850 Å a dissociation continuum increases gradually to a maximum of ~ 120 cm^{-1} at 805 Å. This is followed by a secondary peak with a maximum value of 75 cm^{-1} occurring at 775 Å. The continuum then decreases to 0 at ~ 750 Å. The most prominent dissociation continuum starts at approximately 730 Å and decreases to 90 cm^{-1} at 660 Å. Below 660 Å there is another continuum with a broad maximum at 610 Å, this continuum overlapping the previous one. (73.151)

SPECTROSCOPIC CONSTANTS

State	T_0	ω_e	$x_e\omega_e$	B_e	$\alpha_e \times 10^3$	$D_e \times 10^6$	r_e	Remarks	Bibliography
$e'\,^1\Sigma_u^+$	115767.5							Rydberg	(69.90)
$e\,^1\Pi_u$	115593.6							Rydberg	(69.90)
$z\,^1\Delta_g$	115365.9	(1700)		(1.76)	15.3		(1.16)	Rydberg	(57.27)
$y\,^1\Pi_g$	114166.3	1707.9[a]		1.78[b]			1.16[c]	Rydberg	(57.28)
$x\,^1\Sigma_g^-$	113212.1	1910.0		1.750	22.5	5.88	1.168	Rydberg	(56.26)
$d'\,^1?_u$	111333								(45.16)
$o\,^1\Pi_u$	105682	2020.0	32.28	1.694[b]			1.19[c]	Rydberg	(69.82, 69.81)
$c'\,^1\Sigma_u^+$	104322.4	2046[a]		1.929[b]			1.12[c]	Rydberg	(69.82, 69.81)
$c'\,^1\Pi_u$	104139.2	2410[a]		1.50[b]			1.27[c]	Rydberg	(69.82, 69.81)
$b'\,^1\Sigma_u^+$	103672	746[a]		1.154	4.8		1.444	Rydberg	(69.82, 69.81)
$D\,^3\Sigma_u^+$	103573			1.961[b]		20	1.108[c]	Rydberg	(40.11)

SPECTROSCOPIC CONSTANTS

State	T_o	ω_e	$x_e\omega_e$	B_e	$\alpha_e \times 10^3$	$D_e \times 10^6$	r_e	Remarks	Bibliography
$G\,^3\Pi_u$	103338							Rydberg	(73.166)
$F\,^3\Pi_u$	102854							Rydberg	(73.166)
$b\,^1\Pi_u$	100816	635[a]		1.448[b]	4.8	29	1.230[c]		(69.86, 69.81)
$a''\,^1\Sigma_g^+$	99032							Rydberg	(67.67)
$C'\,^3\Pi_u$	97580			1.0496[b]		10.9	1.508[c]	Rydberg	(63.44)
$E\,^3\Sigma_g^+$	95774.50	2185[a]		1.927[b]		6.0	1.117[c]	Rydberg	(74.188, 54.50)
$C\,^3\Pi_u$	88977.9	2047.18	28.4450	1.82473	18.683	5.80	1.1487		(65.50)
$^5\Sigma^-$	77925	650				0.0011	1.55		(62.41)
$w\,^1\Delta_u$	71698.8	1559.24	11.8874	1.498	16.6	5.53	1.2678		(65.50)
$a\,^1\Pi_g$	68951.2	1694.20	13.9491	1.61688	17.933	5.89	1.2203		(65.50)
$a'\,^1\Sigma_u^-$	67739.3	1530.25	12.0747	1.47988	16.574	5.54	1.2755		(65.50)

N_2

SPECTROSCOPIC CONSTANTS

State	T_o	ω_e	$x_e\omega_e$	B_e	$\alpha_e \times 10^3$	$D_e \times 10^6$	r_e	Remarks	Bibliography
$B'^3\Sigma_u^-$	65852.4	1516.88	12.1810	1.47359	16.861	5.56	1.2782		(65.50)
$W^3\Delta_u$	60555.8	1501.4	11.6				1.28		(71.101, 65.50)
$B^3\Pi_g$	59306.8	1733.39	14.1221	1.6374	17.91	5.84	1.2126		(65.50)
$A^3\Sigma_u^+$	49754.8	1460.52	13.8313	1.45455	18.009	5.77	1.2866		(65.50)
$X^1\Sigma_g^+$	0	2358.03	14.1351	1.9980	17.72	5.74	1.0977		(65.50)

(a) ΔG_0, (b) B_o, (c) r_o

Dissociation energy = 9.76 ± 0.01 eV, 225.07 kcal/mole, 78710 cm^{-1} (63.43, 56.24).

Perturbations and General Information

The $D^3\Sigma_u^+$ state is predissociated by the shallow $C'^3\Pi_u$ state (74.188).

The $b^1\Pi_u$ state is perturbed by the $c^1\Pi_u$ state
The $b'^1\Sigma_u^+$ state is perturbed by the $c'^1\Sigma_u^+$ state $\Bigg\}$ (73.166)

The $o^1\Pi_u$ level is predissociated possibly by the $C'^3\Pi_u$ state (73.166).

The $B^3\Pi_g$ ($v' \sim 12$) and a $^1\Pi_g$ ($v' \sim 6$) levels are predissociated by the $^5\Sigma^+$ level (68.80).

The higher levels of the $C^3\Pi_u$ and $C'^3\Pi_u$ states are predissociated by the $^3\Pi_u$ continuum (69.82).

Perturbations and predissociation have been observed in the y state (57.28).

Lifetimes:

$A^3\Sigma_u^+$	$v' = 0$	$\tau = 1.36 + 0.27$ sec for $\Sigma = 0$ substate levels (69.L2, 69.L3)
		$\tau = 2.70 \mp 0.54$ sec for $\Sigma = 1, -1$ substate levels

$B^3\Pi_g$	$v' = 0$	$\tau = 10 + 2$ μsec	(75.217)
	$v' = 2$	$\tau = 7.0 + 0.4$ μsec	(66.L1)
	$v' = 3$	$\tau = 6.8 \mp 0.3$ μsec	
	$v' = 4$	$\tau = 6.7 \mp 0.7$ μsec	
	$v' = 5$	$\tau = 6.7 \mp 1.0$ μsec	
	$v' = 6$	$\tau = 7.0 \mp 0.7$ μsec	
	$v' = 7$	$\tau = 5.4 \mp 0.8$ μsec	
	$v' = 8$	$\tau = 5.4 \mp 0.8$ μsec	
	$v' = 9$	$\tau = 5.4 \mp 0.5$ μsec	

$W^3\Delta_u$	$v' = 0$	$\tau = 1.668$ msec	(73.167)
	$v' = 1$	$\tau = 2.000$ msec	

$a'^1\Pi_g$	$v' = 0$	$\tau = 0.17$ msec	(65.52)

$C^3\Pi_u$	$v' = 0$	$\tau = 40.4 + 0.5$ nsec	(73.177)
	$v' = 1$	$\tau = 40.6 \mp 0.5$ nsec	

$D^3\Sigma_u^+$	$v' = 0$	$\tau = 14.1$ nsec	(73.182)

Oscillator Strengths:

$$A^3\Sigma_u^+ \leftarrow X^1\Sigma_g^+ \qquad f_{0,0} = 2 \times 10^{-3} \qquad\qquad (66.\text{L1})$$

$$a'\Pi_g \leftarrow X^1\Sigma_g^+ \qquad f_{0,0} = 1.3 \times 10^{-6} \qquad\qquad (67.68)$$

$$f_{1,0} = 3.0 \times 10^{-6}$$

$$f_{2,0} = 4.1 \times 10^{-6}$$

$$C^3\Pi_u \leftarrow X^1\Sigma_g^+ \qquad f_{0,0} = 2.2 \times 10^{-6}$$

$$f_{1,0} = 1.1 \times 10^{-6}$$

$$f_{2,0} = 5.6 \times 10^{-7}$$

$$w^1\Delta_u \leftarrow X^1\Sigma_g^+ \qquad f_{3,0} = (3.5 + 0.18p) \times 10^{-8}$$

$$f_{4,0} = (6.1 + 0.21p) \times 10^{-8} \quad\Big\}\ \text{for pressure p in psi}$$

$$f_{5,0} = (4.0 + 0.26p) \times 10^{-8}$$

Franck-Condon factors for the $C^3\Pi_u - B^3\Pi_g$ (Second Positive) system (65.52):

v'', v'	0	1	2	3	4
0	4.55-1	3.88-1	1.34-1	2.16-2	1.16-3
1	3.31-1	2.29-2	3.35-1	2.52-1	5.66-2
2	1.45-1	2.12-1	2.30-2	2.04-1	3.26-1
3	4.94-2	2.02-1	6.91-2	8.81-2	1.13-1
4	1.45-2	1.09-1	1.69-1	6.56-3	1.16-1
5	3.87-3	4.43-2	1.41-1	1.02-1	2.45-3
6	9.68-4	1.52-2	7.72-2	1.37-1	4.70-2
7	2.31-4	4.68-3	3.32-2	9.93-2	1.09-1
8	5.36-5	1.33-3	1.23-2	5.26-2	1.04-1
9	1.21-5	3.57-4	4.12-3	2.31-2	6.67-2
10	2.61-6	9.15-5	1.27-3	8.95-3	3.40-2

Franck-Condon factors followed by a factor of ten

Franck-Condon factors for the $B^3\Pi_g$ - $A^3\Sigma_g^+$ (First Positive) system (65.52):

v'', v'	0	1	2	3	4	5	6	7	8
0	4.06-1	4.01-1	1.58-1	3.17-2	3.47-3	2.01-4	5.72-6	8.81-8	8.28-11
1	3.27-1	3.71-3	2.85-1	2.77-1	9.18-2	1.41-2	1.07-3	3.70-5	5.14-7
2	1.64-1	1.59-1	6.59-2	1.05-1	3.06-1	1.63-1	3.41-2	3.26-3	1.35-4
3	6.67-2	1.93-1	2.25-2	1.50-1	1.11-2	2.59-1	2.26-2	6.36-2	7.50-3
4	2.44-2	1.29-1	1.22-1	4.67-3	1.53-1	6.94-3	1.76-1	2.68-1	1.01-1
5	8.38-3	6.57-2	1.39-1	4.09-2	4.94-2	1.00-1	5.05-2	9.30-2	2.83-1
6	2.80-3	2.92-2	9.94-2	1.03-1	2.04-3	9.29-2	4.02-2	9.90-2	3.20-2
7	9.26-4	1.20-2	5.66-2	1.08-1	5.13-2	8.81-3	1.04-1	5.00-3	1.26-1
8	3.07-4	4.73-3	2.83-2	7.88-2	8.85-2	1.22-2	3.92-2	8.29-2	2.75-3
9	1.03-4	1.83-3	1.31-2	4.78-2	8.58-2	5.37-2	4.73-5	6.71-2	4.68-2

Franck-Condon factors followed by factor of ten

Franck-Condon factors for the $A^3\Sigma_u^+$ - $X^1\Sigma_g^+$ (Vegard-Kaplan) system (65.52):

v'', v'	0	1	2	3	4	5	6	7	8
0	1.06-3	5.55-3	1.57-2	3.15-2	5.07-2	6.93-2	8.38-2	9.21-2	9.38-2
1	8.41-3	3.27-2	6.65-2	9.31-2	9.91-2	8.35-2	5.57-2	2.78-2	8.41-3
2	3.34-2	8.88-2	1.15-1	8.91-2	4.00-2	5.73-3	1.92-3	1.90-2	3.87-2
3	8.29-2	1.33-1	8.12-2	1.35-2	3.65-3	3.44-2	5.52-2	4.64-2	2.21-2
4	1.44-1	1.09-1	9.45-3	1.74-2	6.05-2	5.16-2	1.45-2	1.60-4	1.52-2
5	1.89-1	3.67-2	1.77-2	7.36-2	3.88-2	4.23-4	1.88-2	4.41-2	3.63-2
6	1.92-1	8.43-5	8.13-2	4.21-2	1.05-2	4.10-2	4.70-2	1.23-2	1.06-3
7	1.55-1	4.26-2	7.92-2	1.22-4	5.28-2	4.21-2	8.71-4	1.90-2	4.11-2
8	1.02-1	1.17-1	1.76-2	4.83-2	5.01-2	5.30-5	3.71-2	4.04-2	5.56-3
9	5.47-2	1.53-1	6.52-3	8.10-2	8.41-4	4.56-2	3.70-2	8.28-6	2.58-2
10	2.46-2	1.32-1	7.06-2	3.10-2	3.70-2	4.73-2	5.09-4	4.04-2	3.06-2

Franck-Condon factors followed by factor of ten

Franck-Condon factors for the $a^1\Pi_g$ - $X^1\Sigma_g^+$ (Lyman-Birge-Hopfield) system (65.52):

v'', v'	0	1	2	3	4	5	6
0	4.43-2	1.18-1	1.73-1	1.85-1	1.60-1	1.20-1	8.08-2
1	1.51-1	1.90-1	9.44-2	1.15-2	6.67-3	4.75-2	8.52-2
2	2.50-1	8.02-2	3.30-3	7.51-2	9.62-2	4.70-2	4.94-3
3	2.53-1	5.84-4	1.08-1	6.81-2	4.43-4	3.47-2	7.32-2
4	1.73-1	9.22-2	8.41-2	4.39-3	7.81-2	5.51-2	2.37-3
5	8.61-2	1.91-1	3.19-4	9.76-2	3.47-2	9.80-3	6.39-2
6	3.22-2	1.76-1	7.30-2	6.18-2	2.05-2	7.84-2	1.24-2
7	9.17-3	9.93-2	1.73-1	1.17-3	9.90-2	5.16-3	4.47-2
8	1.99-3	3.87-2	1.60-1	9.17-2	2.93-2	5.50-2	5.01-2
9	3.37-4	1.10-2	8.76-2	1.71-1	1.64-2	8.17-2	5.19-3
10	4.75-5	2.33-3	3.23-2	1.38-1	1.25-1	3.08-3	8.54-2

Franck-Condon factors followed by a factor of ten

N$_2$

Franck-Condon factors for the $C^3\Pi_u$ - $X^1\Sigma_g^+$ (Tanaka) system (65.52):

v'', v'	0	1	2	3	4
0	5.59-1	3.03-1	1.01-1	2.76-2	7.16-3
1	3.36-1	8.73-2	2.71-1	1.82-1	7.88-2
2	9.03-2	3.64-1	1.82-3	1.30-1	1.82-1
3	1.33-2	1.95-1	2.44-1	7.11-2	2.42-2
4	1.12-3	4.47-2	2.67-1	9.92-2	1.48-1
5	6.28-5	5.84-3	9.52-2	2.80-1	9.47-3
6	6.12-6	5.13-4	1.80-2	1.58-1	2.19-1
7	6.77-7	4.90-5	2.27-3	4.32-2	2.13-1
8	4.83-9	5.17-6	2.79-4	8.06-3	8.76-2
9	1.44-10	1.48-7	4.21-5	1.39-3	2.40-2
10	1.68-8	9.72-8	4.41-6	2.59-4	5.66-3

Franck-Condon factors followed by factor of ten

Franck-Condon factors for the $W^3\Delta_u$ - $X^1\Sigma_g^+$ system (70.94):

v', v''	0	1	2	3	4	5	6	7
0	.1713-2	.1310-1	.4721-1	.1065-6	.1691-0	.2005-0	.1846-0	.1354-0
1	.8568-2	.4711-1	.1107-0	.1384-0	.8733-1	.1401-1	.9521-2	.7826-1
2	.2295-1	.8741-1	.1204-0	.5727-1	.1253-3	.4533-1	.9516-1	.5355-1
3	.4383-1	.1099-0	.7206-1	.1040-2	.4385-1	.7576-1	.1548-1	.1425-1
4	.6680-1	.1025-0	.1818-1	.2143-1	.6970-1	.1385-1	.1912-1	.6786-1
5	.8696-1	.7284-1	.7807-4	.5735-1	.3075-1	.7809-2	.5891-1	.1536-1
6	.1003-0	.3743-1	.1649-1	.5698-1	.5108-3	.4486-1	.2615-1	.8806-2
7	.1050-0	.1106-1	.4202-1	.2820-1	.1368-1	.4519-1	.2614-4	.4465-1
8	.1021-0	.3046-3	.5511-1	.4007-2	.3909-1	.1477-1	.2141-1	.3225-1
9	.9325-1	.3737-2	.4992-1	.1816-2	.4350-1	.1630-4	.4062-1	.2814-2
10	.8129-1	.1581-1	.3329-1	.1623-1	.2659-1	.1279-1	.2831-1	.6968-2

Franck-Condon factors followed by a factor of ten

r-Centroids for the $B'^3\Sigma_u^-$ - $X^1\Sigma_g^+$ system (66.65a):

v', v''	0	1	2	3	4	5	6	7	8
0	1.182	1.199	1.216	1.234	1.252	1.271	1.290	1.310	1.330
1	1.171	1.188	1.205	1.222	1.240	1.258	1.277	1.296	1.316
2	1.161	1.177	1.194	1.211	1.228	1.246	1.264	1.283	1.302
3	1.151	1.167	1.183	1.200	1.217	1.234	1.252	1.270	1.289

Lasing from the First Positive system has been observed (75.217, 67.68a, 63.43a).

Lasing from the Second Positive system has been observed (75.217, 74.206) 74.204, 74.199, 74.197, 74.195, 74.193, 74.191, 74.190, 74.189, 73.168, 73.165, 73.163, 68.74a, 67.68a, 66.64a, 64.47a).

Lasing from the Lyman-Birge-Hopfield system has been observed (73.168).

The two MacFarlane infrared systems have only been seen in lasing (75.217, 66.65, 66.63a, 65.58).

BIBLIOGRAPHY

The references for this molecule are only those from which spectroscopic information has been taken directly and those papers published after 1967. An excellent bibliography is given in reference (68.80) that provides a listing through 1967 as well as a review of reactions of the N$_2$ molecule.

(30. 1)   Hopfield Rydberg Series,
          J. J. Hopfield,
          Phys. Rev. 36, 789

(30. 2)   F. Charola,
          Phys. Z. 31, 457-63

(32. 3)   Vegard-Kaplan System,
          L. Vegard,
          Z. Physik 75, 30-62

(34. 4)   First and Second Goldstein-Kaplan Systems,
          J. Kaplan,
          Phys. Rev. 46, 534

(34. 5)   First and Second Goldstein-Kaplan Systems,
          J. Kaplan,
          Phys. Rev. 46, 534

(34. 6)   First and Second Goldstein-Kaplan Systems,
          J. Kaplan,
          Phys. Rev. 46, 631

(34. 7)   W. W. Lozier,
          Phys. Rev. 45, 752

(34. 8)   Hopfield Rydberg Series,
          R. S. Mulliken,
          Phys. Rev. 46, 144-6

(35. 9)   First and Second Herman-Kaplan and Kaplan Systems,
          J. Kaplan,
          Phys. Rev. 47, 259

(38. 10) Hopfield Rydberg Series,
          T. Takamine, T. Suga, and Y. Tanaka,
          Sci. Papers Inst. Phys. Chem. Res. 34, 854-64

(40. 11)   Fourth Positive System,
           L. Gerö and R. Schmid,
           Z. Physik 116, 589-603

(42. 12)   N. M. Emanuel,
           C. R. Acad. Sci. 36, 145-9

(42. 13)   Worley-Jenkins Rydberg Series,
           Y. Tanaka and T. Takamine,
           Sci. Papers Inst. Phys. Chem. Res. 39, 427-36

(43. 14)   Worley Rydberg Series,
           R. E. Worley,
           Phys. Rev. 64, 207-24

(44. 15)   Gaydon Green System,
           A. G. Gaydon,
           Proc. Phys. Soc. 56, 85-95

(45. 16)   Gaydon-Herman and Herman-Kaplan Systems.   Predissociation of
           the a State,
           R. Herman-Montagne,
           Ann. Phys. 20, 241-91

(50. 17)   Tables of Band Heads,
           R. W. B. Pearse and A. G. Gaydon,
           Identification of Molecular Spectra
           2nd Edition, Chapman & Hall, London

(51. 18)   Herman Infrared System,
           R. Herman,
           C. R. Acad. Sci. 233, 738-40

(52. 19)   Rydberg Series,
           G. L. Weissler, P. Lee, and E. I. Mohr,
           J. Opt. Soc. Am. 42, 84-90

(53. 20)   Worley Rydberg Series,
           R. E. Worley,
           Phys. Rev. 89, 863-4

(53. 21)   Rydberg Series,
           R. E. Worley,
           Phys. Rev. 90, 1131

(53. 22)   Fifth Positive, Gaydon Green, and Herman Infrared Systems,
           P. K. Carroll and N. D. Sayers,
           Proc. Phys. Soc. A 66, 1138-44

(54. 23)   Gaydon Green System,
             A. E. Grün,
             Z. Naturforsch. A 9, 1017-9

(56. 24)   Dissociation Energy,
             L. Brewer and A. W. Searcy,
             Ann. Rev. Phys. Chem. 7, 259-86

(56. 25)   Lyman-Birge-Hopfield System in Emission,
             A. Lofthus,
             Can. J. Phys. 34, 780-9

(56. 26)   Fifth Positive System,
             A. Lofthus,
             J. Chem. Phys. 25, 494-7

(57. 27)   Gaydon-Herman, Z - W Systems,
             A. Lofthus,
             Can. J. Phys. 35, 216-34

(57. 28)   First and Second Kaplan Systems. Perturbations and Predissocia-
             tions in the y State,
             A. Lofthus and R. S. Mulliken,
             J. Chem. Phys. 27, 1010-7

(58. 29)   B' - B System,
             F. LeBlanc, Y. Tanaka, and A. Jursa,
             J. Chem. Phys. 28, 979-81

(59. 30)   a' - X and B' - X Systems,
             M. Ogawa and Y. Tanaka,
             J. Chem. Phys. 30, 1354-5

(59. 31)   Vegard-Kaplan System in Absorption,
             P. G. Wilkinson,
             J. Chem. Phys. 30, 773-6

(59. 32)   a' - X System,
             P. G. Wilkinson and R. S. Mulliken,
             J. Chem. Phys. 31, 674-9

(59. 33)   First and Second Positive Systems. Λ Doubling and Triplet Separa-
             tions in the B Perturbations,
             G. H. Dieke and D. F. Heath,
             U. S. Dept. Com., Office Tech. Serv., PB 146.250

(60.34)   B' - B System Rotational Analysis,
           G. H. Dieke and D. F. Heath,
           J. Chem. Phys. 33, 432-6

(60.35)   a' - X, B' - X Systems,
           M. Ogawa and Y. Tanaka,
           J. Chem. Phys. 32, 754-8

(60.36)   B' - B System. Triplet Separation in the B State,
           P. K. Carroll and H. E. Rubalcava,
           Proc. Phys. Soc. 76, 337-45

(60.37)   First and Second Positive Systems,
           D. F. Heath,
           Los Alamos Sci. Lab. Rept., U.S. At. Energy Comm. LA-2335

(61.38)   Persistent Aurora Spectra,
           Y. Tanaka and A. S. Jursa,
           J. Opt. Soc. Am. 51, 1239-45

(62.39)   Rydberg Series,
           M. Ogawa and Y. Tanaka,
           Can. J. Phys. 40, 1593-607

(62.40)   Vegard-Kaplan Systems,
           N. P. Carleton and C. Papaliolios,
           J. Quant. Spectrosc. Radiative Trans. 2, 241-4

(62.41)   P. K. Carroll,
           "Note on the $^5\Sigma_g^+$ State of N_2,
           J. Chem. Phys. 37, 805-9

(62.42)   B - X System,
           P. G. Wilkinson,
           J. Quant. Spectrosc. Radiative Trans. 2, 343-8

(63.43)   Dissociation and Ionization Energies,
           P. G. Wilkinson,
           Astrophys. J. 138, 778-800

(63.43a)   L. E. S. Mathias and J. T. Parker,
           "Stimulated Emission in the Band Spectrum of Nitrogen,"
           Appl. Phys. Letters 3, 16-18

(63.44)   Goldstein-Kaplan Systems,
           P. K. Carroll,
           Proc. Roy. Soc. A 272, 270-83

(64. 45)   Isotope Displacements and Vibrational Structure,
            D. Mahon-Smith and P. K. Carroll,
            J. Chem. Phys. 41, 1377-82

(64. 46)   Forbidden Absorption System,
            Y. Tanaka, M. Ogawa, and A. S. Jursa,
            J. Chem. Phys. 40, 3690-700

(64. 47)   Far Ultraviolet Spectra,
            S. G. Tilford and P. G. Wilkinson,
            J. Molec. Spectrosc. 12, 231-88

(64. 47a)   H. G. Heard,
            "High-Power Ultraviolet Gas Laser,"
            Bull. Am. Phys. Soc. 9, 65

(64. 48)   Inverse Predissociation,
            S. G. Tilford and P. G. Wilkinson,
            J. Molec. Spectrosc. 12, 347-59

(64. 49)   Second Positive System,
            D. C. Tyte and R. W. Nicholls,
            Identification Atlas of Molecular Spectra
            York Univ. Center for Res. in Exp. Space Sci. and Dept. of Phys.,
            Toronto, Ontario, 2

(65. 50)   Potential Energy Curves,
            W. Benesch, J. T. Vanderslice, S. G. Tilford, and P. G. Wilkinson,
            Astrophys. J. 142, 1227-40

(65. 51)   B' - X System. High Resolution Spectra,
            S. G. Tilford, J. T. Vanderslice, and P. G. Wilkinson,
            Astrophys. J. 141, 1226-65

(65. 52)   R. N. Zare, E. O. Larsson, and R. A. Berg,
            "Franck-Condon Factors for Electronic Band Systems of Molecular
            Nitrogen,"
            J. Molec. Spectrosc. 15, 117-39

(65. 53)   C - X System. High Resolution Study,
            S. G. Tilford, J. T. Vanderslice, and P. G. Wilkinson,
            Astrophys. J. 142, 1203-26

(65. 54)   a' - X System. High Resolution Study,
            S. G. Tilford, P. G. Wilkinson, and J. T. Vanderslice,
            Astrophys. J. 141, 427-43

(65.55) a - X System. High Resolution Study,
J. T. Vanderslice, S. G. Tilford, and P. G. Wilkinson,
Astrophys. J. 141, 395-426

(65.56) Predissociation,
P. K. Carroll and R. S. Mulliken,
J. Chem. Phys. 43, 2170-9

(65.57) Vegard-Kaplan System,
R. E. Miller,
J. Chem. Phys. 43, 1695-701

(65.58) a - a' System. Laser Stimulated Transition,
R. A. MacFarlane,
Phys. Rev. A 140, 1070-1

(65.59) Lyman-Birge-Hopfield System,
D. J. MacEwen,
Thesis, University Western Ontario

(66.60) Photoionization,
K. Codling,
Astrophys. J. 143, 552-8

(66.61) Vacuum Ultraviolet Absorption Spectra,
S. G. Tilford, P. G. Wilkinson, V. B. Franklin, R. H. Naber,
W. Benesch, and J. T. Vanderslice,
Astrophys. J. Suppl. 13, 31-64

(66.62) a' - X System. Franck-Condon Factors,
K. C. Joshi,
J. Atm. Terrestr. Phys. 28, 521-5

(66.63) A State,
R. E. Miller,
J. Molec. Spectrosc. 19, 185-7

(66.63a) R. A. McFarlane,
"Precision Spectroscopy of New Infrared Emission Systems of
Molecular Nitrogen,
IEEE J. Quant. Electronics QE-2, 229-32

(66.64) w - a System. Laser Stimulated Transition,
R. A. MacFarlane,
Phys. Rev. 146, 37-9

(66.64a) V. M. Kaslin and G. C. Petrash,
"Rotational Structure of Ultraviolet Generation of Molecular Nitrogen, "
JETP Letters 3, 55-57

(66. 65)   J. W. Moskowitz, D. Neumann, and M. C. Harrison,
           Molecules and the Solid State
           Academic Press, Inc., New York

(66. 65a) K. C. Joshi,
           "Relative Band Strengths and r-Centroids for the $B'^3\Sigma_u^-$ - $Z^1\Sigma_g^+$
           System of Nitrogen,"
           Nature 212, 1459-60

(67. 66)   Rydberg Series,
           P. K. Carroll and K. Yoshino,
           J. Chem. Phys. 47, 3073-4

(67. 67)   a" - X Quadrupole Transition,
           K. Dressler and B. L. Lutz,
           Phys. Rev. Letters 19, 1219-21

(67. 68)   B. K. Ching, G. R. Cook, and R. A. Becker,
           "Oscillator Strengths of the a, w and C Bands of N_2,"
           J. Quant. Spectrosc. Radiative Trans. 7, 323-30

(67. 68a) T. Kasuya and D. R. Lide, Jr.,
           "Measurements on the Molecular Nitrogen Pulsed Laser,"
           Appl. Optics 6, 66-80

(68. 69)   N. A. Borisevich, M. V. Dubovik, and A. Ya. Smirnov,
           Zhur. Priklad. Spectrosk. 9, 807-11

(68. 70)   J. F. M. Aarts, F. J. DeHeer, and D. A. Vroom,
           "Emission Cross Sections of the First Negative Band System of N_2
           by Electron Impact,"
           Physica 40, 197-206

(68. 71)   P. N. Stanton,
           "Electron Excitation of the First Positive Bands of N_2 and of the
           First Negative and Meinel Bands of N_2^+,"
           Thesis, University of Oklahoma

(68. 72)   S. N. Ghosh, Y. Sahai, and K. K. Bhutani,
           "Spectra of N_2 Excited by Proton Bombardment,"
           Indian J. Phys. 42, 146-7

(68. 73)   H. L. Wu and W. Benesch,
           "Evidence for the $^3\Delta_u \rightarrow B^3\Pi_g$ Transition in N_2,"
           Phys. Rev. 172, 31-5

(68. 74)   Yu. A. Plastinin,
           "Optical Absorption Cross Sections of Two Atom Molecules,"
           Fiz. Gazodin. Ioniz. Khim. Reag. Gazov. 89-125

(68.74a) J.H. Parks, D.R. Rao, and A. Jowan,
"A High Resolution Study of the $C^3\Pi_u$ - $B^3\Pi_g$ (0, 0) Stimulated
Transitions in N$_2$,"
Appl. Phys. Letters 13, 142-4

(68.75) Vegard-Kaplan System,
G. Chandraiah and G.G. Shepherd,
Can. J. Phys. 46, 221-6

(68.76) Oscillator Strengths,
G.M. Lawrence, D.L. Mickey, and K. Dressler,
J. Chem. Phys. 48, 1989-94

(68.77) Spectra of Solid N$_2$,
J.Y. Roncin,
J. Molec. Spectrosc. 26, 105-10

(68.78) Raman Spectra,
J.J. Barrett and N.I. Adams III,
J. Opt. Soc. Am. 58, 311-9

(68.79) W - B System,
H.L. Wu and W. Benesch,
Phys. Rev. 172, 31-5

(68.80) A.N. Wright and C.A. Winkler,
Active Nitrogen
Academic Press, New York

(69.81) High Resolution Absorption Study in the Vacuum Ultraviolet,
P.K. Carroll and C.P. Collins,
Can. J. Phys. 47, 563-90

(69.82) Rydberg States,
K. Dressler,
Can. J. Phys. 47, 547-61

(69.83) Theory of Homogeneous Perturbation,
H. Lefebvre-Brion,
Can. J. Phys. 47, 541-6

(69.84) D.J. Burns, F.R. Simpson, and J.W. McConkey,
"Absolute Cross Sections for Electron Excitation of the Second
Positive Bands of Nitrogen,"
J. Phys. B (Atom. Molec. Phys.) 2, 52-64

(69.85) J.W. McConkey and F.R. Simpson,
"Electron Impact Excitation of the $B^3\Pi_g$ State of N$_2$,"
J. Phys. B (Atom. Molec. Phys.) 2, 923-9

(69.86)   P. N. Stanton and R. M. St. John,
"Electron Excitation of the First Positive Bands of N_2 and of the First Negative and Meinel Bands of N_2^+,"
J. Opt. Soc. Am. <u>59</u>, 252-60

(69.87)   D. H. Stedman and D. W. Setser,
"Energy Pooling by Triplet Nitrogen ($A^3\Sigma_u^+$) Molecules,"
J. Chem. Phys. <u>50</u>, 2256-8

(69.88)   R. S. Freund,
"Molecular-Beam Measurements of the Emission Spectrum and Radiative Lifetime of N_2 in the Metastable $E^3\Sigma_g^+$ State,"
J. Chem. Phys. <u>50</u>, 3734-40

(69.89)   M. R. Katti and H. D. Sharma,
"Evaluation of r-Centroids for the Band Systems of a Few Diatomic Molecules Using the Hulburt-Hirschfelder Potential Function,"
Indian J. Pure Appl. Phys. <u>7</u>, 282-3

(69.90)   J. Geiger and B. Schröder,
"Intensity Perturbations Due to Configuration Interaction Observed in the Electron Energy-Loss Spectrum of N_2,"
J. Chem. Phys. <u>50</u>, 7-16

(70.91)   M. J. Hubin-Franskin and J. E. Collin,
"Electron Impact Excitation by the SF_6 Scavenger Technique I. Nitrogen,"
Int. J. Mass Spectrom. Ion Phys. <u>4</u>, 451-63

(70.92)   K. A. Saum and W. M. Benesch,
"Infrared Electronic Emission Spectrum of Nitrogen,"
Appl. Optics <u>9</u>, 195-200

(70.93)   A. W. Johnson and R. G. Fowler,
"Measured Lifetimes of Rotational and Vibrational Levels of Electronic States of N_2,"
J. Chem. Phys. <u>53</u>, 65-72

(70.94)   K. A. Saum and W. M. Benesch,
$W^3\Delta_u - X^1\Sigma_g^+$ System of N_2,"
Phys. Rev. A <u>2</u>, 1655-9

(70.95)   D. C. Cartwright,
"Total Cross Sections for the Excitation of the Triplet States in Molecular Nitrogen,"
Phys. Rev. A <u>2</u>, 1331-48

(70.96)   H. H. Michels,
          "Identification of Two Low-Lying Non-Rydberg States of the
          Nitrogen Molecule,"
          J. Chem. Phys. 53, 841-2

(70.97)   D. W. Setser, D. H. Stedman, and J. A. Coxon,
          "Chemical Applications of Metastable Argon Atoms. IV. Excitation
          and Relaxation of Triplet States of N$_2$,"
          J. Chem. Phys. 53, 1004-20

(70.98)   K. L. Wray, E. V. Feldman, and P. F. Lewis,
          "Shock Tube Study of the Effect of Vibrational Energy of N$_2$ on the
          Kinetics of the O + N$_2$ → NO + N Reaction,"
          J. Chem. Phys. 53, 4131-6

(70.99)   D. C. Jain,
          "A Study of Some Potential Energy Functions for Diatomic Molecules,"
          Int. J. Quantum Chem. 4, 579-86

(70.100) A. A. Gribanova, V. V. Kokhaenko, N. A. Prilezhaeva, and
          L. A. Sigaenko,
          "On the Question of the Intensity Distribution in the Rotational
          Structure and Electronic-Vibrational Bands of Nitrogen,"
          Izvest. Vyssh. Ucheb. Zaved. Fiz. 9, 17-21

(71.101) W. M. Benesch and K. A. Saum,
          "The W$^3\Delta_u$ State of Molecular Nitrogen,"
          J. Phys. B (Atom. Molec. Phys.) 4, 732-8

(71.102) M. F. Golde and B. A. Thrush,
          "Vacuum Ultraviolet Emission by Active Nitrogen,"
          Chem. Phys. Letters 8, 375-7

(71.103) J. M. Calo and R. C. Axtmann,
          "Vibrational Relaxation and Electronic Quenching of the C $^3\Pi_u$
          (v' = 1) State of Nitrogen,"
          J. Chem. Phys. 54, 1332-41

(71.104) T. T. Kassal and E. S. Fishburne,
          "Energy Transfer Between N$_2$ and Electronically Excited Inert Gas
          Atoms and Ions,"
          J. Chem. Phys. 54, 1363-8

(71.105) W. E. Sharp,
          "Rocket-Borne Spectroscopic Measurements in the Ultraviolet
          Aurora: Nitrogen Vegard-Kaplan Bands,"
          J. Geophys. Res. 76, 987-1005

(72. 106)   W. L. Borst,
"Excitation of Several Important Metastable States of N$_2$ by Electron Impact,"
Phys. Rev. A 5, 648-56

(72. 107)   S. N. Ghosh and S. K. Gupta,
"Effect of Molecular Oxygen and Temperature on the N$_2$ First Positive Bands in the Nitrogen Afterglow,"
Indian J. Phys. 46, 18-27

(72. 108)   J. H. Moore, Jr.,
"Electronic Excitation of N$_2$ and Dissociative Excitation of O$_2$ by Proton Impact,"
J. Geophys. Res. 77, 5567-72

(72. 109)   N. Bjorna,
"Density-Distribution Analysis of a Constrained SCF-LCAO-MO Wave Function for N$_2$,"
Phys. Norvegica 6, 81-5

(72. 110)   J. W. Dreyer and D. Perner,
"The Deactivation of N$_2$B$^3\Pi_g$, v = 0-2 and N$_2$a'$^1\Sigma_u^-$, v = 0 by Nitrogen,"
Chem. Phys. Letters 16, 169-73

(72. 111)   R. J. McNeal, M. E. Whitson, Jr., and G. R. Cook,
"Quenching of Vibrationally Excited N$_2$ by Atomic Oxygen,"
Chem. Phys. Letters 16, 507-10

(72. 112)   D. G. Truhlar,
"Test of Massey's Method for Calculating the Static Potential for Electron Scattering by N$_2$,"
Chem. Phys. Letters 15, 486-9

(72. 113)   D. G. Truhlar, F. A. Van-Catledge, and T. H. Dunning,
"Ab Initio and Semiempirical Calculations of the Static Potential for Electron Scattering Off the Nitrogen Molecule,"
J. Chem. Phys. 57, 4788-99

(72. 114)   S. J. Young and K. P. Horn,
"Measurements of Temperatures of Vibrationally Excited N$_2$,"
J. Chem. Phys. 57, 4835-46

(72. 115)   R. J. McNeal, M. E. Whitson, Jr., and G. R. Cook,
"Photoionization of Vibrationally Excited N$_2$. II. Quenching by CO$_2$ and N$_2$O,"
J. Chem. Phys. 57, 4752-8

(72.116)   L. S. Polak, D. I. Slovetskii, and A. S. Sokolov,
"Predissociation and Quenching Probabilities for the Vibrational
Levels of the $B^3\Pi_g$ State of Molecular Nitrogen,"
Opt. Spectrosc. 32, 247-51

(72.117)   A. A. Konkov and A. V. Vorontsov,
"Experimental Study of Nitrogen Infrared Radiation,"
Opt. Spectrosc. 32, 348-50

(72.118)   J. B. Tellinghuisen, C. A. Winkler, C. G. Freeman, M. J. McEwan,
and L. F. Phillips,
Quenching Rates for N_2^+, N_2O^+, and CO_2^+ Emission Bands Excited
by 58.4 nm Irradiation of N_2, N_2O, and CO_2,"
J. Chem. Soc. Faraday Trans. II 68, 833-8

(72.119)   P. G. Burke and N. Chandra,
"Electron-Molecule Interactions   III.   A Pseudo-Potential Method
for e^--N_2 Scattering,"
J. Phys. B (Atom. Molec. Phys.) 5, 1696-1711

(72.120)   R. G. Gann, F. Kaufman, and M. A. Biondi,
"Interferometric Study of the Chemiluminescent Excitation of
Sodium by Active Nitrogen,"
Chem. Phys. Letters 16, 380-4

(72.121)   G. C. Berend, R. L. Thommarson, and S. W. Benson,
"Vibration-Vibration Energy Exchange in N_2 With O_2 and HCl
Collision Partners,"
J. Chem. Phys. 57, 3601-4

(72.122)   D. G. Truhlar,
"Vibrational Matrix Elements of the Quadrupole Moment Functions
of H_2, N_2 and CO,"
Int. J. Quant. Chem. 6, 975-88

(72.123)   D. Spence, J. L. Mauer, and G. J. Schulz,
"Measurement of Total Inelastic Cross Sections for Electron
Impact in N_2 and CO_2,"
J. Chem. Phys. 57, 5516-21

(72.124)   N. Bjorna,
"Methods for Solving Constrained SCF-LCAO-MO Equations.
Application to a Calculation on N_2,"
Molec. Phys. 24, 1-10

(72.125)   E. Kisker,
"Optical Measurement of Electron-Impact Excitation of Xenon and Molecular Nitrogen in the Threshold Region With High Energy Resolution,"
Z. Physik 257, 51-61

(72.126)   R. F. Holland and W. B. Maier II,
"Emission From Long-Lived States of N_2^+. A New Interpretation of $N_2^+ + N_2 \rightarrow N_3^+ + N$,"
J. Chem. Phys. 57, 4497-8

(72.127)   P. C. Cosby and T. F. Moran,
"Product Internal State Distributions From Interactions of Metastable Ar With N_2,"
J. Chem. Phys. 57, 4111-5

(72.128)   W. Von Niessen,
"Density Localization of Atomic and Molecular Orbitals,"
Theoret. Chim. Acta 27, 9-23

(72.129)   R. Simonaitis and J. Heicklen,
"The $H_g(^1P_1)$ Sensitized Photolysis of N_2 and CO,"
J. Photochem. 1, 181-97

(72.130)   R. Carbonneau and P. Marmet,
"Autoionizing Levels of N_2 Converging to the $A^2\Pi_u$ and $B^2\Sigma_u^+$ Limits,"
Int. J. Mass Spectrom. Ion Phys. 10, 143-55

(72.131)   G. Hartmann,
"Relative Populations of the $C^3\Pi_u$ and $B^3\Sigma_u^+$ States of Nitrogen Created by the Passing of a Predisruptive Discharge Through Flames,"
C. R. Acad. Sci. B 275, 311-4

(72.132)   M. Leoni and K. Dressler,
"Deperturbation of the Worley-Jenkins Rydberg Series of N_2,"
Helv. Phys. Acta 45, 959-961

(72.133)   S. Chung and C. C. Lin,
"Excitation of the Electronic States of the Nitrogen Molecule by Electron Impact,"
Phys. Rev. A 6, 988-1002

(72.134)   D. C. Cartwright, W. Williams, and S. Trajmar,
"The IR Emission Spectrum of N_2 Excited Under Auroral Conditions,"
Ann. Geophys. 28, 403-7

(72. 135)   N. P. Danilevskii, L. I. Popova, A. G. Koval, N. I. Fedorova, V. T. Koppe, and Ya. M. Fogel, "The Infrared Spectrum of Nitrogen Excited by Fast Electrons, " Ann. Geophys. <u>28</u>, 409-14

(72. 136)   A. Sharma and J. C. Joshi, "The Mechanism of Vacuum Ultraviolet Emission From the Lewis-Rayleigh Nitrogen Afterglow, " J. Quant. Spectrosc. Radiative Trans. <u>12</u>, 1641-6

(72. 137)   B. A. Thrush and A. H. Wild, "Excitation of Species in Active Nitrogen, Part 1. – Mercury Hg(6^{3}P$_1$), " J. Chem. Soc. Faraday Trans. II <u>68</u>, 2023-30

(73. 138)   R. S. Hickman and L. Liang, "Comment on 'Rotational Temperature Measurement in Nitrogen', " Rev. Sci. Instrum. <u>44</u>, 246

(73. 139)   V. I. Kleimenov, Yu. V. Chizhov, and F. I. Vilesov, "Applicability of the Franck-Condon Principle to the Autoionization of the N$_2$ Molecule, " Opt. Spectrosc. <u>34</u>, 590

(73. 140)   P. Millet, Y. Salamero, H. Brunet, J. Galy, and D. Blanc, "De-excitation of N$_2$ (C^{3}Π$_u$; v' = 0 and 1) Levels in Mixtures of Oxygen and Nitrogen, " J. Chem. Phys. <u>58</u>, 5839-41

(73. 141)   S. A. Lawton and F. M. J. Pichanick, "Resonances in the Metastable Excitation of Molecular Nitrogen, " Phys. Rev. A <u>7</u>, 1004-7

(73. 142)   A. I. Lyutui and L. D. Meinikov, "Excitation Temperature for Active Nitrogen, " Opt. Spectrosc. <u>34</u>, 385-6

(73. 143)   M. Outred and M. E. Pillow, "Some Observations on the Decay of the Lewis-Rayleigh Afterglow in a Nitrogen-Oxygen Mixture, " J. Phys. B (Atom. Molec. Phys.) <u>6</u>, 2701-6

(73. 144)   L. S. Polak and D. I. Slovetskii, "Electron-Impact Vibrational Excitation Cross Sections for Molecular Nitrogen, " High Temp. <u>10</u>, 575-6

(73. 155)   W. L. Borst and S. L. Chang,
"Excitation of Metastable $N_2(A\ ^3\Sigma_u^+)$ Vibrational Levels by
Electron Impact,"
J. Chem. Phys. $\underline{59}$, 5830-6

(73. 156)   A. Crowe and J. W. McConkey,
"Dissociative Ionization by Electron Impact II. N^+ and N^{++}
From N_2,"
J. Phys. B (Atom. Molec. Phys.) $\underline{6}$, 2108-17

(73. 157)   J. M. Hoffman, G. J. Lockwood, and G. H. Miller,
"Emission Cross Sections for the N_2 Second Positive (0, 0)
Transition for H, H^+, He, and He^+ Impact,"
Phys. Rev. A $\underline{7}$, 118-25

(73. 158)   A. Chutjian, D. C. Cartwright, and S. Trajmar,
"Excitation of the $W^3\Delta_u$, $w^1\Delta_u$, $B'^3\Sigma_u^-$, and $a'^1\Sigma_u^-$ States of N_2 by
Electron Impact,"
Phys. Rev. Letters $\underline{30}$, 195-8

(73. 159)   L. Klynning,
"On the 6895.5Å and 8937.0Å Bands of N_2,"
Spectrosc. Letters $\underline{6}$, 291-2

(73. 160)   H. A. Hyatt, J. M. Cherlow, W. R. Fenner, and S. P. S. Porto,
"Cross Section for the Raman Effect in Molecular Nitrogen Gas,"
J. Opt. Soc. Am. $\underline{63}$, 1604-6

(73. 161)   K. C. Smyth, J. A. Schiavone, and R. S. Freund,
"Dissociative Excitation of N_2 by Electron Impact: Translational
Spectroscopy of Long-Lived High-Rydberg Fragment Atoms,"
J. Chem. Phys. $\underline{59}$, 5225-41

(73. 162)   A. I. Dashchenko, I. P. Zapesochnyi, and A. I. Imre,
"Excitation Cross Sections of Bands of the First Negative System
of Nitrogen in Electron-Ion Collisions,"
Opt. Spectrosk. $\underline{35}$, 970-2

(73. 163)   B. W. Woodward, V. J. Ehlers, and W. C. Lineberger,
"A Reliable, Repetitively Pulsed, High-Power Nitrogen Laser,"
Rev. Sci. Instrum. $\underline{44}$, 882-7

(73. 164)   L. Y. Nelson, G. J. Mullaney, and S. R. Byron,
"Superfluorescence in N_2 and H_2 Electron-Beam-Stabilized
Discharges,"
Appl. Phys. Letters $\underline{22}$, 79-80

(73. 165) I. Nagata and Y. Kimura,
"A Compact High-Power Nitrogen Laser,"
J. Phys. E (Sci. Instrum.) $\underline{6}$, 1103-5

(73. 166) G. Joyez, R. I. Hall, J. Reinhardt, and J. Mazeau,
"Low Energy Electron Spectroscopy of N_2 in the 11.8-13.8 eV
Energy Range,"
J. Electron. Spectrosc. Rel. Phenom. $\underline{2}$, 183-90

(73. 167) R. Covey, K. A. Saum, and W. Benesch,
"Transition Probabilities for the $W^3\Delta_u$ - $B^3\Pi_g$ System of Molecular
Nitrogen,"
J. Opt. Soc. Am. $\underline{63}$, 592-6

(73. 168) A. A. Tagliaferri, M. Gallardo, C. A. Massone, and
M. Garavaglia,
"UV Stimulated Emission From N_2 and NO,"
Phys. Letters $\underline{45A}$, 211-2

(73. 169) G. N. Hays and H. J. Oskam,
"Population of $N_2(B^3\Pi_g)$ by $N_2(A\ ^3\Sigma_u^+)$ During the Nitrogen
Afterglow,"
J. Chem. Phys. $\underline{59}$, 1507-16

(73. 170) P. J. Hicks, J. Comer, and F. H. Read,
"Autoionizing Transitions in N_2 and H_2 Produced by Electron
Impact,"
J. Phys. B (Atom. Molec. Phys.) $\underline{6}$, L65-L69

(73. 171) W. E. Jones and M. Rujimethabhas,
"Reaction of Active Nitrogen With Tetrachloroethylene,"
Can. J. Chem. $\underline{51}$, 3680-3

(73. 172) J. Rose, T. Shibuya, and V. McKoy,
"Application of the Equations-of-Motion Method to the Excited
States of N_2, CO, and C_2H_4,"
J. Chem. Phys. $\underline{58}$, 74-83

(73. 173) L. S. Cederbaum, G. Hohlneicher, and W. Von Niessen,
"On the Breakdown of the Koopmans' Theorem for Nitrogen,"
Chem. Phys. Letters $\underline{18}$, 503-8

(73. 174) V. I. Egorov, Yu. M. Gershenzon, V. B. Rosenshtein, and
S. Ya. Umanskii,
"On the Mechanism of Heterogeneous Relaxation of Vibrationally
Excited Nitrogen Molecules,"
Chem. Phys. Letters $\underline{20}$, 77-80

(73. 145)   L. Veseth,
"On the Calculation of Molecular Parameters for Triplet States
in Diatomic Molecules.   The $G^3\Delta_g$ and $H^3\Phi_u$ States of N_2,"
Molec. Phys. 26, 101-7

(73. 146)   W. S. Watson, J. Lang, and D. T. Stewart,
"Photoabsorption Coefficients of Molecular Nitrogen in the
300-700 Å Region,"
J. Phys. B (Atom. Molec. Phys.) 6, L148-L151

(73. 147)   J. S. Briggs and M. R. Hayns,
"Molecular Orbital Calculations for Close Atomic Collisions:
The N_2 System,"
J. Phys. B (Atom. Molec. Phys.) 6, 514-20

(73. 148)   J. Comer and M. Harrison,
"Observation of the Effects of Rotational Transitions in the
Resonant Scattering of Electrons From N_2,"
J. Phys. B (Atom. Molec. Phys.) 6, L70-L72

(73. 149)   R. W. Dreyfus and R. T. Hodgson,
"Relativistic Electron-Beam Pumped uv Gas Lasers,"
J. Vac. Sci. Tech. 10, 1033-6

(73. 150)   J. W. Dreyer and D. Perner,
"Deactivation of N_2 ($A^3\Sigma_u^+$, $v = 0$-7) by Ground State Nitrogen,
Ethane, and Ethylene Measured by Kinetic Absorption Spectroscopy,
J. Chem. Phys. 58, 1195-1201

(73. 151)   G. R. Cook, M. Ogawa, and R. W. Carlson,
"Photodissociation Continuums of N_2 and O_2,"
J. Geophys. Res. 78, 1663-7

(73. 152)   L. C. Lee, R. W. Carlson, D. L. Judge, and M. Ogawa,
"The Absorption Cross Sections of N_2, O_2, CO, NO, CO_2, N_2O,
CH_4, C_2H_4, C_2H_6 and C_4H_{10} From 180 to 700 Å,"
J. Quant. Spectrosc. Radiative Trans. 13, 1023-31

(73. 153)   D. DeSantis, A. Lurio, T. A. Miller, and R. S. Freund,
"Radio-Frequency Spectrum of Metastable $N_2(A^3\Sigma_u^+)$.   II.   Fine
Structure, Magnetic Hyperfine Structure, and Electric Quadrupole
Constants in the Lowest 13 Vibrational Levels,"
J. Chem. Phys. 58, 4625-65

(73. 154)   G. N. Hays and H. J. Oskam,
"Reaction Rate Constant for $2N_2(A\ ^3\Sigma_u^+) \rightarrow N_2(C\ ^3\Pi_u) + N_2(X\ ^1\Sigma_g^+,$
$v' > 0$),"
J. Chem. Phys. 59, 6088-91

(73. 175)   L. Kurzweg, G. T. Egbert, and D. J. Burns,
             "Contribution of the Metastable $E^3\Sigma_g^+$ State to the Population of
             the $C^3\Pi_u$ State of N_2 Following Electron-Impact Excitation, "
             Phys. Rev. A 7, 1966-71

(73. 176)   O. V. Ravodina, T. N. Popova, A. A. Eliseev, and S. S. Smolyakov,
             "Excitation of the $B^3\Pi_g$ State of Molecular Nitrogen in a Glow
             Discharge, "
             Opt. Spectrosc. 34, 243-5

(73. 177)   L. W. Dotchin and E. L. Chupp,
             "Radiative Lifetimes and Pressure Dependence of the Relaxation
             Rates of Some Vibronic Levels in N_2^+, N_2, CO^+, and CO, "
             J. Chem. Phys. 59, 3960-7

(73. 178)   M. Cazjkowski, L. Krause, and G. M. Skardis,
             "Quenching of Mercury-Sensitized Fluorescence in Sodium
             Induced in Collisions With N_2 Molecules, "
             Can. J. Phys. 51, 1582-9

(73. 179)   M. J. W. Boness and G. J. Schulz,
             "Excitation of High Vibrational States of N_2 and CO via Shape
             Resonances, "
             Phys. Rev. A 8, 2883-6

(73. 180)   V. A. Kosinov and P. A. Skovorodko,
             "Calculation of the Franck-Condon Factors for the Processes
             Involved in the Excitation of the Molecular Bands of the First
             Negative System of Nitrogen and Carbon Monoxide, "
             Opt. Spectrosc. 34, 344

(73. 181)   E. A. Andreev,
             "Energy Transfer in N_2-Alkali Collisions, "
             Chem. Phys. Letters 23, 516-7

(73. 182)   L. Kurzweg, G. T. Egbert, and D. J. Burns,
             "Lifetime of the D $^3\Sigma_u^+$ State of N_2, "
             J. Chem. Phys. 59, 2641-45

(73. 183)   A. Corney,
             "Measurements of Transition Probabilities of Forbidden Lines of
             Neutral Atoms and Molecules, "
             Nuc. Instrum. Methods 110, 151-66

(73. 184)   A. Pochat, M. Doritch, and J. Pereese,
             "Measurement of the Absolute Lifetimes of Helium, Neon,
             Molecular Nitrogen and Carbon Monoxide, "
             J. Chim. Phys. 70, 936-40

(73. 185) F. P. Billingsley II,
"Quadrupole Moment of CO, N_2 and NO^+,"
J. Chem. Phys. 60, 2767-72

(74. 186) J. M. Hoffman, G. J. Lockwood, and G. H. Miller,
"Emission Cross Sections for $N_2^+(3914 Å)$ for F^+, Ne^+, and Na^+
Ions Incident on N_2 Gas,"
Phys. Rev. A 9, 187-91

(74. 187) V. N. Lisitin, P. L. Chapovski, and A. A. Chernenko,
"Monochromatization of Emission Frequency of the Infrared
Nitrogen Laser,"
Kvant. Elekt. 1, 341-5

(74. 188) P. K. Carroll and A. P. Doheny,
"The Herman-Kaplan System of N_2: High Resolution Studies,"
J. Molec. Spectrosc. 50, 257-65

(74. 189) S. K. Searles and G. A. Hart,
"Laser Emission at 3577 and 3805 Å in Electron-Beam-Pumped
Ar-N_2 Mixtures,"
Appl. Phys. Letters 25, 79-82

(74. 190) E. R. Ault, M. L. Bhaumik, and N. T. Olson,
"High-Power Ar-N_2 Transfer Laser at 3577 Å,"
J. Quant. Elect. QE-10, 624-6

(74. 191) B. Godard,
"A Simple High-Power Large-Efficiency N_2 Ultraviolet Laser,"
J. Quant. Elect. QE-10, 147-53

(74. 192) R. A. Young,
"New Theory of Active Nitrogen,"
J. Chem. Phys. 60, 5050-3

(74. 193) D. A. McArthur and J. W. Poukey,
"Theory of Plasma Electron Contribution to the Electron-Beam-
Excited Nitrogen Laser,"
Phys. Rev. Letters 32, 89-92

(74. 194) W. Brennen, R. V. Gutowski, and E. C. Shane,
"Vibrational Distributions of $N_2(A^3\Sigma_u^+)$ in the Nitrogen Afterflow,"
Chem. Phys. Letters 27, 138-140

(74. 195) H. Fischer, R. Girnus, and F. Rühl,
"Low Threshold Coaxial N_2 Laser With a Resonator,"
Appl. Optics 13, 1759-60

(74.196) M. Imami and W. L. Borst,
"Electron Excitation of the (0,0) Second Positive Band of Nitrogen
From Threshold to 1000 eV,"
J. Chem. Phys. 61, 1115-7

(74.197) R. T. Brown and D. C. Smith,
"Optically Pumped Electric-Discharge uv Laser,"
Appl. Phys. Letters 24, 236-8

(74.198) D. H. Winicur and J. L. Fraites,
"Electronic-Energy Exchange Cross Sections for Ar*(^{3}P) and N$_2$
(X$^1\Sigma_g^+$),"
J. Chem. Phys. 71, 1548-53

(74.199) H. E. B. Andersson and R. C. Tobin,
"Electrical Breakdown and Pumping in an Axial-Field Nitrogen
Laser,"
Physica Scripta 9, 7-14

(74.200) G. C. Lie and E. Clementi,
"Study of the Electronic Structure of Molecules. XXII. Correla-
tion Energy Corrections as a Function of the Hartree-Fock Type
Density and its Application to the Homonuclear Diatomic Molecules
of the Second Row Atoms,"
J. Chem. Phys. 60, 1288-96

(74.201) K. J. Miller and A. E. S. Green,
"Energy Levels and Potential Energy Curves for H$_2$, N$_2$, and O$_2$
With an Independent Particle Model,"
J. Chem. Phys. 60, 2617-26

(74.202) G. C. Baldwin,
"Nitrogen Total Cross Section for Electrons Below 2.0 eV,"
Phys. Rev. A 9, 1225-9

(74.203) T. G. Slanger and G. Black,
"Electronic-to-Vibrational Energy Transfer Efficiency in the
O(^{1}D)-N$_2$ and O(^{1}D)-CO Systems,"
J. Chem. Phys. 60, 468-77

(74.204) S. N. Suchard, L. Galvan, and D. G. Sutton,
"Quasi-CW Laser Emission From the Second Positive Band of
Nitrogen,"
Appl. Phys. Lett. 26, 521-3

(74.205) S. K. Searles,
"Superfluorescent Laser Emission From Electron-Beam-Pumped
Ar-N$_2$ Mixtures,"
Appl. Phys. Letters 25, 735-7

(74.206) J. I. Levatter and S. C. Lin,
"High-Power Generation From a Parallel-Plates-Driven Pulsed Nitrogen Laser,"
Appl. Phys. Letters 25, 703-5

(74.207) T. D. Nguyen, N. Sadeghi, and J. C. Pebay-Peyroula,
"Energy Transfer Reaction Between Metastable Xenon Atoms and Nitrogen Molecules: Excitation of the $N_2(B\ ^3\Pi_g$, $v' \leq 5)$ States in the Afterglow,"
Chem. Phys. Letters 29, 242-6

(74.208) V. N. Ishchenko, V. N. Lisitsyn, A. M. Razhev, and V. N. Starinskii,
"Superradiance on the 2^+ and 1^- Bands of Nitrogen in a Discharge at Pressures Above 10 atm,"
JETP Letters 19, 233-4

(74.209) V. Hasson, D. Preussler, J. Klimek, and H. M. von Bergmann,
"Transverse Double-Discharge High-Pressure Glow Excitation of uv Lasing Action in Molecular Nitrogen,"
Appl. Phys. Letters 25, 654-6

(74.210) F. Albugues, A. Birot, D. Blanc, H. Brunet, J. Galy, and P. Millet,
"Destruction of the Levels $C^3\Pi_u$ ($v' = 0$, $v' = 1$) of Nitrogen by O_2, CO_2, CH_4, and H_2O,"
J. Chem. Phys. 61, 2695-9

(74.211) J. H. Birely,
"Formation of $N_2^+ B^2\Sigma_u^+$ and $N_2 C^3\Pi_u$ in Collisions of H^+ and H With N_2,"
Phys. Rev. A 10, 550-62

(74.212) P. K. Carroll and A. P. Doheny,
"The Herman-Kaplan System of N_2: High Resolution Studies,"
J. Molec. Spectrosc. 50, 257-65

(74.213) D. C. Cartwright and T. H. Dunning, Jr.,
"Vibrational Matrix Elements of the Quadrupole Moment of N_2 $(x\ ^1\Sigma_g^+)$,"
J. Phys. B (Atom. Molec. Phys.) 7, 1776-81

(74.214) J. W. Dreyer, D. Perner, and C. R. Roy,
"Rate Constants for the Quenching of $N_2 (A\ ^3\Sigma_u^+$, $v_A = 0$-8) by CO, CO_2, NH_3, NO, and O_2,"
J. Chem. Phys. 61, 3164-9

(74. 215)   K. J. Rajan,
"The 1-0 Band of the $b^1\Pi_u$ -$a^1\Pi_g$ Transition and the 1-10 and
2-12 Bands of the Fifth Positive System of N_2. Rotational
Analyses,"
Proc. Roy. Irish Acad. A 74, 17-23

(75. 216)   C. S. Willett and D. M. Litynski,
"Power Increase of N_2 uv and ir Lasers by Addition of SF_6,"
Appl. Phys. Letters 26, 118-21

(75. 217)   S. N. Suchard, R. F. Heidner III, and D. G. Sutton,
Behavior of the First and Second Positive Emission in the
N_2/SF_6 Laser,"
IEEE J. Quant. Electron. QE-11, 908-16

(n. p. 218)   S. N. Suchard and D. G. Sutton,
Private Communication

(66. L1)   M. Jeunehomme
"Transition Moment of the First Positive Band System of
Nitrogen,"
J. Chem. Phys. 45, 1805-11

(69. L2)   D. E. Shemansky, N. P. Carleton
"Lifetime of the N_2 Vegard-Kaplan System,"
J. Chem. Phys. 51, 682-8

(69. L3)   D. E. Shemansky
"N_2 Vegard-Kaplan System in Absorption,"
J. Chem. Phys. 51, 689-700

Na$_2$

Methods of Production and Experimental Technique

Absorption.

Emission from a discharge in Na$_2$ vapor, heat pipe.

Fluorescence.

BAND SYSTEMS

	System	Transition	Sources	Wavelength Limits	Degrading	Characteristic Bands, λ	Remarks	Bibliography
	I	$A\,^1\Sigma_u^+ \rightleftharpoons X\,^1\Sigma_g^+$	Absorption, discharge, fluorescence	8000-6000	R			(70.44, 33.20, 29.13)
	II	$B\,^1\Pi_u \rightleftharpoons X\,^1\Sigma_g^+$	Absorption, discharge, fluorescence	5040-4560	R			(69.41, 32.17, 28.10)
	III	$C\,^1\Pi_u \rightleftharpoons X\,^1\Sigma_g^+$	Absorption, discharge	3600-3200	R	3338.8(5,0) 3326.3(6,0)		(50.34, 49.33)
	IV	$D\,^1\Pi_u \leftarrow X\,^1\Sigma_g^+$	Absorption	3325-3030	R			(50.34)
	V	$E \rightleftharpoons ?$	Absorption, discharge	3120-2880	R	2945.5(7,0)		(47.31)
	VI	?	Absorption, discharge	3050-2500	R	2750, 2735		(47.31)

I. $\underline{A\,^1\Sigma_u^+ \rightleftharpoons X\,^1\Sigma_g^+ \text{ System}}$

Most intense band heads in absorption, λ (33.20, 29.13):

(v', v'')	(4, 2)	(4, 1)	(5, 1)	(6, 0)	(7, 0)	(8, 0)	(9, 0)
λ	6751.2	6679.7	6561.5	6513.2	6465.8	6418.4	6374.2

II. $\underline{B\,^1\Pi_u \rightleftharpoons X\,^1\Sigma_g^+ \text{ System}}$

Most intense band heads in absorption, λ (32.17, 28.10):

(v', v'')	(0, 3)	(0, 2)	(0, 1)	(1, 1)	(0, 0)	(1, 0)	(2, 0)	(3, 0)
λ	5040.4	5001.4	4962.8	4932.6	4924.2	4894.5	4865.5	4837.2

III. $\underline{C\,^1\Pi_u \rightleftharpoons X\,^1\Sigma_g^+ \text{ System}}$

Most intense band heads, λ (absorption intensity, emission intensity) (50.34, 49.33):

(v', v'')	(5, 1)	(4, 0)	(5, 0)	(6, 0)	(7, 0)	(9, 0)	(10, 0)
λ	3356.5	3351.5	3338.8	3326.3	3314.0	3290.0	3278.4
Absorption intensity	8	7	10	10	10	9	8
Emission intensity	4	4	5	4	4	4	4

IV. $\underline{D\,^1\Pi_u \leftarrow X\,^1\Sigma_g^+ \text{ System}}$

Most intense band heads, λ (Intensity) (50.34):

(v', v'')	(1, 2)	(3, 3)	(2, 2)	(1, 1)	(0, 0)	(2, 1)	(1, 0)	(2, 0)
λ	3151.6	3145.2	3140.0	3135.7	3131.2	3125.1	3120.5	3109.5
(Intensity)	2	2	2	2	2	2	2	2

Na_2

V.   <u>E ⇌ ? System</u>

Most intense band heads, λ (absorption intensity, emission intensity) (47.31):

λ	2983.1	2959.6	2945.5	2936.2	2932.5	2928.6	2927.6
Absorption intensity	6	6	10	8	6	6	8
Emission intensity	0	0	0	4	2	2	4

VI.   <u>3050-2500Å Bands</u>

Possibly four fragmentary systems (4-7), preliminary vibrational analysis, λ (Intensity) (47.31):

(v', v'')	(0, 3)	(1, 3)	(0, 1)	(2, 1)	(0, 0)	(1, 0)	(2, 0)	(8, 0)	(0, 6)
λ	2986.4	2977.0	2958.6	2948.2	2944.0	2935.6	2970.6	2750.0	2735.(
Intensity	6	5	5	6	5	8	5	5	5
System	4	4	4	4	4	4	4	5	6

Na$_2$

SPECTROSCOPIC CONSTANTS

State	T_e	ω_e	$x_e\omega_e$	B_e	$\alpha_e \times 10^4$	$D_e \times 10^6$	r_e	Remarks	Bibliography
$D\,^1\Pi_u$	33486.9	111.93	0.573	0.1152	11.0				(60.35)
$C\,^1\Pi_u$	29384.8	119.53	0.782	0.1185	9.6				(60.35, 32.17)
$B\,^1\Pi_u$	20319.596	124.065	0.6863	0.125829	8.6754	0.3614	3.41398	(a)	(69.41, 32.17)
$A\,^1\Sigma_u^+$	14680.4	117.6	0.38	0.1107	5.4		3.64		(29.13)
$b\Pi(0_u^+)$ (b)	<14680.4	~145		~0.14					(33.21)
$X\,^1\Sigma_g^+$	0	159.126	0.7262	0.154853	8.5637	0.6552	3.07745	(c)	(69.41, 33.20)

(a) $y_e\omega_e = -5.441 \times 10^{-3}$, $z_e\omega_e = -1.15 \times 10^{-4}$, $\gamma_e = -1.535 \times 10^{-5}$; (b) calculated by deperturbation analysis of $A\,^1\Sigma_u^+$; (c) $y_e\omega_e = -9.145 \times 10^{-3}$, $z_e\omega_e = -5.02 \times 10^{-5}$, $\gamma_e = -7.646 \times 10^{-6}$.

Dissociation energy = 0.75 ± 0.03 eV, 17.3 kcal/mole, 6049 cm^{-1}.

Na$_2$

Perturbations and General Information

Gyromagnetic ratio (g_j) = 0.03892 nuclear magnetons (64.36).

$A\,^1\Sigma_u^+$ state is perturbed by the $b\Pi\left(0_u^+\right)$ state (33.21).

Radiative lifetimes:

$$A\,^1\Sigma_u^+,\quad \tau_r = 10^{-7} - 10^{-6}\ \text{sec (70.44)}$$

$$B\,^1\Pi_u,\quad \tau_r = 6.41\ \text{nsec (69.43)}$$

Average polarizability $(736°\text{K})$ = $30 \times 10^{-24}\ \text{cm}^3$ (74.55).

Transition moment for $B\,^1\Pi_u \rightarrow X\,^1\Sigma_g^+$ system (74.56):

$$D = 6.8 + 0.5r \quad 2.6\text{Å} \le r \le 5.0\text{Å}$$

Potential energy curves — RKR potential (69.40):

State	v	E(v)cm^{-1}	r_{min}(Å)	r_{max}(Å)
$T_e = 0.0$ $\qquad X\,^1\Sigma_g^+$	0	79.4	2.9481	3.2200
	1	237.2	2.8593	3.3320
	2	393.5	2.8014	3.4141
	3	548.3	2.7563	3.4841
	4	701.6	2.7187	3.5475
	5	853.4	2.6864	3.6065
	6	1003.6	2.6581	3.6624
	7	1152.3	2.6327	3.7163
	8	1299.3	2.6099	3.7686
	9	1444.9	2.5893	3.8196
	10	1588.8	2.5705	3.8699
	11	1731.0	2.5533	2.9195
	12	1871.7	2.5375	3.9687
	13	2010.7	2.5231	4.0176
	14	2148.0	2.5100	4.0665
	15	2283.6	2.4979	4.1153

	State	v	E(v)cm^{-1}	r$_{min}$(A)	r$_{max}$(A)
T$_e$ = 14680.4 cm^{-1}	A $^1\Sigma_u^+$	0	58.7	3.4875	3.8037
		1	175.5	3.3839	3.9330
		2	291.6	3.3159	4.0268
		3	406.9	3.2626	4.1060
		4	521.5	3.2179	4.1769
		5	635.3	3.1789	4.2421
		6	748.3	3.1442	4.3032
		7	860.6	3.1128	4.3612
		8	972.1	3.0839	4.4168
		9	1082.9	3.0573	4.4703
		10	1192.9	3.0324	4.5222
		11	1302.1	3.0091	4.5728
		12	1410.6	2.9871	4.6221
		13	1518.3	2.9663	4.6704
		14	1625.3	2.9466	4.7178
		15	1731.5	2.9278	4.7645
T$_e$ = 20319.596	B $^1\Pi_u$	0	61.7	3.2663	3.5747
		1	184.2	3.1678	3.7044
		2	305.4	3.1038	3.7998
		3	425.1	3.0539	3.8814
		4	543.4	3.0122	3.9553
		5	660.2	2.9759	4.0242
		6	775.4	2.9435	4.0895
		7	889.0	2.9141	4.1523
		8	1000.9	2.8870	4.2132
		9	1111.1	2.8618	4.2727
		10	1219.5	2.8381	4.3313
		11	1326.0	2.8157	4.3892
		12	1430.6	2.7943	4.4467
		13	1533.2	2.7737	5.5039
		14	1633.9	2.7539	4.5612
		15	1732.4	2.7347	4.6186

Franck-Condon factors – RKR potential (69.41):

$$B^1\Pi_u - X^1\Sigma_g^+$$

v', v''	0	1	2	3	4	5	6	7	8
0	6.55-1	1.61-1	2.13-1	2.00-1	1.51-1	9.77-2	5.61-2	2.94-2	1.44-2
1	1.93-1	1.92-1	5.53-2	2.47-4	4.52-2	1.03-1	1.23-1	1.07-1	7.68-2
2	2.69-1	4.05-2	3.19-2	1.15-1	7.40-2	8.32-3	9.19-3	5.28-2	8.78-2
3	2.35-1	1.67-2	1.30-1	2.68-2	1.69-2	8.17-2	6.84-2	1.56-2	1.46-3
4	1.43-1	1.36-1	4.39-2	3.84-2	9.15-2	1.41-2	1.53-2	6.60-2	5.99-2
5	6.44-2	1.96-1	9.62-3	1.08-1	2.86-3	5.45-2	6.53-2	5.72-3	1.66-2
6	2.22-2	1.49-1	1.16-1	2.24-2	6.54-2	5.29-2	3.20-3	6.06-2	4.46-2
7	5.95-3	7.41-2	1.77-1	2.28-2	8.28-2	6.65-3	7.60-2	1.46-2	1.68-2
8	1.26-3	2.63-2	1.33-1	1.31-1	2.55-3	8.92-3	9.94-3	4.16-2	5.20-2

Franck-Condon factor followed by a factor of ten

BIBLIOGRAPHY

(09. 1)   A, B ⇄ X Systems,
          R. W. Wood and F. E. Hackett,
          Astrophys. J. 30, 339-72

(11. 2)   A, B ⇄ X Systems,
          R. W. Wood and R. G. Galt,
          Astrophys. J. 33, 72-80

(24. 3)   B ⇄ X System, Incorrect Vibrational Analysis,
          H. G. Smith,
          Proc Roy. Soc. A 106, 400-15

(27. 4)   A ⇄ X System, Vibrational and Rotational Analysis, B ⇄ X System,
          Incorrect Rotational Analysis,
          W. R. Fredrickson and W. W. Watson,
          Phys. Rev. 30, 429-38

(27. 5)   B ⇄ X System,
          F. W. Loomis,
          Phys. Rev. 29, 607

(27. 6)   A ⇄ X System, Excitation Mechanism,
          H. Schüler,
          Z. Physik 43, 474-9

(28. 7)   B ⇄ X System,
          F. W. Loomis,
          Phys. Rev. 31, 323-32

(28. 8)   B ⇄ X System,
          F. W. Loomis,
          Phys. Rev. 31, 705

(28. 9)   A ⇄ X System,
          F. W. Loomis and S. W. Nile, Jr.,
          Phys. Rev. 32, 873-9

(28. 10) B ⇄ X System, Vibrational Analysis,
          F. W. Loomis and R. W. Wood,
          Phys. Rev. 32, 223-36

(28. 11) B ⇄ X System,
          D. S. Villars,
          Proc. Nat. Acad. Sci. 14, 508-11

(28. 12)   Ultraviolet Systems,
           J. M. Walter and S. Barratt,
           Proc. Roy. Soc. A 119, 257-75

(29. 13)   A ⇄ X System, Vibrational and Rotational Analysis, B ⇄ System,
           Incorrect Rotational Analysis,
           W. R. Fredrickson,
           Phys. Rev. 34, 207-12

(30. 14)   Ultraviolet Systems, Incorrect Analysis,
           W. Weizel and M. Kulp,
           Ann. Phys. 4, 971-84

(30. 15)   E. Hutchisson,
           Phys. Rev. 36, 410-20

(32. 16)   A → X System,
           Y. Uchida,
           Japan J. Phys. 8, 25-50

(32. 17)   B ⇄ X System, Vibrational Analysis,
           F. W. Loomis and R. E. Nusbaum,
           Phys. Rev. 40, 380-6

(32. 18)   A → X and Ultraviolet Systems,
           M. Kimura and Y. Uchida,
           Sci. Papers Inst. Phys. Chem. Res. 18, 109-18

(32. 19)   Ultraviolet Systems,
           H. Kuhn,
           Z. Physik 76, 782-92

(33. 20)   A ⇄ X System, Vibrational Analysis,
           W. R. Fredrickson and C. R. Stannard,
           Phys. Rev. 44, 632-7

(33. 21)   A ⇄ X System, Deperturbation Analysis,
           V. Kondratjew and L. Polak,
           Phys. Z. Sowjetunion 4, 764-86

(37. 22)   Ultraviolet Systems,
           H. Yoshinaga,
           Proc. Phys. -Math. Soc. Japan 19, 847-59

(37. 23)   Ultraviolet Systems,
           H. Yoshinaga,
           Proc. Phys. -Math. Soc. Japan 19, 1073-83

(37.24)   C. H. D. Clark,
Trans. Faraday Soc. 33, 1398-401

(38.25)   Fluorescence,
P. Pringsheim,
Physica, Pays-Bas 5, 489-94

(40.26)   Theory,
G. B. B. M. Sutherland,
J. Chem. Phys. 8, 161-4

(40.27)   Theory,
R. F. Barrow,
Trans. Faraday Soc. 36, 624-5

(40.28)   Theory,
C. H. D. Clark,
Trans. Faraday Soc. 36, 370-6

(41.29)   Theory,
H. M. Hulbert and J. O. Hirschfelder,
J. Chem. Phys. 9, 61-9

(47.30)   Preliminary Note to (49.33),
R. W. B. Pearse and S. P. Sinha,
J. Phys. Soc. 160, 159

(47.31)   Ultraviolet Systems, C ⇄ X System,
S. P. Sinha,
Proc. Phys. Soc. A 59, 610-21

(49.32)   S. P. Sinha,
Indian J. Phys. 23, 229-36

(49.33)   C ⇄ X System, Vibrational Analysis,
S. P. Sinha,
Proc. Phys. Soc. A 62, 124-30

(50.34)   G. S. Chang,
Chinese J. Phys. 7, 377-82

(60.35)   D, C - X Systems,
R. F. Barrow, N. Travis, and C. V. Wright,
Nature 187, 141-2

(64.36)   Magnetic Moment,
R. A. Brooks, C. H. Anderson, and N. F. Ramsey,
Phys. Rev. 136, 62-8

(65. 37)   RKR Potential Curves, Dissociation Energy,
R. B. Singh and D. K. Rai,
Indian J. Pure Appl. Phys. 3, 475-8

(66. 38)   L. Szasz and G. McGinn,
"Atomic and Molecular Calculations With the Pseudopotential Method.
I.   The Binding Energy and Equilibrium Internuclear Distance of the
Na$_2$ Molecule,"
J. Chem. Phys. 45, 2898-2912

(68. 39)   L. Szasz and G. McGinn,
"Atomic and Molecular Calculations With the Pseudopotential Method.
III.   The Theory of Li$_2$, Na$_2$, K$_2$, LiH, NaH, and KH,"
J. Chem. Phys. 48, 2997-3008

(69. 40)   D. C. Jain and R. C. Sahni,
"Reduced Potential Energy Curves of Some Electronic States of
Alkali Molecules,"
Trans. Faraday Soc. 65, 897-903

(69. 41)   W. Demtröder, M. McClintock, and R. N. Zare,
"Spectroscopy of Na$_2$ Using Laser-Induced Fluorescence,"
J. Chem. Phys. 51, 5495-5508

(69. 42)   P. L. Goodfriend,
"Estimation of Spectroscopic Trends by a Perturbation Method Using
Hellmann Psuedopotentials,"
J. Molec. Spectrosc. 30, 111-5

(69. 43)   M. McClintock, W. Demtröder, and R. N. Zare,
"Level-Crossing Studies of Na$_2$ Using Laser-Induced Fluorescence,"
J. Chem. Phys. 51, 5509-21

(70. 44)   S. E. Johnson, K. Sakurai, and H. P. Broida,
"Fluorescence of Na$_2$ Induced by a Helium-Neon Laser at 632. 8 and
640. 1 nm,"
J. Chem. Phys. 52, 6441-2

(70. 45)   A. C. Roach and P. Baybutt,
"Potential Curves of Alkali Diatomic Molecules and the Origins of
Bonding Anomalies,"
Chem. Phys. Letters 7, 7-10

(70. 46)   G. Baumgartner, W. Demtröder, and M. Stock,
"Lifetime-Measurements of Alkali-Molecules Excited by Different
Laserlines,"
Z. Physik 232, 462-72

(71.47) K. Bergmann and W. Demtröder,
"Inelastic Collision Cross Section of Excited Molecules. I. Rotational Energy Transfer Within the B$^1\Pi_u$-State of Na$_2$ Induced by Collisions with He,"
Z. Physik 243, 1-13

(71.48) W. S. Struve, T. Kitagawa, and D. R. Herschbach,
"Chemiluminescence in Molecular Beams: Electronic Excitation in Reactions of Cl Atoms With Na$_2$ and K$_2$ Molecules,"
J. Chem. Phys. 54, 2759-61

(71.49) P. P. Sorokin and J. R. Lankard,
"Emission Spectra of Alkali-Metal Molecules Observed With a Heat-Pipe Discharge Tube,"
J. Chem. Phys. 55, 3810-3

(72.50) A. C. Roach,
"Theoretical Ground State and Excited State Potential Energy Curves for Alkali Diatomic Molecules,"
J. Molec. Spectrosc. 42, 27-37

(72.51) K. Bergmann and W. Demtröder,
"Inelastic Collision Cross Section of Excited Molecules. II. Asymmetries in the Cross Section for Rotational Transitions in the Na$_2$ (B$^1\Pi_u$) State,"
J. Phys. B 5, 1386-95

(72.52) K. Bergmann and W. Demtröder,
"Inelastic Cross Sections of Excited Molecules. III. Absolute Cross Sections for Rotational and Vibrational Transitions in the Na$_2$(B$^1\Pi_u$) State,"
J. Phys. B 5, 2098-106

(72.53) K. Bergmann, H. Klar, and W. Schlecht,
"Asymmetries in Collision-Induced Rotational Transitions,"
Chem. Phys. Letters 12, 522-5

(74.54) R. H. Callender, J. I. Gersten, R. W. Leigh, and J. L. Yang,
"Dependence of Transition Moment on Internuclear Separation in Na$_2$,"
Phys. Rev. Letters 32, 917-20

(74.55) R. W. Molof, T. M. Miller, H. L. Schwartz, B. Bederson, and J. T. Park,
"Measurements of the Average Electric Dipole Polarizabilities of the Alkali Dimers,"
J. Chem. Phys. 61, 1816-22

Na$_2$

(74. 56)   M. M. Hessel, E. W. Smith, and R. E. Drullinger,
          "Transition Dipole Moment of Na$_2$ and Its Variation With Internuclear
          Distance,"
          Phys. Rev. Letters 33, 1251-4

Nd$_2$

Spectroscopic Constants

Dissociation energy = 0.82 ± 0.30 eV, 19 kcal/mole, 6614 cm^{-1} (72.1).

Nd_2

BIBLIOGRAPHY

(72. 1)   A. Kant and S. Lin,
           "Dissociation Energies of the Homonuclear Diatomic Molecules of
           the Rare Earths,"
           Monatshefte für Chemie 103, 757-63

Ne$_2$

Methods of Production and Experimental Technique

Absorption.

Discharge.

BAND SYSTEMS

	System	Transition	Sources	Wavelength Limits	Degrading	Band Head, $\nu_{0,0}$	Remarks	Bibliography
	I	$A(0_u^+) \leftarrow X^1\Sigma_g^+(0_g^+)$	Absorption	747-745	V			(72.5)
	II	$B(0_u^+) \leftarrow X^1\Sigma_g^+$	Absorption	737-736	V			(72.5)
	III	$C(1_u) \leftarrow X^1\Sigma_g^+$	Absorption	639-630				(72.5)
	IV	$D(0_u^+) \leftarrow X^1\Sigma_g^+$	Absorption	631-629				(72.5)
	V	$E(0_u^+) \leftarrow X^1\Sigma_g^+$	Absorption	628-626				(72.5)
	VI	$F(0_u^-)? \leftarrow X^1\Sigma_g^+$	Absorption	629-627				(72.5)
	VII	$G(0_u^+) \leftarrow X^1\Sigma_g^+$	Absorption	624-619				(72.5)
	VIII	$H(0_u^+) \leftarrow X^1\Sigma_g^+$	Absorption	624-619				(72.5)
	IX	$I(0_u^+)? \leftarrow X^1\Sigma_g^+$	Absorption	618-615				(72.5)
	X	$J(1_u) \leftarrow X^1\Sigma_g^+$	Absorption	609-603				(72.5)
	XI	$K(0_u^+) \leftarrow X^1\Sigma_g^+$	Absorption	604-602				(72.5)

Ne$_2$

BAND SYSTEMS

	System	Transition	Sources	Wavelength Limits	Degrading	Band Head, $\nu_{0,0}$	Remarks	Bibliography
	XII	$L\left(0_u^+, 0_u^-\right) \leftarrow X^1\Sigma_g^+$	Absorption	601-600				(72.5)

I. $\underline{A\left(0_u^+\right) \leftarrow X^1\Sigma_g^+\left(0_g^+\right) \text{System}}$

Band heads, λ (Intensity) (72.5):

(v', v'')	(v,0)	(v-1,0)	(v-2,0)	(v-3,0)
λ	745.11	745.34	745.85	746.83
(Intensity)	10	3	4	0

II. $\underline{B\left(0_u^+\right) \leftarrow X^1\Sigma_g^+ \text{System}}$

Band heads, λ (Intensity) (72.5):

(v', v'')	(v,0)	(v,1)	(v-1,0)	(v-1,1)
λ	736.18	736.25	736.49	736.57
(Intensity)	10	8	3	1

III. $\underline{C(1_u) \leftarrow X^1\Sigma_g^+ \text{System}}$

Band heads, λ (Intensity) (72.5):

(v', v'')	(v,0)	(v-1,0)	(v-2,0)	(v-3,0)	(v-4,0)	(v-5,0)
λ	630.98	631.49	632.05	632.71	633.45	634.26
(Intensity)	10	9	8	6	4	2

VI. $\underline{D\left(0_u^+\right) \leftarrow X^1\Sigma_g^+ \text{System}}$

Band heads, λ (Intensity) (72.5):

(v', v'')	(v,0)	(v-1,0)	(v-2,0)
λ	629.87	630.06	630.27
(Intensity)	4	6	10

V. $\underline{E\left(0_u^+\right) \leftarrow X^1\Sigma_g^+ \text{System}}$

Band heads, λ (Intensity) (72.5):

(v', v'')	(v,0)	(v-1,0)	(v-2,0)	(v-3,0)
λ	626.92	627.03	627.23	627.46
(Intensity)	2	5	6	10

VI. $F\left(0_u^-\right)? \leftarrow X^1\Sigma_g^+$ System

Band heads, λ (Intensity) (72.5):

(v', v'')	$(v,0)$	$(v-1,0)$	$(v-2,0)$	$(v-2,1)$	$(v-3,0)$
λ	619.26	619.62	620.07	620.13	620.61
(Intensity)	10	7	6	2	5

VII. $G\left(0_u^+\right) \leftarrow X^1\Sigma_g^+$ System

Band heads, λ (Intensity) (72.5):

(v', v'')	$(v,0)$	$(v-1,0)$	$(v-2,0)$	$(v-3,0)$
λ	619.42	619.80	620.28	620.82
(Intensity)	10	7	4	2

VIII. $H\left(0_u^+\right) \leftarrow X^1\Sigma_g^+$ System

Band heads, λ (Intensity) (72.5):

(v', v'')	$(v,0)$	$(v-1,0)$	$(v-2,0)$	$(v-3,0)$
λ	619.42	619.80	620.28	620.82
(Intensity)	10	7	4	2

IX. $I\left(0_u^+\right)? \leftarrow X^1\Sigma_g^+$ System

Band heads, λ (Intensity) (72.5):

(v', v'')	$(v,0)$	$(v-1,0)$	$(v-2,0)$	$(v-3,0)$
λ	616.30	616.53	616.81	617.06
(Intensity)	10	5	8	3

X. $J(1_u) \leftarrow X^1\Sigma_g^+$ System

Band heads, λ (Intensity) (72.5):

(v', v'')	$(v-1,0)$	$(v-2,0)$	$(v-3,0)$	$(v-4,0)$
λ	603.57	603.85	604.28	604.74
(Intensity)	10	8	7	7

XI. $\underline{K\left(0_u^+\right) \leftarrow X^1\Sigma_g^+ \text{ System}}$

Band heads, λ (Intensity) (72.5):

(v', v'')	$(v,0)$	$(v-1,0)$	$(v-2,0)$	$(v-3,0)$	$(v-4,0)$
λ	602.88	602.90	602.97	603.08	603.23
(Intensity)	6	4	5	6	10

SPECTROSCOPIC CONSTANTS

State	T_e	ω_e	$x_e\omega_e$	B_e	$\alpha_e \times 10^3$	$D_e \times 10^6$	r_e	Remarks	Bibliography
$X^1\Sigma_g^+$	0	31.3	6.84	0.20	60		2.91		(72.5)

Dissociation energy = 3.74×10^{-3} eV, 10.6 cal/mole, 30.2 cm^{-1} (72.5).

Perturbations and General Information

Radiative lifetimes — calculated (74.15):

$$1_u\left(^3P_2\right) \to X^1\Sigma_g^+ \qquad \tau = 11.9 \ \mu sec$$

$$0_u^+\left(^3P_1\right) \to X^1\Sigma_g^+ \qquad \tau = 2.8 \ nsec$$

$$0_u^+\left(^1P_1\right) \to X^1\Sigma_g^+ \qquad \tau = 1.2 \ nsec$$

BIBLIOGRAPHY

(58. 1)   Observation of Ultraviolet Continuum,
           Y. Tanaka, A. S. Jursa, and F. J. LeBlanc,
           J. Opt. Soc. Am. 48, 304-8

(67. 2)   J. F. Prince and W. W. Robertson,
           "Visible Continua in Xenon, Krypton, and Neon,"
           J. Chem. Phys. 46, 3309-13

(67. 3)   Theory,
           T. L. Gilbert and A. C. Wahl,
           J. Chem. Phys. 47, 3425-38

(67. 4)   P. G. Wilkinson,
           The Mechanism of the Argon Emission Continuum in the Vacuum
           Ultraviolet,"
           Can. J. Phys. 45, 1715-27

(72.5)   Y. Tanaka and K. Yoshino,
           "Absorption Spectra of Ne$_2$ and HeNe Molecules in the Vacuum-uv
           Region,"
           J. Chem. Phys. 57, 2964-76

(72. 6)   D. D. Konowalow, P. Weinberger, J. L. Calais, and J. W. D. Connolly,
           "Self-Consistent-Field X α Cluster Calculations for the Ground State
           Ne$_2$ Molecule,"
           Chem. Phys. Letters 16, 81-5

(73. 7)   G. C. Maitland,
           "The Determination of the Intermolecular Potential Energy Function
           of Neon From Spectroscopic, Equilibrium and Transport Data,"
           Molec. Phys. 26, 513-28

(73. 8)   J. M. Farrar, Y. T. Lee, V. V. Goldman, and M. L. Klein,
           "Neon Interatomic Potentials From Scattering Data and Crystalline
           Properties,"
           Chem. Phys. Letters 19, 359-62

(73. 9)   M. L. Klein,
           "Comments on the Interatomic Potential of Ne$_2$,"
           Chem. Phys. Letters 18, 203-4

(73. 10)   Y. Tanaka, K. Yoshino, and D. E. Freeman,
           "On the Determination of the Ground State Potential Energy of Ne$_2$
           From its Vacuum Ultraviolet Spectrum,"
           J. Chem. Phys. 59, 564-5

(74. 11) W. J. Stevens, A. C. Wahl, M. A. Gardner, and A. M. Karo,
"Ab Initio Calculation of the Neon-Neon $^1\Sigma_g^+$ Potential at Intermediate Separations,"
J. Chem. Phys. 60, 2195-7

(74. 12) R. J. LeRoy, M. L. Klein, and I. J. McGee,
"On the Dissociation Energy and Interaction Potential of Ground-State Ne$_2$,"
Molec. Phys. 28, 587-91

(74. 13) A. Conway and J. N. Murrell,
"The Exchange Energy Between Neon Atoms,"
Molec. Phys. 27, 873-8

(74. 14) J. S. Cohen and B. Schneider,
"Ground and Excited States of Ne$_2$ and Ne$_2^+$. I. Potential Curves With and Without Spin-Orbit Coupling,"
J. Chem. Phys. 61, 3230-9

(74. 15) B. Schneider and J. S. Cohen,
"Ground and Excited States of Ne$_2$ and Ne$_2^+$. II. Spectroscopic Properties and Radiative Lifetimes,"
J. Chem. Phys. 61, 3240-3

Ni$_2$

Ni$_2$

Spectroscopic Constants

Dissociation energy = 2.37 ± 0.22 eV, 54.5 kcal/mole, 19100 cm^{-1} (64.1).

BIBLIOGRAPHY

(64. 1)   Dissociation Energy,
          A. Kant,
          J. Chem. Phys. 41, 1872-6

O_2

Methods of Production and Experimental Technique

Absorption: in high frequency discharges, pulsed discharges, ac discharges, flash photolysis.

Emission: all types of discharges, flames, explosions, luminescence.

In astrophysics.

Ground state studied by microwave spectroscopy.

BAND SYSTEMS

	System	Transition	Sources	Wavelength Limits	Degrading	Band Head, $\nu_{0,0}$	Remarks	Bibliography
Infrared atmospheric	I	$a^1\Delta_g \rightleftharpoons X^3\Sigma_g^-$	Absorption, emission	15800-9240	R	7882.39		(72.73, 62.39, 59.32, 58.29, 47.14, 33.6)
Atmospheric	II	$b^1\Sigma_g^+ \rightleftharpoons X^3\Sigma_g^-$	Absorption, emission	9970-5380	R	13120.9085		(72.73, 69.57, 64.41, 61.36, 50.18, 49.17)
Noxon	III	$b^1\Sigma_g^+ \rightarrow a^1\Delta_g$	Discharge	19080		5240 (head)	Only a single band	(69.57)
Herzberg II	IV	$c^1\Sigma_u^- \rightleftharpoons X^3\Sigma_g^-$	Absorption, luminescence	4790-4490 2715-2540	R	32664.1 (calculated)		(68.49, 53.22)
Herzberg III, High pressure	V	$C^3\Delta_u \leftarrow X^3\Sigma_g^-$	Absorption at high pressure	2630-2570 2924-2440	R	34319 (head)		(53.22, 39.11, 34.8, 32.5, 28.1)

	System	Transition	Sources	Wavelength Limits	Degrading	Band Head, $v_{0,0}$	Remarks	Bibliography
Chamberlain	VI	$C^3\Delta_u \to a^1\Delta_g$	Luminescence	4380-3700	R			(58.27)
Herzberg I	VII	$A^3\Sigma_u^+ \rightleftharpoons X^3\Sigma_g^-$	Absorption, luminescence	4880-2430	R	35007.15 (calculated)		(60.33, 59.31, 57.26, 55.25)
Schumann-Runge	VIII	$B^3\Sigma_u^- \rightleftharpoons X^3\Sigma_g^-$	All sources	5350-1750 1750-1300	R Continuum	49358.15		(72.73, 68.54, 68.52, 66.45, 64.43, 64.42, 61.35, 59.30, 54.24, 54.23, 50.19)
	IX	$\alpha^1\Sigma_u^+ \leftarrow b^1\Sigma_g^+$		1585-1538	V	63141.5		(68.48)
		$\alpha^1\Sigma_u^+ \leftarrow X^3\Sigma_g^-$		1280-1196	V			(69.58, 68.48)
		$\beta^3\Sigma_u^+ \leftarrow X^3\Sigma_g^-$	Absorption	1294-1181	V			(69.58, 68.48, 52.21)
		$^1\Delta_u \leftarrow a^1\Delta_g$		1243.8 (only a single band)		80396.0		(68.48)

	System	Transition	Sources	Wavelength Limits	Degrading	Band Head, $v_{0,0}$	Remarks	Bibliography
Rydberg Series	IX (cont)	$^1\Pi_u \leftarrow a^1\Delta_g$		1229.0 (only a single band)		81362.5		(68.48)
		$^3\Sigma_u^+ \leftarrow X^3\Sigma_g^-$		1144.6 (only a single band)	V	87369.1		(69.58)
	X	$X^2\Pi_g(0_2^+) \leftarrow X^3\Sigma_g^-$		1290-1180	V			(61.38, 52.21)
		$b^4\Sigma_g^-(0_2^+) \leftarrow X^3\Sigma_g^-$	Absorption	730-660	R			(68.51, 33.9)
		$B^2\Sigma_g^-(0_2^+) \leftarrow X^3\Sigma_g^-$		650-600	R			(68.51, 68.50, 42.12)
		$c^4\Sigma_u^-(0_2^+) \leftarrow X^3\Sigma_g^-$		595-510				(69.61)
	XI	Many bands that are unclassified or whose identification is doubtful						(68.51, 68.48, 67.47, 61.37, 54.24, 52.21, 48.16, 43.13)

I. $a\,^1\Delta_g \rightleftarrows X\,^3\Sigma_g^-$ System (Infrared Atmospheric)

Band origins, λ (58.29, 47.14, 33.6):

(v', v'')	(0, 1)	(0, 0)	(1, 0)	(2, 0)
λ	(15800)	1263.0	10674.1	(9240)

II. $b\,^1\Sigma_g^+ \rightleftarrows X\,^3\Sigma_g^-$ System (Atmospheric)

Band heads in emission, λ (69.57, 64.41, 61.36, 50.18, 49.17):

v', v''	0	1	2	3	4
0	7593.73	(8623)	(9970)		
1	6867.2	7683.85	8697.8		
2	6276.6	6953	7779.03		
3			7043	7879.17	
4				7141	7987

III. $b\,^1\Sigma_g^+ \rightarrow a\,^1\Delta_g$ System (Noxon)

Only a single band, Q branch (69.57):
$$\lambda(0, 0)\,|\,19080$$

IV. $c\,^1\Sigma_u^- \rightleftarrows X\,^3\Sigma_g^-$ System (Herzberg II)

Band origins (calculated), λ (68.49):

v', v''	0	1	2	3	4	5	6	7	8
0	3060.6	3213.7	3380.3	3562.0	3761.2	3980.3	4222.4	4491.2[a]	4791.5[a]
1	2990.3	3136.3	3294.7	3467.2	3655.6	3862.2	4089.7	4341.5	4621.4
2	2925.5	3065.1	3216.2	3380.4	3559.2	3754.8	3969.5	4206.2	4468.5
3	2865.8	2999.7	3144.3	3301.0	3471.3	3657.1	3860.5	4084.0	4330.8
4	2811.0	2939.6	3078.4	3228.4	3391.1	3568.2	3761.6	3973.5	4206.7
5	2760.6	2884.6	3018.1	3162.1	3318.1	3487.5	3671.9	3873.6	4094.9
6	2714.5[b]	2834.2	2963.0	3101.8	3251.7	3414.2	3590.8	3783.4	3994.2
7	2672.3[b]	2788.3	2912.9	3046.9	3191.4	3347.8	3517.4	3702.0	3903.6
8	2634.0[b]	2746.6	2867.4	2997.1	3136.9	3287.8	3451.3	3628.8	3822.4
9	2599.2[b]	2708.9	2826.2	2952.2	3087.7	3233.9	3391.9	3563.2	3749.6
10	2568.0[b]	2674.9	2789.3	2911.9	3043.7	3185.6	3338.8	3504.7	3684.9

[a] Observed in luminescence, [b] observed in absorption (53.22)

V. $C^3\Delta_u \leftarrow X^3\Sigma_g^-$ System (Herzberg III, High Pressure Bands)

Herzberg III

Two fragments with three heads have been observed (53.24).
Vibrational numbering is uncertain.

(v', v'')	$F_2(6,0)$	$F_3(6,0)$	(5,0)
λ	2589.14	2579.39	2620.71

High Pressure Bands (diffuse)

Maxima in absorption (no heads), λ (39.11).
Vibrational numbering is uncertain.

(v',v'')	(0,0)	(1,0)	(2,0)	(3,0)	(4,0)	(5,0)	(6,0)	(7,0)	(8,0)	(9,0)
λ	2924	2855	2795	2739.8	2689.8	2642.7	2598.8	2555.9	2525.4	2497.4
	2913	2842	2783.9	2729.9	2679.3	2632.7	2590.3	2553.5	2517	2488.7
	2904	2832	2769.1	2720.7	2671.6	2626	2582.4	2537	2510	2482

VI. $C^3\Delta_u \rightarrow a^1\Delta_g$ System (Chamberlain)

27 weak bands have been observed, but the identification is uncertain.
Vibrational numbering of the lower state is uncertain.

Possible band heads, λ (53.24):

v', v''	0	1	$^3\Delta_1$ 2	3	4	5
0						
1			4135			
2						
3			3887	4114		
4					4244	
5			3698		4127	4378
6				3813	4031	

$$\mathbf{{}^3\Delta_2}$$

v', v''	0	1	2	3	4	5
0						
1			4107			
2						
3			3866	4090		
4					4221	
5					4107	
6				3792	4009	4240

$$\mathbf{{}^3\Delta_3}$$

v', v''	0	1	2	3	4	5
0						
1			4086			
2						
3			3844	4071	4317	
4						
5				3861	4086	4326
6				3771	3985	4215

VII. $A\,^3\Sigma_u^+ \rightleftharpoons X\,^3\Sigma_g^-$ System (Herzberg I)

Band heads in emission, λ (Intensity) (59.31):

v', v''	0	1	2	3	4	5	6	7	8
0							3840 (5)	4064 (5)	4309 (7)
1					(3366.5) (2)	3542 (8)	3734 (8)	3938 (7)	4170 (6)
2					3285 (7)	3453 (8)	3633 (8)	3829 (8)	4044 (2)
3			2931 (1)	3066 (5)	3211 (10)	3370 (10)	3542 (8)	(3726.1) (2)	(3842.2) (2)
4			2873 (2)	3002 (5)	3142 (7)	3292 (4)	3459 (2)	(3634.6) (2)	
5			2820 (3)	2945 (5)	3080	(3225.0) (2)		(3552.5) (4)	(3737.7) (4)
6			2775 (3)	2895 (6)	3026 (2)		(3315.7) (2)	(3479.3) (4)	3657 (2)
7		2622 (3)	2734 (5)	2850 (5)			(3257.1) (4)	(3414.7) (4)	
8		2588 (2)	2696 (4)						

VIII. $B\,^3\Sigma_u^- \rightleftharpoons X\,^3\Sigma_g^-$ System (Schumann-Runge)

Band origins in absorption, λ (68.54, 66.45, 64.43, 64.42, 59.30, 54.23, 50.19):

v', v''	0	1	2	3	4	5	6	7	8
0	2026.01								
1	1998.17							2522.67	2614.67
2	1971.97	2034.29				2316.82	2396.80	2481.02	2569.95
3	1947.33	2008.11				2282.89	2360.52	2442.25	
4	1924.19	1983.60		2110.91	2179.36	2251.21	2326.53		
5	1902.23	1960.58	2021.28	2084.93	2151.61	2221.53	2295.03		
6	1882.43	1939.25	1998.63	2060.84	2125.94	2194.20			
7	1863.72	1919.37	1977.57	2038.35	2102.05	2168.78			
8	1846.51	1901.14	1958.21	2017.84	2080.22	2145.54			
9	1830.76	1884.47	1940.47	1999.05	2060.27	2124.31			
10	1816.50	1869.37	1924.48	1982.02	2042.23	2105.05			

IX.   Partial Systems

$\alpha\,^1\Sigma_u^+ \leftarrow b\,^1\Sigma_g^+$ System
___

Band heads, λ (68.48):

(v', v'')	(0, 0)	(1, 1)	(1, 0)
λ	1583.9	1571.9	1537.9

$\alpha\,^1\Sigma_u^+ \leftarrow X\,^3\Sigma_g^-$ System
___

Band heads, λ (69.58, 68.48):

(v', v'')	(1, 0)	(2, 0)	(3, 0)	(4, 0)
λ	1279.5	1250.0	1222.1	1196.4

$\beta\,^3\Sigma_u^+ \leftarrow X\,^3\Sigma_g^-$ System
___

Band origins, λ (69.58, 68.48):

(v', v'')	(2, 0)	(3, 0)
λ	1262.18	1233.47

$^1\Delta_u \leftarrow a\,^1\Delta_g$ System
___

Band head, λ (68.48):

(v', v'')	(0, 0)
λ	1243.8

$^1\Pi_u \leftarrow a\,^1\Delta_g$ System
___

Band head, λ (68.48):

(v', v'')	(0, 0)
λ	1229.0

$^3\Sigma_u^+ \leftarrow X\,^3\Sigma_g^-$ System

Double headed bands with 3 branches.   Band head, λ (69.58):

$$
\begin{array}{cc}
(v', v'') & (0, 0) \\
\lambda & 1144.6
\end{array}
$$

X.   ## Rydberg Series

$X\,^2\Pi_g\left(0_2^+\right) \leftarrow X\,^3\Sigma_g^-$ System

Single progression of doublets.   Classification is doubtful (61.38, 52.21).

$b\,^4\Sigma_g^-\left(0_g^+\right) \leftarrow X\,^3\Sigma_g^-$ System

Many progressions with the proposed configuration $\cdots np\,\sigma_u\ ^3\Sigma_u$ have been observed (68.38, 62.40, 35.9).

Band head formula: $\nu = 146568 - \dfrac{R}{(n-1.679)^2}$ $(n = 5 \cdots \infty)$

Another weak, diffuse series has been observed with a proposed configuration of $np\,\pi_u\ ^3\Pi_u$ (68.38).

$B\,^2\Sigma_g^-\left(0_2^+\right) \leftarrow X\,^3\Sigma_g^-$ System

Bands with simple heads (68.51, 68.50, 42.12).

Band head formula: $\nu = 163602 - \dfrac{R}{(n-0.658)^2}$ $(n = 4 \cdots \infty)$

$c\,^4\Sigma_u^-\!\left(0_2^+\right) \leftarrow X\,^3\Sigma_g^-$ System

Several series have been observed (69.61).

$\underline{\Pi\ \text{Series}}$ – probably excited to the nd $\pi_g\ ^3\Pi_u$ Rydberg state.

Band head formula: $\nu = 198\,125 - \dfrac{R}{(n-1.559)^2}$ $(n = 4 \cdots \infty)$

$\underline{\Sigma\ \text{Series}}$ – probably excited to the ns $\sigma_g\ ^3\Sigma_u^-$ Rydberg state.

Band head formula: $\nu = 198\,125 - \dfrac{R}{(n-0.955)^2}$ $(n = 4 \cdots \infty)$

O_2

SPECTROSCOPIC CONSTANTS

State	T_e	ω_e	$x_e\omega_e$	B_e	$\alpha_e \times 10^2$	$D_e \times 10^6$	r_e	Remarks	Bibliography
$^1\Pi_u$	89244.9[a]			(1.451)					(68.48)
$^1\Delta_u$	88278.4[a]			(1.446)					(68.48)
$^3\Sigma_u^+$	87369.1[a]			(1.706)	(2800)				(69.58)
$\alpha\,^1\Sigma_u^+$	76089	(1927)	(19)	1.699	1.6				(69.58, 68.48)
$\beta\,^3\Sigma_u^+$	75263	(1957)	(19.7)	(1.7)	(2)				(69.58, 68.48, 52.21)
$B\,^3\Sigma_u^-$	49794.33	709.058	10.6141	0.818975	1.19225		1.60428	(b, g)	(70.63, 66.45, 54.23, 34.7)
$A\,^3\Sigma_u^+$	35398.70	799.08	12.16	0.91053	1.416	4.79	1.52153	(c, h)	(54.24, 52.20)
$C\,^3\Delta_{u,i}$	34735	(750)	(14)				(1.5)	(d, i)	(53.22, 39.11, 32.5)
$c\,^1\Sigma_u^-$	33058.4	794.29	12.736	0.9155	1.391	(10.5)	1.5174	(d, i)	(68.49, 53.22)
$b\,^1\Sigma_g^+$	13195.314	1432.66	13.9336	1.4004796	1.8169303	5.356	1.22684	(e, j)	(n.p. 125, 48.15)
$a\,^1\Delta_g$	7918.11	(1509.3)	(12.9)	1.4263	1.71	(4.97)	1.21569		(47.14)

SPECTROSCOPIC CONSTANTS

State	T_e	ω_e	$x_e \omega_e$	B_e	$\alpha_e \times 10^2$	$D_e \times 10^6$	r_e	Remarks	Bibliography
$X^3\Sigma_g^-$	0	1580.19	11.981	1.445622	1.593268		1.20754	(f, k)	(n.p. 125, 66.45, 54.23, 34.7)

(a) T_o; (b) $y_e \omega_e = -0.059212435$, $z_e \omega_e = -0.023974994$; (c) $y_e \omega_e = -0.550$; (d) $y_e \omega_e = -0.2444$, $z_e \omega_e = 0.00055$;

(e) $y_e \omega_e = -0.0143$; (f) $y_e \omega_e = 0.047474736$, $z_e \omega_e = -0.00012727481$; (g) $\gamma_e = -6.30472 \times 10^{-4}$;

(h) $\gamma_e = -9.7 \times 10^{-4}$, $\beta_e = 3.0 \times 10^{-7}$; (i) $\gamma_e = -7.40 \times 10^{-4}$; (j) $\gamma_e = -4.2941920 \times 10^{-5}$, $\beta_e = 0.077$;

(k) $\gamma_e = 6.406456 \times 10^{-5}$

Dissociation energy = 5.12 ± 0.0019 eV, 117.97 kcal/mole, 41260 cm^{-1} (54.23).

Perturbations and General Information

Ionization potential (I_p) to $X^2\Pi_{g,i}\left(0_2^+\right)$ = 12.059 ± 0.001 eV (68.53, 66.44).

$A^3\Sigma_u^+$ - $X^3\Sigma_g$ has a strong perturbation in the (11,0) band for N > 11 (52.20).

$B^3\Sigma_u^-$ state is perturbed at v = 16, J = 8 and v = 19, J = 8 (54.23).

$B^3\Sigma_u^-$ state is predissociated, probably by a repulsive $^3\Pi_u$ state. The predissociation is characterized by an onset at v = 2 and broadening at v = 4, 8, and 11, with a minimum at v = 9. The interpretation of the predissociation is in question (72.73, 70.62, 69.60, 69.59, 61.36, 59.30, 58.28, 36.10).

Vibrational Raman effect has been observed (60.33, 30.3, 29.2).

Rotational Raman effect has been observed (74.114, 60.33, 30.3).

Potential energy curves — RKR potentials (72.73 and references cited therein):

State	v	$V(cm^{-1})$	$r_{min}(\text{Å})$	$r_{max}(\text{Å})$
$X^3\Sigma_g^-$	0	787.3818	1.1590417	1.2626908
	1	2343.7613	1.1272513	1.3078976
$T_e = 0\ cm^{-1}$	2	3876.57	1.10700	1.34170
	3	5386.03	1.09146	1.37093
	4	6872.34	1.07864	1.39759
	5	8335.65	1.06767	1.42257
	6	9776.11	1.0580	1.4464
	7	11193.80	1.0494	1.4693
	8	12588.82	1.0417	1.4917
	9	13961.18	1.0346	1.5136
	10	15310.91	1.0280	1.5351
$a^1\Delta_g$	0	751.658	1.16619	1.27228
	1	2235.158	1.13396	1.31904
$T_e = 7918.11\ cm^{-1}$	2	3692.86	1.11353	1.35422
	3	5124.76	1.0979	1.3848
$b^1\Sigma_g^+$	0	712.9766	1.176241	1.285186
	1	2117.7290	1.143442	1.333696
$T_e = 13195.314\ cm^{-1}$	2	3494.4855	1.122734	1.370428
	3	4843.1603	1.106952	1.402561

State	v	$V(cm^{-1})$	$r_{min}(\overset{\circ}{A})$	$r_{max}(\overset{\circ}{A})$
$A^3\Sigma_u^+$	0	395.8	1.454	1.600
	1	1168.7	1.411	1.668
$T_e = 35398.70\ cm^{-1}$	2	1912.5	1.385	1.722
	3	2623.5	1.366	1.772
	4	3298.9	1.350	1.822
	5	3934.9	1.337	1.872
	6	4527.2	1.326	1.925
	7	5070.0	1.317	1.982
	8	5555.6	1.310	2.050
	9	5973.4	1.304	2.131
	10	6309.1	1.298	2.245
$B^3\Sigma_u^-$	0	351.204	1.53266	1.68771
	1	1038.736	1.48649	1.75876
$T_e = 49794.33\ cm^{-1}$	2	1703.961	1.45776	1.81426
	3	2345.774	1.43623	1.86450
	4	2962.845	1.41889	1.91257
	5	3553.643	1.40434	1.96005
	6	4118.425	1.39181	2.00806
	7	4649.207	1.38084	2.05761
	8	5149.746	1.37117	2.10976
	9	5615.548	1.36264	2.16578
	10	6043.932	1.35518	2.22722
	11	6432.167	1.34876	2.29602

Radiative lifetimes, Einstein coefficients and oscillator strengths:

Transition	Band	τ(sec)	$A_{v'}$(sec^{-1})	$A_{v'v''}$(sec^{-1})	Absorption f-Value	Reference
$a^1\Delta_g - X^3\Sigma_g^-$	0-0	3.88(10^3)		2.58(10^{-4})	4.15(10^{-12})	(68.55)
$b^1\Sigma_g^+ - X_3\Sigma_g^-$	0-0			0.085	2.47(10^{-10})	(67.46)
	1-0			(0.0069)		(32.4)
	2-0			(0.1636)10^{-3}		(68.56)
	1-1			0.0704		(68.56)
$b^1\Sigma_g^+ - a^1\Delta_g$	0-0	(1 - 10^3)		1.5(10^{-3})		(61.34)
$A^3\Sigma_u^+ - X^3\Sigma_g^-$						(67.46, 64.41, 62.40)
	7-0				1.24(10^{-10})	(70.64)
$c^1\Sigma_u^- - X^3\Sigma_g^-$		> 10^{-3}	~ 10^{-4}			(62.40)
						(64.41)
$C^3\Delta_u - X^3\Sigma_g^-$		> 10^{-3}	≤ 10^{-5}			(62.40)
						(64.41)

Absolute f-values for the $B^3\Sigma_u^- - X^3\Sigma_g^-$ bands (72.73 and references cited therein):

v', v"	0	1	2
0	3.45-10		
1	3.90-9		
2	2.38-8	5.35-7	
3	9.90-8	2.08-6	
4	3.21-7	6.15-6	
5	8.52-7	1.53-5	
6	1.91-6	3.15-5	2.13-4
7	3.81-6	5.78-5	3.39-4
8	6.68-6	9.40-5	5.46-4
9	1.06-5	1.38-4	9.87-4
10	1.57-5	1.91-4	1.03-3
11	2.09-5	2.38-4	1.04-3
12	2.53-5	2.73-4	1.22-3
13	2.88-5	2.93-4	1.04-3
14	3.03-5	2.95-4	
15	2.92-5	2.77-4	
16	2.59-5	2.42-4	
17	2.23-5	2.01-4	
18	1.83-5		
19	1.44-5		

f-value followed by a factor of ten

Franck-Condon factors — RKR potentials (n.p. 125, 72.73):

$a^1\Delta_g - X^3\Sigma_g^-$

v', v"	0	1	2	3	4
0	9.869-1	1.297-2	1.260-4		
1	1.303-2	9.586-1	2.791-2	4.296-4	1.735-6
2	6.795-5	2.814-2	9.258-1	4.497-2	9.802-4
3	2.591-4	4.548-2	8.881-1	6.423-2	1.867-3

Franck-Condon factors followed by a factor of ten

O_2

$b^1\Sigma_g^+ - X^3\Sigma_g^-$

v', v''	0	1	2	3	4	5	6
0	9.308-1	6.660-2	2.523-3	5.648-5			
1	6.647-2	7.928-1	1.322-1	8.284-3	2.736-4	6.417-6	
2	2.639-3	1.315-1	6.527-1	1.943-1	1.802-2	8.232-4	2.512-5
3	6.911-5	8.753-3	1.924-1	5.144-1	2.499-1	3.240-2	1.968-3

Franck-Condon factors followed by a factor of ten

$A^3\Sigma_u^+ - X^3\Sigma_g^-$

v', v''	6	7	8	9	10	11	12
0	4.260-2	7.935-2	1.214-1	1.546-1	1.654-1	1.495-1	1.140-1
1	8.985-2	1.052-1	8.298-2	3.500-2	1.510-3	1.512-2	6.765-2
2	8.158-2	4.457-2	4.492-3	1.049-2	5.486-2	7.589-2	4.343-2
3	3.593-2	1.434-3	1.700-2	5.478-2	4.681-2	6.822-3	9.761-3
4	3.900-3	1.162-2	4.595-2	3.559-2	1.700-3	1.847-2	5.157-2

Franck-Condon factors followed by a factor of ten

$B^3\Sigma_u^- - X^3\Sigma_g^-$

v', v''	12	13	14	15	16	17
0	1.192-1	1.443-1	1.514-1	1.378-1	1.087-1	7.417-2
1	6.350-2	2.328-2	3.441-4	1.553-2	6.165-2	1.087-1
2	5.507-5	1.853-2	5.696-2	6.930-2	3.928-2	3.934-3
3	3.150-2	5.446-2	3.492-2	2.823-3	1.221-2	5.283-2
4	4.503-2	1.900-2	1.904-4	2.620-2	4.910-2	2.392-2
5	1.579-2	6.553-4	2.667-2	3.844-2	9.388-3	5.262-3

Franck-Condon factors followed by a factor of ten

$b^1\Sigma_g^+ - a^1\Delta_g$

v', v''	0	1	2	3
0	9.770-1	2.283-2	2.136-4	
1	2.267-2	9.290-1	4.760-2	7.217-4
2	3.628-4	4.694-2	8.768-1	7.430-2
3	4.426-6	1.213-3	7.266-2	8.202-1

Franck-Condon factors followed by a factor of ten

BIBLIOGRAPHY

References for this molecule are only those from which spectroscopic information has been taken directly and those papers published after 1970. The reason for excluding the publications from before 1971 is the reference (72.73) in our Bibliography. This publication provides a Bibliography through the beginning of 1971, as well as an excellent critical review of the O$_2$ molecule.

(28. 1)   Triplet Systems at High Pressure,
           D. S. Villars,
           Proc. Nat. Acad. Sci. 14, 508-11

(29. 2)   Vibration-Rotation Raman Effect,
           F. Rasetti,
           Phys. Rev. 34, 367-71

(30. 3)   Vibration-Rotation and Rotation, Raman Effect,
           F. Rasetti,
           Z. Physik 61, 598-601

(32. 4)   A - X System,
           G. Herzberg,
           Naturwissenschaften 20, 577

(32. 5)   C - X System, 2300-2900Å Bands,
           W. Finkelnburg and W. Steiner,
           Z. Physik 79, 69-88

(33. 6)   a - X System,
           J. W. Ellis and H. O. Kneser,
           Z. Physik 86, 583-91

(34. 7)   B - X System, a - X System Rotational Constant,
           J. Curry and G. Herzberg,
           Ann. Phys. 19, 800-18

(34. 8)   Triplet System,
           W. Finkelnburg,
           Z. Physik 90, 1-10

(35. 9)   b - X Rydberg Series and Unclassified Rydberg Series,
           W. C. Price and G. Collins,
           Phys. Rev. 48, 714-9

O_2

(36. 10)   Predissociation,
P. J. Flory,
J. Chem. Phys. **4**, 23-7

(39. 11)   C, A - X Systems,
L. Herman,
Ann. Phys. **11**, 548-611

(42. 12)   b, B - X Rydberg Series, Vibrational Structure,
Y. Tanaka and T. Takamine,
Sci. Papers Inst. Phys. Chem. Res. **39**, 437-46

(43. 13)   Unclassified Rydberg Series,
N. L. Singh and L. Lal,
Sci. and Culture **9**, 89

(47. 14)   a - X System, Spectroscopic Constants,
L. Herzberg and G. Herzberg,
Astrophys. J. **105**, 353-9

(48. 15)   b - X System, Spectroscopic Constants of X State,
H. D. Babcock and L. Herzberg,
Astrophys. J. **108**, 167-90

(48. 16)   B - X System,
L. Lal,
Nature **161**, 477-8

(49. 17)   b - X System,
R. C. Herman, H. S. Hopfield, G. A. Hornbeck, and S. Silverman,
J. Chem. Phys. **17**, 220-1

(50. 18)   b - X System,
A. B. Meinel,
Astrophys. J. **112**, 464-8

(50. 19)   B - X System at High Temperature,
A. Herczog and K. Wieland,
Helv. Phys. Acta **23**, 432-6

(52. 20)   A - X System, Spectroscopic Constants,
G. Herzberg,
Can. J. Phys. **30**, 185-210

(52. 21) α, β - X Rydberg Systems,
Y. Tanaka,
J. Chem. Phys. **20**, 1728-33

(53.22)   c, C - X Systems, C - a System,
G. Herzberg,
Can. J. Phys. 31, 657-69

(54.23)   B - X System, a - X System, Spectroscopic Constants,
P. Brix and G. Herzberg,
Can. J. Phys. 32, 110-35

(54.24)   A, c - X System,
H. P. Broida and A. G. Gaydon,
Proc. Roy. Soc. A 222, 181-95

(55.25)   A - X System,
J. W. Chamberlain,
Astrophys. J. 121, 277-86

(57.26)   A - X System,
C. A. Barth and J. Kaplan,
J. Chem. Phys. 26, 506-10

(58.27)   C → a System, Spectroscopic Constants,
J. W. Chamberlain,
Astrophys. J. 128, 713-7

(58.28)   New B - X System Bands in the Ultraviolet, Predissociation,
D. Rakotoarijimy, S. Weniger, and H. Grenat,
C. R. Acad. Sci. 246, 2883-6

(58.29)   a - X System,
A. V. Jones and A. W. Harrison,
J. Atm. Terrestr. Phys. 13, 45-60

(59.30)   B - X System, Predissociation,
P. K. Carroll,
Astrophys. J. 129, 794-800

(59.31)   A - X System,
C. A. Barth and J. Kaplan,
J. Molec. Spectrosc. 3, 583-7

(59.32)   Infrared Systems,
J. Connes and H. P. Gush,
J. Phys. Radium 20, 915-7

(60.33)   Raman Spectrum,
A. Weber and E. A. MacGinnis,
J. Molec. Spectrosc. 4, 195-200

O_2

(61.34)   b - a System, Spectroscopic Constants,
          J. F. Noxon,
          Can. J. Phys. 39, 1110-9

(61.35)   B - X System,
          G. R. Broert and R. W. Nicholls,
          J. Atm. Terrestr. Phys. 21, 21

(61.36)   b, B - X Systems, Predissociation,
          L. Herman, R. Herman, and D. Rakotoarijimy,
          J. Phys. Radium 22, 1-8

(61.37)   Unclassified Rydberg Series,
          J. Byrne,
          Proc. Phys. Soc. 78, 1074-5

(61.38)   Rydberg Series,
          W. R. Jarmain,
          Yerkes Obs., Univ. Chicago, Sci. Report No. 35

(62.39)   b - X System, Spectroscopic Constants,
          A. Landau, E. J. Allin, and H. L. Welsh,
          Spectrochim. Acta 18, 1-19

(62.40)   3000-12400Å Atlas,
          V. I. Krossovsky, N. N. Shefov, and V. I. Yaim,
          Planet. Space Sci. 9, 883-915

(64.41)   b - X System,
          J. A. Curcio, L. F. Drummeter, and G. L. Knestrick,
          Appl. Opt. 3, 1401-9

(64.42)   B - X System,
          A. M. Bass and D. Garvin,
          J. Chem. Phys. 40, 1772-3

(64.43)   B - X System,
          R. V. Fitzsimmons and E. J. Blair,
          J. Chem. Phys. 40, 451-8

(66.44)   Ionization Potential,
          J. A. R. Samson and R. B. Cairns,
          J. Opt. Soc. Am. 56, 769-75

(66.45)   a, B - X Systems,
          M. Ogawa,
          Science Light 15, 97-114

(67.46)   Line Shapes,
          D. E. Burch and D. A. Gryvnak,
          Appl. Opt. **8**, 1493-9

(67.47)   830-900Å Bands,
          R. E. Huffman, J. C. Larrabee, and Y. Tanaka,
          J. Chem. Phys. **46**, 2213-33

(68.48) α-b, α-X, B-X, $^1\Delta_u$-a$^1\Delta_g$, $^1\Pi_u$-a$^1\Delta_g$ Systems and X-X Rydberg
          System,
          F. Alberti, R. A. Ashby, and A. E. Douglas,
          Can. J. Phys. **46**, 337-42

(68.49)   c-X System,
          R. K. Dhumwad and N. A. Narasimham,
          Can. J. Phys. **46**, 1254-5

(68.50)   B-X Rydberg System,
          M. Ogawa,
          Can. J. Phys. **46**, 312-3

(68.51)   b, B-X Rydberg Systems, Unclassified Bands < 600Å, Ionization
          Potential,
          K. Yoshino and Y. Tanaka,
          J. Chem. Phys. **48**, 4859-67

(68.52)   B-X System Between 300 to 900°K,
          R. D. Hudson and V. L. Carter,
          J. Opt. Soc. Am. **58**, 1621-9

(68.53)   Ionization Potential,
          G. L. Bhale and P. R. Rao,
          Proc. Indian Acad. Sci. A **67**, 350-7

(68.54)   B-X System,
          M. Ogawa and H. C. Chang,
          Science Light **17**, 45-56

(68.55)   Radiative Lifetime a$^1\Delta_g$,
          R. M. Badger, A. D. Wright, and R. F. Whitlock,
          J. Chem. Phys. **43**, 3341-63

(68.56)   Radiative Lifetime b$^1\Sigma_g^+$,
          G. F. Sitnik and A. I. Khlystov,
          Izv. Atmos. Ocean. Phys. **4**, 1120-2

(69.57)   b - X System,
G. Herzberg, A. Lagerqvist, and B. J. MacKenzie,
Can. J. Phys. **47**, 1889-97

(69.58) α, β, $^3\Sigma_u^+$ - X Systems, X - X Rydberg System, Spectroscopic
Constants,
M. Ogawa and K. R. Yamawaki,
Can. J. Phys. **47**, 1805-11

(69.59)   Predissociation,
I. Riess and Y. Ben-Aryeh,
J. Quant. Spectrosc. Radiative Trans. **9**, 1463-8

(69.60)   B - X System, Predissociation,
J. N. Murrell and J. M. Taylor,
Molec. Phys. **16**, 609-21

(69.61)   C - X Rydberg Series,
J. Hoeft, F. J. Lovas, E. Tiemann, and T. Törring,
Z. Naturforsch. A **24**, 1843-4

(70.62)   Potential Curves, Predissociation,
D. V. K. Rao and P. T. Rao,
J. Phys. B **3**, 430-7

(70.63)   B - X System,
M. Ackerman and F. Biaume,
J. Molec. Spectrosc. **35**, 73-82

(70.64)   A - X System,
V. Hasson, R. W. Nicholls, and V. Degen,
J. Phys. B **3**, 1192-4

(71.65)   T. A. Miller,
"Rotational Moment, Rotational g Factor, Electronic Orbital g
Factor, and Anisotropy of the Magnetic Susceptibility of $^1\Delta$ O_2,"
J. Chem. Phys. **54**, 330-7

(71.66)   N. Jonathan, A. Morris, K. J. Ross, and D. J. Smith,
"High Resolution Vacuum Ultraviolet Photoelectron Spectra of
Transient Species: $O_2(^1\Delta_g)$ and Previously Unobserved States
of O_2^+,"
J. Chem. Phys. **54**, 4954-5

(71.67)   F. D. Findlay and D. R. Snelling,
"Collisional Deactivation of $O_2(^1\Delta_g)$,"
J. Chem. Phys. **55**, 545-51

(71. 68)   K. H. Becker, W. Grath, and U. Schurath,
"The Quenching of Metastable $O_2(^1\Delta_g)$ and $O_2(^1\Sigma_g^+)$ Molecules,"
Chem. Phys. Letters 8, 259-62

(71. 69)   D. W. McCullough and W. D. McGrath,
"The Collisional Deactivation of $O(^1D)$ Atoms by Molecular Oxygen,"
Chem. Phys. Letters 8, 353-7

(71. 70)   D. R. Snelling and M. Gauthier,
"Efficiency of $O_2(^1\Sigma_g^+)$ Formation by $O(^1D)+O_2$,"
Chem. Phys. Letters 9, 254-6

(71. 71)   S. Durmaz and J. N. Murrell,
"The Effect of Rotations on the Predissociation Probabilities of
Diatomic Molecular Spectra,"
Molec. Phys. 21, 209-16

(71. 72)   K. A. Dick and G. G. Sivjee,
"O_2 Herzberg I Bands in the Night Airglow: Covariation With OI,"
J. Geo. Res. 76, 6987-9

(72. 73)   P. H. Krupenie,
"The Spectrum of Molecular Oxygen,"
J. Phys. Chem. Ref. Data 1, 423-534

(72. 74)   U. Mingelgrin, R. G. Gordon, L. Frenkel, and T. E. Sullivan,
"Microwave Spectrum of Compressed O_2-Foreign Gas Mixtures in
the 48-41 GHz Region,"
J. Chem. Phys. 57, 2923-31

(72. 75)   K. Furukawa and E. A. Ogryzlo,
"A Redetermination of the Rate Constants for the Quenching of
Gaseous $O_2(^1\Delta_g)$ by Aliphatic Amines,"
J. Photochem. 1, 163-9

(72. 76)   D. W. McCullough and W. D. McGrath,
"Electronic-Vibrational Energy Transfer in the Reaction of $O(^1D)$
Atoms With Molecular Oxygen,"
J. Photochem. 1, 241-53

(72. 77)   S. D. Peyerimhoff and R. J. Buenker,
"Comparison of Various CI Treatments for the Description of
Potential Curves for the Lowest Three States of O_2,"
Chem. Phys. Letters 16, 235-43

(72. 78)   R. H. Pritchard, M. L. Sink, J. D. Allen, and C. W. Kern,
"Theoretical Studies of Fine Structure in the Ground State of O_2,"
Chem. Phys. Letters 17, 157-9

O$_2$

(72.79) D. C. Cartwright, S. Trajmar, and W. Williams,
"The Excitation of O$_2$ in Auroras,"
Ann. Geophys. 28, 397-401

(72.80) H. C. Chang and M. Ogawa,
"Rotational Analysis of a High-Resolution Absorption Band of O$_2$
at 1161Å,"
J. Molec. Spectrosc. 44, 405-6

(72.81) A. Konkov and A. V. Vorontsov,
"Experimental Study of Infrared Radiation of Oxygen,"
Opt. Spectrosc. 32, 243-6

(72.82) P. B. Merkel and D. R. Kearns,
"Radiationless Decay of Singlet Molecular Oxygen in Solution. An
Experimental and Theoretical Study of Electronic-to-Vibrational
Energy Transfer,"
J. Am. Chem. Soc. 94, 7244-53

(72.83) P. Gerber,
"Hyperfine Structure From the Electron Spin Resonance Spectrum
of Gas Phase Oxygen,"
Helv. Phys. Acta 45, 655-82

(72.84) P. S. Julienne and M. Krauss,
"Excitation of O$_2$ $^1\Delta_g$ by Electron Impact,"
J. Res. Nat. Bur. Stand. 76A, 661-3

(72.85) J. F. Noxon and A. E. Johanson,
"Changes in Thermospheric Molecular Oxygen Abundance Inferred
From Twilight 6300Å Airglow,"
Planet. Space Sci. 20, 2125-51

(72.86) J. H. Moore, Jr.,
"Electronic Excitation of N$_2$ and Dissociative Excitation of O$_2$ by
Proton Impact,"
J. Geo. Res. 77, 5567-72

(72.87) K. R. Yamawaki,
"Absorption Spectrum of O$_2$ in the a$^1\Delta_g$ Metastable State in the
Region From 1090 to 1700Å,"
Thesis, University of Southern California

(73.88) F. Koike and T. Watanabe,
"On the Mechanism of Electron Attachment by O$_2$,"
J. Phys. Soc. Japan 34, 1022-8

(73.89) F. Koike,
"Resonant Vibrational Excitation of O_2 by Slow Electron Impact,"
J. Phys. Soc. Japan **35**, 1166-70

(73.90) G. R. Cook, M. Ogawa, and R. W. Carlson,
"Photodissociation Continuums of N_2 and O_2,"
J. Geo. Res. **78**, 1663-7

(73.91) L. C. Lee, R. W. Carlson, D. L. Judge, and M. Ogawa,
"The Absorption Cross Sections of N_2, O_2, CO, NO, CO_2, N_2O,
CH_4, C_2H_4, C_2H_6 and C_4H_{10} From 180 to 700Å,"
J. Quant. Spectrosc. Rad. Trans. **13**, 1023-31

(73.92) D. L. Albritton, W. J. Harrop, and A. L. Schmeltekopf,
"Calculation of Centrifugal Distortion Constants for Diatomic
Molecules From RKR Potentials,"
J. Molec. Spectrosc. **46**, 25-36

(73.93) D. L. Albritton, W. J. Harrop, A. L. Schmeltekopf, and R. N. Zare,
"Resolution of the Discrepancies Concerning the Optical and Micro-
wave Values for B_0 and D_0 of the $X\,^3\Sigma_g^-$ State of O_2,"
J. Molec. Spectrosc. **46**, 103-18

(73.94) M. Gauthier and D. R. Snelling,
"Possible Production of $O_2(^1\Delta_g)$ and $O_2(^1\Sigma_g^+)$ in the Reaction of NO
With O_3,"
Chem. Phys. Letters **20**, 178-81

(73.95) C. Schmidt and H. I. Schiff,
"Reactions of $O_2(^1\Delta_g)$ With Atomic Nitrogen and Hydrogen,"
Chem. Phys. Letters **23**, 339-42

(73.96) V. D. Galkin,
"Line Shifts in the A Oxygen Band as a Function of the Pressure,"
Opt. Spectrosc. **35**, 630-3

(73.97) V. D. Galkin, L. N. Zhukova, and L. A. Mitrofanova,
"Line Intensities and Halfwidths in the A and B Bands of the Red
Atmospheric Band System of O_2,"
Opt. Spectrosc. **33**, 462-5

(73.98) D. C. Cartwright, W. J. Hunt, W. Williams, S. Trajmar, and
W. A. Goddard III,
"Theoretical and Experimental (Electron-Impact) Studies of the
Low-Lying Rydberg States in O_2,"
Phys. Rev. A **8**, 2436-48

(73.99)   R. J. Collins and D. Husain,
           "A Kinetic Study of Vibrationally Excited $O_2(a^1\Delta_g, v=1)$ by Time-
           Resolved Absorption Spectroscopy in the Vacuum Ultra-Violet,"
           J. Photochem. 1, 481-90

(73.100)   W. S. Watson, J. Lang, and D. T. Stewart,
           "Photoabsorption Coefficients of Molecular Oxygen in the 400-600Å
           Region,"
           Phys. Letters 44A, 293-4

(73.101)   J. A. Kinsinger and J. W. Taylor,
           "Autoionization and the Photoelectron Spectra of Oxygen,"
           Int. J. Mass Spect. Ion Phys. 11, 461-74

(73.102)   S. F. Wong, M. J. W. Boness, and G. J. Schulz,
           "Vibrational Excitation of O_2 by Electron Impact Above 4 eV,"
           Phys. Rev. Letters 31, 969-72

(73.103)   R. J. Collins, D. Husain, and R. J. Donovan,
           "Kinetic and Spectroscopic Studies of $O_2(a^1\Delta_g)$ by Time-Resolved
           Absorption Spectroscopy in the Vacuum Ultra-Violet,"
           J. Chem. Soc. Faraday Trans. II 69, 145-57

(73.104)   J. G. Parker and D. N. Ritke,
           "Vibrational Relaxation Times of Oxygen in the Pressure Range
           10-110 atm,"
           J. Chem. Phys. 58, 314-23

(73.105)   J. A. Hall,
           "Comment on the Spin-Orbit Contribution to the Zero-Field
           Splitting of the Oxygen Molecule,"
           J. Chem. Phys. 58, 410-2

(73.106)   T. J. Cook, B. R. Zegarski, W. H. Breckenridge, and T. A. Miller,
           "Gas Phase EPR of Vibrationally Excited O_2,"
           J. Chem. Phys. 58, 1548-52

(73.107)   T. C. Frankiewicz and R. S. Berry,
           "Production of Metastable Singlet O_2 Photosensitized by NO_2,"
           J. Chem. Phys. 58, 1787-95

(73.108)   I. T. N. Jones and K. D. Bayes,
           "Formation of $O_2(a^1\Delta_g)$ by Electronic Energy Transfer in Mixture
           of NO_2 and O_2,"
           J. Chem. Phys. 59, 3119-24

(73. 109) J. G. Parker and D. N. Ritke,
"Collisional Deactivation of Vibrationally Excited Singlet Molecular Oxygen,"
J. Chem. Phys. 59, 3713-22

(73. 110) P. D. Burrow,
"Dissociative Attachment From the $O_2(a^1\Delta_g)$ State,"
J. Chem. Phys. 59, 4922-31

(73. 111) C. Long and D. R. Kearns,
"Selection Rules for the Intermolecular Enhancement of Spin Forbidden Transitions in Molecular Oxygen,"
J. Chem. Phys. 59, 5729-36

(74. 112) J. A. Davidson and E. A. Ogryzlo,
"The Quenching of $O_2(^1\Sigma_g^+)$ by Aliphatic Hydrocarbons,"
Can. J. Chem. 52, 240-5

(74. 113) D. R. Snelling,
"The Ultraviolet Flash Photolysis of Ozone and the Reactions of $O(^1D)$ and $O_2(^1\Sigma_g^+)$,"
Can. J. Chem. 52, 257-70

(74. 114) C. M. Penney, R. L. St. Peters, and M. Lapp,
"Absolute Rotational Raman Cross Sections for N_2, O_2, and CO_2,"
J. Opt. Soc. Am. 64, 712-6

(74. 115) D. L. Huestis, G. Black, S. A. Edelstein,
"Fluorescence and Quenching of $O_2(^1\Delta_g)$ and $[O_2(^1\Delta_g)]_2$ in Liquid Oxygen,"
J. Chem. Phys. 60, 4471-4

(74. 116) W. P. West, T. B. Cook, F. B. Dunning, R. D. Rundel, and R. F. Stebbings,
"Chemiionization of O_2 Molecules by Helium Metastable Atoms,"
J. Chem. Phys. 60, 5126-7

(74. 117) L. C. Lee, R. W. Carlson, D. L. Judge, and M. Ogawa,
"Vacuum Ultraviolet Fluorescence From Photodissociation Fragments of O_2 and N_2,"
J. Chem. Phys. 61, 3261-9

(74. 118) J. G. Parker and D. N. Ritke,
"On the Mechanism for Collisional Deactivation of Vibrationally Excited Singlet Molecular Oxygen,"
J. Chem. Phys. 61, 3408-13

(74.119) L. Veseth and A. Lofthus,
"Fine Structure and Centrifugal Distortion in the Electronic and
Microwave Spectra of O_2 and SO,"
Molec. Phys. 27, 511-19

(74.120) L. P. Giver, R. W. Boese, and J. H. Miller,
"Intensity Measurements, Self-Broadening Coefficients, and
Rotational Intensity Distribution for Lines of the Oxygen B Band
at 6880 A,"
J. Quant. Spectrosc. Rad. Trans. 14, 793-802

(74.121) E. A. Ogryzlo and B. A. Thrush,
"The Vibrational Excitation of H_2O and CO_2 by $O_2(^1\Sigma_g^+)$,"
Chem. Phys. Letters 24, 314-6

(74.122) T. J. Cook and T. A. Miller,
"Production of $^1\Delta_g$ O_2 From Microwave Discharges in CO_2, NO_2
and SO_2,"
Chem. Phys. Letters 25, 396-8

(74.123) I. B. C. Matheson, J. Lee, B. S. Yamanashi, and M. L. Wolbarsht,
"Observation of the Singlet Oxygen Dimal Emission From Neodymium
Laser Pumped Oxygen in Gas Phase and in 1,1,2-Trichlorotrifluoro-
ethane Solution,"
Chem. Phys. Letters 27, 355-8

(74.124) L. E. Khvorostovskaya and V. A. Yankovskii,
"Mechanism of Ozone Formation in the Molecular Oxygen Glow
Discharge,"
Opt. Spectrosk. 37, 26-30

(n. p. 125) D. L. Albritton, A. L. Schmeltekopf, and R. N. Zare,
"Diatomic Intensity Factors,"
(to be published by Harper and Row)

P$_2$

Methods of Production and Experimental Technique

Absorption in phosphorus vapor, flash photolysis of PH$_3$.

Emission from a discharge of He or H$_2$ with phosphorus, discharge in PH$_3$ or microwave discharge in PCl$_3$.

Fluorescence.

BAND SYSTEMS

System	Transition	Sources	Wavelength Limits	Degrading	Characteristic Bands, λ	Remarks	Bibliography
I	$A^1\Pi_g \rightarrow X^1\Sigma_g^+$	Emission	3110–2850	R	2970(0, 1)		(73.42, 58.21)
II	$C^1\Sigma_u^+ \rightleftharpoons X^1\Sigma_g^+$	Emission	3500–2000	R	2953.6(6, 22) 2757.1(4, 17) 2456.9(3, 10)		(67.31, 67.30, 66.24, 64.23, 61.22, 50.18, 50.17, 49.16, 46.14, 43.12, 43.11, 40.10, 35.9, 33.8, 32.7, 32.6, 32.5, 32.4, 31.3, 30.2, 07.1)
		Absorption	2300–1800	R	2108.1(3, 1)		
III	$E^1\Pi_u \rightleftharpoons X^1\Sigma_g^+$	Absorption, emission	1750–1600	R	1705.5(0, 1) 1728.1(0, 2)		(66.24, 55.20, 55.19)
IV	$G^1\Sigma_u^+ \rightleftharpoons X^1\Sigma_g^+$	Absorption, emission	1530–1480	R	1508.7(0, 0)		(66.24, 55.19)
V	$I^1\Pi_u \leftarrow X^1\Sigma_g^+$	Absorption	1480–1460	R	1460.7(0, 0)		(66.24)
VI	$K^1\Pi_u \leftarrow X^1\Sigma_g^+$	Absorption	1400–1320	R	1384.0(0, 0)		(66.24)
VII	$M^1\Sigma_u^+ \leftarrow X^1\Sigma_g^+$	Absorption	1350–1300	R	1355.1(0, 0)		(66.24)
VIII	$N^1\Sigma_u^+ \leftarrow X^1\Sigma_g^+$	Absorption	1310–1290	R	1294.5(0, 0)		(66.24)

BAND SYSTEMS

	System	Transition	Sources	Wavelength Limits	Degrading	Characteristic Bands, λ	Remarks	Bibliography
	IX	$Q^1\Pi_u \leftarrow X^1\Sigma_g^+$	Absorption	~ 1250	R	1253.5(0, 0)		(66.24)
	X	$S^1\Sigma_u^+ \leftarrow X^1\Sigma_g^+$	Absorption	~ 1227	R	1227.6		(66.24)
	XI	$b^3\Sigma_u^- \rightarrow X^1\Sigma_g^+$	Emission	4400-3500	R	3720.1(0, 2) 3828.8(0, 3)		(74.44, 67.29)
	XII	$B^1\Pi_u \rightarrow A^1\Pi_g$	Emission	6674-6270		6414.0(0, 0)		(72.42, 71.40)
	XIII	$c(^3\Pi_u) \rightarrow b(^3\Pi_g)$	Emission	10050-7700	V	8622.0(4, 2) 8738.9(4, 2) 8829.2(4, 2)		(68.32, 67.29, 67.27, 64.23)

I. $\underline{A\,^1\Pi_g \rightarrow X\,^1\Sigma_g^+ \text{ System}}$

Band heads, λ (58.21):

(v', v'')	(0, 3)	(0, 2)	(0, 1)	(0, 0)	(1, 0)
λ	3112.43	3039.29	2969.84	2902.99	2852.23

II. $\underline{C\,^1\Sigma_u^+ \rightleftharpoons X\,^1\Sigma_g^+ \text{ System}}$

Band heads, λ

v',v''	0	1	2	3	4	5	6	7
0	2136.58				2286.36	2326.5	2367.6	2409.9
1	2115.23	2150.0	2186.4		2261.6	2301.0		
2	2094.38	2128.6		2164.3			2315.97	2356.3
3	2074.66	2108.1			2143.0	2253.24	2291.8	
4	2055.32	2088.3		2157.35		2122.6	2267.86	
5	2036.55	2069.0			2172.2		2245.4	2283.2
6	2018.08	2050.0					2223.0	
7	2000.26					2165.9		
8	1983.52					2145.31	2180.43	2216.1
9	1966.61		2027.52		2092.21		2159.89	2195.01
10	1950.15		2009.80			2073.56		

III. $\underline{E\,^1\Pi_u \rightleftharpoons X\,^1\Sigma_g^+ \text{ System}}$

Band heads, λ (66.24, 55.19):

v', v''	0	1	2	3	4
0	1683.22	1705.47	1728.14	1751.23	
1	1663.76		1709.6	1732.24	1755.10
2	1644.92				
3	1626.65				
4	1608.89				

IV. $\underline{\text{G}^1\Sigma_u^+ \rightleftharpoons \text{X}^1\Sigma_g^+ \text{ System}}$

Band heads, λ (66.24, 55.19):

v', v''	0	1	2	3	4	5
0	1508.68	1526.50				
1	1493.30	1510.75	1528.45			
2	1478.39	1495.51	1512.85	1530.54		
3		1480.74	1497.77	1515.07		
4				1500.12	1517.33	
5					1502.53	1519.67
6						

V. $\underline{\text{I}^1\Pi_u \leftarrow \text{X}^1\Sigma_g^+ \text{ System}}$

Band heads, λ (66.24):

(v', v'')	(0, 1)	(0, 0)
λ	1477.42	1460.69

VI. $\underline{\text{K}^1\Pi_u \leftarrow \text{X}^1\Sigma_g^+ \text{ System}}$

Band heads, λ (66.24):

(v', v'')	(0, 1)	(0, 0)	(1, 0)	(2, 0)	(3, 0)	(4, 0)
λ	1398.98	1383.98	1370.67	1357.81	1345.17	1333.16

VII. $\underline{\text{M}^1\Sigma_u^+ \leftarrow \text{X}^1\Sigma_g^+ \text{ System}}$

Band heads, λ (66.24):

(v', v'')	(0, 0)	(1, 0)	(2, 0)	(3, 0)	(4, 0)
λ	1355.06	1342.82	1330.92	1319.33	1308.04

VIII. $\underline{\text{N}^1\Sigma_u^+ \leftarrow \text{X}^1\Sigma_g^+ \text{ System}}$

Band heads, λ (66.24):

(v, v'')	(1, 2)	(0, 1)	(2, 2)	(1, 1)	(0, 0)	(1, 0)	(2, 1)
λ	1309.82	1307.54	1299.96	1296.78	1294.47	1283.88	1287.09

IX. $Q\,^1\Pi_u \leftarrow X\,^1\Sigma_g^+$ System

Band heads, λ (66.24):

(v', v'')	(1, 1)	(0, 0)
λ	1255.94	1253.45

XI. $b'\,^3\Sigma_u^- \to X\,^1\Sigma_g^+$ System

Band heads, λ (74.44, 67.29):

v', v''	0	1	2	3	4
0		3617.9	3721.5	3830.4	3944.9
1		3541.0	3640.2		
2					3767.2

XII. $B\,^1\Pi_g \to A\,^1\Pi_u$ System

Band heads, λ (73.42):

(v', v'')	(0, 2)	(1, 2)	(0, 0)	(1, 0)
λ	6674.0	6517.8	6414.0	6269.7

XIII. $c\left(^3\Pi_u\right) \to b\left(^3\Pi_g\right)$ System

Band heads, λ (67.29, 67.27, 64.23):

v', v''	λ			v', v''	λ		
0, 0	10047.5	9934.7	9784.9	2, 0	8924.6	8829.2	8716.8
1, 0	9449.2	9345.8	9218.3	3, 1	8875.4	8786.5	8673.6
2, 1	9389.7	9289.5	9159.1	4, 2	8829.2	8738.9	8622.0
3, 2	9325.8	9218.3	9105.0	5, 3	8786.5	8693.3	8585.7
4, 3	9269.6	9159.2	9047.3	6, 4	8738.9	8648.1	8537.4

SPECTROSCOPIC CONSTANTS

State	T_e	ω_e	$x_e\omega_e$	B_e	$\alpha_e \times 10^3$	$D_e \times 10^6$	r_e	Remarks	Bibliography
S$^1\Sigma_u^+$	81843.6[a]	-	-	0.2783[d]	-	-	1.978[e]		(66.24)
Q$^1\Pi_u$	80169.2[a]	618[b]	-	-	-	-	-		(66.24)
N$^1\Sigma_u^+$	77286.8	701.2	(29.70)	0.29845	5.11	3.1	1.910		(66.24)
M$^1\Sigma_u^+$	73845.7	678.5	3.0	0.2786	1.6	-	1.977		(66.24)
K$^1\Pi_u$	72288.5	713	5.5	0.2704[d]	-	-	2.006[e]		(66.24)
I$^1\Pi_u$	68849.4	-	-	0.2541[d]	-	2.5	2.070[e]		(66.24)
G$^1\Sigma_u^+$	66313.43	694.12	4.18	0.2973	1.95	2.25	1.913		(66.24, 55.19)
E$^1\Pi_u$	59446.28	700.66	2.92	0.2807[d]	-	1.84	1.969[e]		(66.24, 55.19)
B$^1\Pi_g$	50223.30	391.3	16.2	0.2300	6.0	3.3	2.176		(73.42)
C$^1\Sigma_u^+$	46941.33	473.93	2.340	0.24211	1.75	2.57	2.1204	(c)	(66.24)
A$^1\Pi_g$	34515.34	618.78	2.92	0.2752	1.70	2.2	1.9889		(73.42, 58.21)

SPECTROSCOPIC CONSTANTS

State	T_e	ω_e	$x_e\omega_e$	B_e	$\alpha_e \times 10^3$	$D_e \times 10^6$	r_e	Remarks	Bibliography
c($^3\Pi_u$)	$10180 + x_1$ $10038 + x_2$ $9915 + x_3$	$640^{(f)}$	4.0						(67.29)
b'$^3\Sigma_u^-$	28507.74	604.48	2.2	0.2583	1.4	1.6			(74.44, 73.43, 67.29)
b($^3\Pi_g$)	x_1, x_2, x_3	562	3.6						(67.29)
X$^1\Sigma_g^+$	0	780.89	2.820	0.30356	1.43	1.88	1.8937	(g)	(73.43, 67.31, 66.24)

(a) $T_e + G'(0)$; (b) $\Delta G_{1/2}$; (c) $y_e\omega_e = 0.0066$ cm^{-1}; (d) B_o; (e) r_o; (f) v uncertain;

(g) $y_e\omega_e = -0.005511$ cm^{-1}

Dissociation energy = 5.04 ± 0.11 eV, 147.5 kcal/mole, 40651 cm^{-1} (68.34).

Perturbations and General Information

Many of the vibrational levels of the C$^1\Sigma_u^+$ state are strongly perturbed
  (50.18, 50.17, 32.4).

Many of the levels of the E$^1\Pi_u$ state are perturbed (66.24).

Predissociation of the C$^1\Sigma_u^+$ state, by a $^3\Sigma_u^+$ state, is observed at v = 10,
  J = 58 and v = 11, J = 34. A second predissociation is observed at v = 19
  (66.24).

A region of diffuse absorption at 1425Å probably belongs to the I - X system.

Levels of the K$^1\Pi_u$ state are diffuse (maximum at v = 3, 4), probably due to
  predissociation.

Potential energy curves – RKRV potentials (70.36):

State	v	U+T$_e$(cm^{-1})	r$_{min}$(Å)	r$_{max}$(Å)
E$^1\Pi_u$	0	59795.9	1.914	2.025
	1	60490.9	1.879	2.073
	2	61179.3	1.854	2.106
	3	61862.2	1.836	2.135
	4	62540.8	1.821	2.160
G$^1\Sigma_u^+$	0	66659.4	1.860	1.972
	1	67341.8	1.825	2.020
	2	68016.9	1.800	2.054
	3	68683.4	1.782	2.084
	4	69341.3	1.767	2.111
	5	69990.7	1.754	2.136
	6	70631.2	1.742	2.160
K$^1\Pi_u$	0	72643.6	1.966	2.076
	1	73345.3	1.932	2.125
	2	74036.0	1.911	2.161
	3	74728.1	1.895	2.193
	4	75398.1	1.864	2.194
	5	76078.2	1.857	2.232
M$^1\Sigma_u^+$	0	74184.2	1.922	2.035
	1	74856.4	1.886	2.083
	2	75523.5	1.862	2.118
	3	76182.5	1.842	2.146
	4	76836.6	1.828	2.173

State	v	U+T$_e$(cm^{-1})	r$_{min}$(Å)	r$_{max}$(Å)
N$^1\Sigma_u^+$	0	77286.8	1.858	1.972
	1	78264.0	1.828	2.032
	2	78844.0	1.805	2.079

BIBLIOGRAPHY

(07. 1)   Emission,
          P. Geuter,
          Z. Wiss. Phot. 5, 33-60

(30. 2)   Predissociation,
          G. Herzberg,
          Nature 126, 239-40

(31. 3)   Absorption and Fluorescence,
          A. Jakovleva,
          Z. Physik 69, 548-63

(32. 4)   Emission,
          G. Herzberg,
          Ann. Phys. 15, 677-706

(32. 5)   F. A. Jenkins and M. F. Ashley,
          Nature 129, 829-30

(32. 6)   G. Herzberg,
          Phys. Rev. 40, 313-4

(32. 7)   F. A. Jenkins and M. F. Ashley,
          Phys. Rev. 39, 552

(33. 8)   Emission
          M. F. Ashley,
          Phys. Rev. 44, 919-26

(35. 9)   Nuclear Spin,
          F. A. Jenkins,
          Phys. Rev. 47, 783

(40. 10) Absorption,
          G. Herzberg, L. Herzberg, and G. G. Milne,
          Can. J. Res. 18, 139-43

(43. 11) Emission,
          K. N. Rao,
          Indian J. Phys. 17, 135-40

(43. 12) Nuclear Spin,
          K. N. Rao,
          Indian J. Phys. 17, 149-52

(44. 13)   R. F. Barrow,
           Proc. Phys. Soc. 56, 211-2

(46. 14)   Emission,
           E. J. Marais,
           Phys. Rev. 70, 499-510

(48. 15)   Emission,
           K. Sreeramamurty,
           Current Sci. 17, 119-20

(49. 16)   L. Gerö and C. Fonô,
           J. Chem. Phys. 17, 345-6

(50. 17)   Emission,
           E. J. Marais and H. Verleger,
           Phys. Rev. 80, 429-31

(50. 18)   Perturbations,
           S. M. Naude and H. Verleger,
           Phys. Rev. 80, 432-5

(55. 19)   Emission,
           K. Dressler,
           Helv. Phys. Acta 28, 563-90

(55. 20)   K. Dressler and E. Miescher,
           Proc. Phys. Soc. 68, 542-4

(58. 21)   Emission,
           A. E. Douglas and K. S. Rao,
           Can. J. Phys. 36, 565-70

(61. 22)   Flash Photolysis,
           R. W. G. Norrish and G. A. Oldershaw,
           Proc. Roy. Soc. 262, 1-9

(64. 23)   Emission,
           H. Guenebaut, B. Pascat, and J. Brion,
           C. R. Acad. Sci. 259, 3545-8

(66. 24)   Absorption,
           F. Creutzberg,
           Can. J. Phys. 44, 1583-92

(66. 25)   Dissociation Energy,
           K. A. Gingerich,
           J. Chem. Phys. 44, 1717-18

(66. 26)   R. B. Singh and D. K. Rai,
           "Potential Curves for Some Diatomic Molecules: P$_2$, PN, SiN, NBr,
           BaO, BeF, SiF and SnF,"
           Indian J. Pure Appl. Phys. 4, 102-5

(67. 27)   Emission,
           J. Brion, J. Malicet, and H. Guenebaut,
           C. R. Acad. Sci. 264, 622-5

(67. 28)   Theory,
           D. B. Boyd and W. N. Lipscomb,
           J. Chem. Phys. 46, 910-9

(67. 29)   Emission,
           A. Weber, S. P. S. Porto, L. E. Cheesman, and J. J. Barrett,
           J. Opt. Soc. Am. 57, 19-28

(67. 30)   Flash Photolysis,
           N. Basco and K. K. Yee,
           Nature 216, 998-9

(67. 31)   Emission,
           M. N. Dixit,
           Proc. Indian Acad. Sci. A 66, 325-41

(68. 32)   Emission,
           J. Brion, J. Malicet, and H. Guenebaut,
           C. R. Acad. Sci. 266, 82-3

(68. 33)   F. Jenč,
           "Ground State Reduced Potential Curves (RPC) of BeF, CS, SiN, P$_2$,
           SiS and GeO,"
           Spectrochimica Acta 24A, 259-64

(68. 34)   K. A. Gingerich,
           "Gaseous Phosphorus Compounds. III. Mass Spectrometric Study
           of the Reaction Between Diatomic Nitrogen and Phosphorus Vapor
           and Dissociation Energy of Phosphorus Mononitride and Diatomic
           Phosphorus,"
           J. Phys. Chem. 73, 2734-41

(69. 35)   G. A. Ozin,
           "Gas-Phase Raman Spectroscopy of Phosphorus, Arsenic, and
           Saturated Sulphur Vapours,"
           J. Chem. Soc. D 1969, 1325-7

(70.36)   T. V. R. Rao and S. V. J. Lakshman,
"Potential Energy Curves, r-Centroids and Franck-Condon Factors
for the Bands of P$_2$ Molecule,"
Indian J. Pure Appl. Phys. **8**, 617-20

(70.37)   I. R. Beattie, G. A. Ozin, and R. O. Perry,
"The Gas-Phase Raman Spectra of P$_4$, P$_2$, As$_4$, and As$_2$. The
Resonance Fluorescence Spectrum of ^{80}Se$_2$. Resonance Fluorescence-
Raman Effects in the Gas-Phase Spectra of Sulphur and I$_2$. The Effect
of Pressure on the Depolarization Ratios for I$_2$,"
J. Chem. Soc. A **12**, 2071-4

(70.38)   R. D. Verma and H. P. Broida,
"Spectral Study of the Phosphorus Glow,"
Can. J. Phys. **48**, 2991-5

(71.39)   B. Rai, J. Singh, and D. K. Rai,
"Dissociation Energies of S$_2$, SO, Te$_2$, SeO and P$_2$,"
Israel J. Chem. **9**, 563-8

(71.40)   J. Brion, J. Malicet, and J. Mongin,
"Rotational Analysis of Two Bands of the New Visible Electronic
System of the P$_2$ Radical,"
C. R. Acad. Sci. B **272**, 127-30

(73.41)   J. Kordis and K. A. Gingerich,
"Mass Spectroscopic Investigation of the Equilibrium Dissociation of
Gaseous Sb$_2$, Sb$_3$, Sb$_4$, SbP, SbP$_3$ and P$_2$,"
J. Chem. Phys. **58**, 5141-9

(73.42)   J. Malicet, J. Brion, and H. Guenebaut,
"Analysis of Four Bands From a New $^1\Pi$ - $^1\Pi$ Transition of the P$_2$
Radical,"
C. R. Acad. Sci. C **276**, 991-4

(73.43)   J. Brion and J. Malicet,
"Rotational Analysis of the (1, 1) Band of the a $^3\Sigma_u^-$ - X$^1\Sigma_g^+$ Transition
in the P$_2$ Radical,"
C. R. Acad. Sci. C **278**, 223-6

(74.44)   J. Brion, J. Malicet, and H. Guenebaut,
"Emission Spectra of the P$_2$ Radical: Study of the b'$^3\Sigma_u^-$ - X$^1\Sigma_g^+$
Transition,"
Can. J. Phys. **52**, 2143-4

Pb$_2$

Methods of Production and Experimental Technique

Absorption.

Thermal emission.

Laser-induced fluorescence.

BAND SYSTEMS

	System	Transition	Sources	Wavelength Limits	Degrading	Characteristic Bands, λ	Remarks	Bibliography
	I	A $\rightleftharpoons$ X	Absorption, fluorescence	7000-6200	R			(72.9, 67.8)
	II	B $\rightleftharpoons$ X	Absorption, fluorescence	5270-4200	R			(72.9, 67.8, 35.4)
	III	C $\leftarrow$ X	Absorption	3000-2830	R			(n.p. 10, 67.8)
	IV	D $\leftarrow$ X	Absorption	2780-2620	R			(n.p. 10, 67.8)
	V	E $\leftarrow$ X	Absorption	2600-2460				(n.p. 10)
	VI	F $\leftarrow$ X	Absorption	2450-2300	R			(n.p. 10, 67.8)
	VII	G $\leftarrow$ X	Absorption	2167-2136				(n.p. 10)

II.   <u>B $\rightleftharpoons$ X System</u>

Band heads, λ (72.9):

(v', v'')	(3, 2)	(3, 1)	(3, 0)	(4, 1)	(4, 0)	(5, 0)
λ	5058.30	5030.56	5002.95	4991.79	4964.50	4927.56

III.   <u>C $\leftarrow$ X System</u>

Most intense band heads, λ (n.p. 10):

λ	3003.1	2942.3	2931.3	2920.4	2911.0	2901.0
Intensity	10	4	5	6	7	7

V.   <u>E $\leftarrow$ X System</u>

Most intense ultraviolet system, with several bands converging (n.p. 10).

VI.   <u>F $\leftarrow$ X System</u>

Most intense band heads, λ (Intensity) (n.p. 10):

λ	2435.7	2430.4	2417.4	2410.1	2403.4	2397.0	2390.7
Intensity	9	10	7	6	6	5	5

SPECTROSCOPIC CONSTANTS

State	T_e	ω_e	$x_e \omega_e$	B_e	$\alpha_e \times 10^3$	$D_e \times 10^6$	r_e	Remarks	Bibliography
B	19490.3	161.64	1.036					(a)	(72.9)
A	14465.5	162.4	0.4						(72.9)
X	0	119.1	0.35						(72.9)

(a) $y_e \omega_e = 0.0055$ cm^{-1}

Dissociation energy = 0.8 ± 0.2 eV, 18.5 kcal/mole, 6450 cm^{-1}.

BIBLIOGRAPHY

(32. 1)   Attributed to Pb$_2$, possibly PbS,
          M. Domaniewska-Kruger,
          Acta Phys. Polon. 1, 357-62

(33. 2)   Attributed to Pb$_2$, possibly PbS,
          W. Kloskowska,
          Acta Phys. Polon. 2, 239-44

(35. 3)   Ultraviolet, Low Resolution,
          N. V. Kremenevskii,
          C. R. Acad. Sci. (USSR) 3, 251-2

(35. 4)   Absorption and Emission,
          E. N. Shawhan,
          Phys. Rev. 48, 343-6

(39. 5)   Attributed to Pb$_2$, possibly N$_2$,
          L. Natanson,
          Acta Phys. Polon. 7, 275-8

(57. 6)   Dissociation Energy,
          J. Drowart and R. E. Honig,
          J. Phys. Chem. 61, 980-5

(62. 7)   Absorption,
          J. G. Kay, N. A. Kuebler, and L. S. Nelson,
          Nature 194, 671

(67. 8)   Absorption, Vibrational Analysis,
          S. Weniger,
          J. Phys. 28, 595-601

(72. 9)   S. E. Johnson, D. Cannell, J. Lunacek, and H. P. Broida,
          "New Molecular Constants for the Ground Electronic State of Pb$_2$, "
          J. Chem. Phys. 56, 5723-5

(n.p. 10) Ultraviolet Systems,
          B. Eisler and R. F. Barrow,
          Unpublished

Pd$_2$

Pd$_2$

Spectroscopic Constants

Dissociation energy = 1.13 ± 0.21 eV, 26 kcal/mole, 9114 cm^{-1} (69.3).

BIBLIOGRAPHY

(62. 1)   Dissociation Energy,
           M. Ackerman, F. E. Stafford, and G. Verhaegen,
           J. Chem. Phys. 36, 1560-2

(67. 2)   Observation,
           K. A. Gingerich,
           Naturwisschaften 54, 43

(69. 3)   S. Lin, B. Strauss, and A. Kant,
           "Dissociation Energy of Pd$_2$,"
           J. Chem. Phys. 51, 2282-3

Po_2

Po_2

Methods of Production and Experimental Technique

Emission from an electrodeless discharge.

Band Systems

Emission, degrading R, has been observed in the region 5130-3600Å.

SPECTROSCOPIC CONSTANTS

State	T_e	ω_e	$x_e\omega_e$	B_e	$\alpha_e \times 10^3$	$D_e \times 10^6$	r_e	Remarks	Bibliography
$\left(0_u^+\right)$	25149.3	108.532	0.4417						
$X\left(0_g^+\right)$	0	155.715	0.3353					(a)	

(a) $y_e\omega_e = -0.0003226$ cm^{-1}

Dissociation energy = 1.89 ± 0.1 eV, 43.5 kcal/mole, 15244 cm^{-1}.

BIBLIOGRAPHY

(57. 1)   Vibrational Analysis,
           G. W. Charles, D. J. Hunt, G. Pish, and D. L. Timma,
           J. Opt. Soc. Am. 47, 291-7

Pr$_2$

Spectroscopic Constants

Dissociation energy = 1.30 $\pm$ 0.30 eV, 30 kcal/mole, 10490 cm^{-1} (72.1).

BIBLIOGRAPHY

(72. 1)   A. Kant and S. Lin,
"Dissociation Energies of the Homonuclear Diatomic Molecules of
the Rare Earths,"
Monatshefte für Chemie 103, 757-63

Rb$_2$

Methods of Production and Experimental Technique

Absorption.

Emission from a discharge in Rb vapor, from a discharge in a heat pipe.

Laser-induced fluorescence.

BAND SYSTEMS

System	Transition	Sources	Wavelength Limits	Degrading	Characteristic Bands, λ	Remarks	Bibliography
I	$A\,^1\Sigma_u^+ \rightleftharpoons X\,^1\Sigma_g^+$	Absorption, discharge	11000-8400	R	Max. $\sim$ 10500		(71.20, 34.8)
II	$B\,^1\Pi_u \rightleftharpoons X\,^1\Sigma_g^+$	Absorption, discharge	7350-6400	R	6824.2(1,1) 6797.8(1,0)		(71.20, 36.10)
III	$C\,^1\Pi_u \rightleftharpoons X\,^1\Sigma_g^+$	Absorption, laser-induced fluorescence	5030-4690	R	4746.5(10,2)		(71.20, 37.11)
IV	$D \leftarrow X\,^1\Sigma_g^+$	Absorption	4550-4220	R	4326.8(10,1) 4288.2(14,0)		(37.11)
V	$? \rightarrow X\,^1\Sigma_g^+$	Laser-induced fluorescence	6100-5400			Quasi-continuum	(71.20)
VI	Bands associated with resonance lines (Van der Waals molecules)						(35.7, 32.6)

Rb$_2$

I. $\underline{A\,^1\Sigma_u^+ \rightleftharpoons X\,^1\Sigma_g^+\text{ System}}$

Bands are fragmentary, not analyzed (71.20, 34.8):

$\lambda\,|\,10500\,|\,9033\,|\,8989\,|\,8941\,|\,8897\,|\,8852\,|\,8807\,|\,8762$

II. $\underline{B\,^1\Pi_u \rightleftharpoons X\,^1\Sigma_g^+\text{ System}}$

Band heads of ^{85}Rb$_2$ of greatest intensity, λ (Intensity) (36.10):

(v', v'')	(1, 0)	(2, 0)	(3, 0)	(4, 0)	(6, 1)	(5, 0)
λ	6797.8	6775.7	6754.5	6734.0	6718.1	6713.2
(Intensity)	10	10	10	10	5	6

III. $\underline{C\,^1\Pi_u \rightleftharpoons X\,^1\Sigma_g^+\text{ System}}$

Band heads of greatest intensity, λ (Intensity) (71.20, 37.11):

(v', v'')	(2, 1)	(3, 0)	(4, 0)	(6, 1)	(9, 2)	(8, 1)	(10, 2)
λ	4797.1	4775.8	4767.7	4764.6	4754.1	4749.0	4746.5
(Intensity)	9	8	8	8	8	9	10

IV. $\underline{D \leftarrow X\,^1\Sigma_g^+\text{ System}}$

Band heads of greatest intensity, λ (Intensity) (37.11):

(v', v'')	(7, 2)	(8, 2)	(8, 1)	(9, 1)	(10, 1)	(11, 1)	(11, 0)
λ	4359.3	4351.9	4341.1	4333.8	4326.8	4319.7	4309.2
(Intensity)	8	8	9	9	10	9	9

Rb$_2$

SPECTROSCOPIC CONSTANTS

State	T_e	ω_e	$x_e\omega_e$	B_e	$\alpha_e \times 10^3$	$D_e \times 10^6$	r_e	Remarks	Bibliography
D	22777.5	40.42	0.745					(e)	(37.11)
C $^1\Pi_u$	20835.1	36.46	0.124						(37.11)
B $^1\Pi_u$	14662.1	48.05 (a)	0.191 (c)						(36.10)
A $^1\Sigma_u^+$	~11500	-	-						(34.8)
X $^1\Sigma_g^+$	0	57.31 (b)	0.105 (d)	~0.02					(71.20, 37.11, 36.10)

(a) $\omega_e = 47.78$ for ^{85}Rb^{87}Rb, (b) $\omega_e = 56.98$ for ^{85}Rb^{87}Rb, (c) $x_e\omega_e = 9.188$ for ^{85}Rb^{87}Rb,

(d) $x_e\omega_e = 0.103$ for ^{85}Rb^{87}Rb, (e) $y_e\omega_e = -0.00144$

Dissociation energy = 0.47 ± 0.05 eV, 10.8 kcal/mole, 3790 cm^{-1}.

Rb_2

Perturbations and General Information

Radiation in the region 6100-5400Å due to transfer from the C state into an unidentified state followed by transitions to high-lying and continuum levels of the ground state (71.20).

Predissociation of the C state caused by crossing of A state (71.20).

Radiative lifetimes (70.17):

$$B^1\Pi_u \quad - \quad \tau_r \sim 16 \text{ nsec}$$

$$C^1\Pi_u \quad - \quad \tau_r \sim 61 \text{ nsec}$$

Potential energy curves – empirical (71.20)

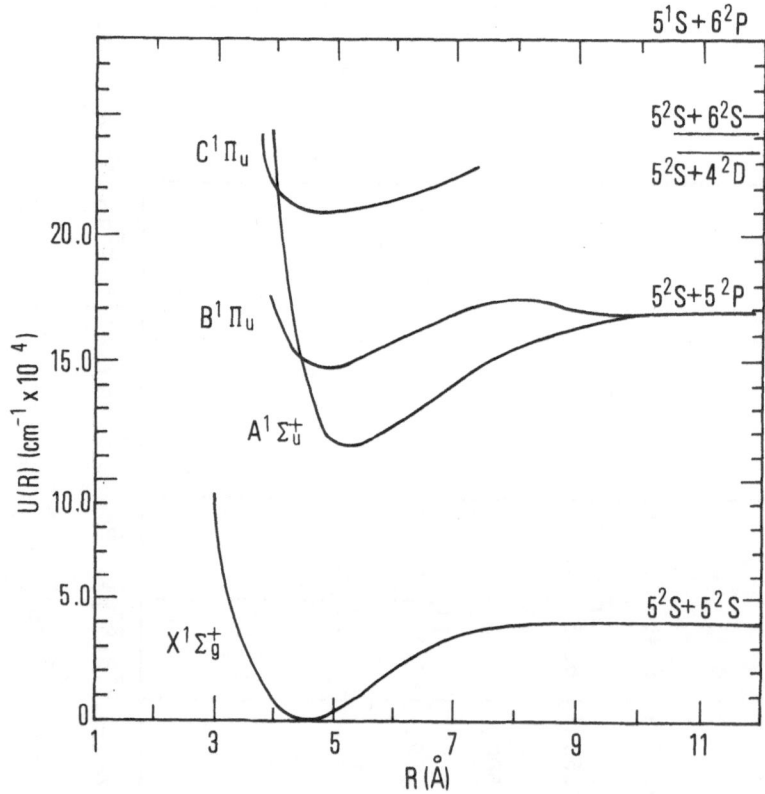

Average electric dipole polarizability (534°K) $68 \pm 7 \times 10^{-24}$ cm^3 (74.25).

BIBLIOGRAPHY

(10. 1)   Low Dispersion Investigation,
T. S. Carter,
Z. Physik 11, 632-3

(11. 2)   Low Dispersion Investigation,
P. V. Bevan,
Proc. Roy. Soc. A 85, 58-76

(12. 3)   Low Dispersion Investigation,
L. Dunager,
Radium 99, 218-23

(23. 4)   Low Dispersion Investigation,
J. C. MacLennan and D. S. Ainslie,
Proc. Roy. Soc. A 103, 304-14

(28. 5)   Low Dispersion Investigation,
J. M. Walter and S. Barratt,
Proc. Roy. Soc. A 119, 257-75

(32. 6)   Bands Associated With Resonance Lines,
S. Datta and B. Chakravarty,
Indian J. Phys. 7, 273-82

(34. 7)   E. Matuyama,
Nature 133, 567-8

(34. 8)   D ← X System, Incorrect Analysis,
E. Matuyama,
Sci. Rep. Tohoku Univ. 23, 296-307

(35. 9)   Bands Associated With Resonance Lines,
N. Tsi-Ze and C. Shin-Piaw,
J. Phys. Radium 6, 203-8

(36.10)   B ⇌ X System,
P. Kusch,
Phys. Rev. 49, 218-22

(37.11)   C, D ← X Systems,
H. Yoshimaga,
Proc. Phys. -Math. Soc. 19, 847-59

(37. 12)   Comparative Tables,
           C. H. D. Clark,
           Trans. Faraday Soc. 33, 1390-4

(37. 13)   Comparative Tables,
           C. H. D. Clark,
           Trans. Faraday Soc. 33, 1398-401

(37. 14)   Comparative Tables,
           C. H. D. Clark and C. W. Scaife,
           Trans. Faraday Soc. 33, 1394-8

(40. 15)   Comparative Tables,
           R. F. Barrow,
           Trans. Faraday Soc. 36, 624-5

(68. 16)   D. M. Creek and G. V. Marr,
           "Some Ultraviolet Cross-Section Measurements on Molecular
           Alkali-Metal Vapours,"
           J. Quant. Spectrosc. Radiative Trans. 8, 1431-6

(70. 17)   G. Baumgartner, W. Demtröder, and M. Stock,
           "Lifetime-Measurements of Alkali-Molecules Excited by Different
           Laserlines,"
           Z. Physik 232 462-72

(71. 18)   R. J. Gordon, Y. T. Lee, and D. R. Herschbach,
           "Supersonic Molecular Beams of Alkali Dimers,"
           J. Chem. Phys. 54, 2393-409

(71. 19)   Y. T. Lee, R. J. Gordon, and D. R. Herschbach,
           "Molecular Beam Kinetics: Reactions of H and D atoms With
           Diatomic Alkali Molecules,"
           J. Chem. Phys. 54, 2410-23

(71. 20)   P. P. Sorokin and J. R. Lankard,
           "Emission Spectra of Alkali-Metal Molecules Observed With a
           Heat-Pipe Discharge Tube,"
           J. Chem. Phys. 55, 3810-3

(73. 21)   N. N. Kostin, V. A. Khodovoy, V. V. Khromov, and N. A. Chigir,
           "Optical Pumping and Dissociation of the Rb$_2$ Molecule by a Pulse
           Laser,"
           Zh. Eksp. Teor. Fiz. 14, 589-92

(73. 22)   N. N. Kostin and V. A. Khodovoy,
           "Information on the Molecular Band Absorption of Rb$_2$ and Cs$_2$
           Using Pulse Laser Radiation,"
           Izvest. Akcad. Nauk 37, 2083-8

(74.23) R. C. Oldenborg, J. L. Gole, and R. N. Zare,
"Chemiluminescent Spectra of Alkali-Halogen Reactions,"
J. Chem. Phys. 60, 4032-42

(74.24) J. M. Brom, Jr. and H. P. Broida,
"Laser Photoluminescence and Photopredissociation of Rb$_2$,"
J. Chem. Phys. 61, 982-7

(74.25) R. W. Molof, T. M. Miller, H. L. Schwartz, B. Bederson, and
J. T. Park,
"Measurements of the Average Electric Dipole Polarizabilities of
the Alkali Dimers,"
J. Chem. Phys. 61, 1816-22

S_2

Methods of Production and Experimental Technique

Absorption: at elevated temperatures, in matrices, after flash photolysis.

Emission: high frequency discharge, microwave discharge, flames.

Fluorescence: excited by OH*, laser-induced.

BAND SYSTEMS

System	Transition	Sources	Wavelength Limits	Degrading	Band Head, $\nu_{0,0}$	Remarks	Bibliography
I	$b\,^1\Sigma_g^+ \rightarrow X\,^3\Sigma_g^-$	Photolysis	11055-10920				(72.110)
II	$B\,^3\Sigma_u^- \rightleftharpoons X\,^3\Sigma_g^-$	Absorption, discharge, fluorescence	7110-2400	R	31689	(a)	(72.104, 68.90, 63.73, 62.69, 60.67, 53.61, 48.54)
III	$C\,^3\Sigma_u^- \leftarrow X\,^3\Sigma_g^-$	Absorption	1870-1650	V	55633.3		(65.83, 48.56, 48.55, 34.26)
IV	$C'\,^3\Sigma_u^- \rightarrow X\,^3\Sigma_g^-$	Microwave	1860-1760	V	56983.6	(b)	(62.71)
V	$D\,^3\Pi_u \leftarrow X\,^3\Sigma_g^-$	Absorption	1750-1650	V	58750	(b)	(65.83, 48.55, 34.26)
VI	$B'\,^3\Pi_{g,i} \rightarrow A\,^3\Sigma_u^+$	Discharge, microwaves	8083-7434	V	13447.7	(c)	(66.86, 64.76, 62.69, 35.28)
VII	$B'\,^3\Pi_{g,i} \rightarrow A'\,^3\Delta_{u,i}$	Discharge, microwaves	7761-6984	V	$^3\Pi_1 - {}^3\Delta_2$ -14144.7 $^3\Pi_2 - {}^3\Delta_3$ -14318.0	(c)	(64.76, 62.69, 35.28)
VIII	$f\,^1\Delta_u \rightleftharpoons a\,^1\Delta_g$	Absorption, discharge	3350-2400	R	36743		(70.103, 69.100, 64.78, 64.77, 64.76, 63.75)

BAND SYSTEMS

System	Transition	Sources	Wavelength Limits	Degrading	Band Head, $v_{0,0}$	Remarks	Bibliography
IX	$g\,^1\Delta_u \rightarrow a\,^1\Delta_g$	Discharge, microwaves	2130-1880	V	52244.7		(68.93, 62.71)
X	$h\,^1\Sigma_u^+ \rightarrow b\,^1\Sigma_g^+$	Discharge, microwaves	2130-1760	V	51401.3	(b)	(68.93, 67.89, 65.83, 62.71)
XI	$i \rightarrow b\,^1\Sigma_g^+$	Discharge, microwaves	2130-1760	V	55448.3	(b)	(68.93, 65.83, 62.71)
XII	$e\,^1\Pi_g \rightarrow c\,^1\Sigma_u$	Discharge	7430-7152	V	13452		(62.69)
XIII	?	Microwaves	1850-1780	V	56077.7	(b)	(67.89)

(a) Numerous perturbations and predissociations. Several bands possess secondary heads.

(b) Analysis is uncertain.

(c) Predissociates.

I. $\underline{b^1\Sigma_g^+ \to X^3\Sigma_g^- \text{ System}}$

Observed in laser emission only (75.L117, 72.110).

$$\lambda \,|\, 11055 \,|\, 10975 \,|\, 10920$$

II. $\underline{B^3\Sigma_u^- \rightleftharpoons X^3\Sigma_g^- \text{ System}}$

Band heads of $^{32}S_2$, λ (Intensity) (36.30, 31.22, 31.21):

v', v''	0	1	2	3	4	5	6
0				3387.0(1)	3469.6(2)	3555.8(3)	3645.2(5)[a]
1			3259.9(2)	3336.7(2)	3417.0(4)	3500.5(5)	3587.4(5)
2		3143.7(1)	3216.1(2)	3290.7(3)	3369.6(4)	3451.0(2)	
3	3033.1(1)	3101.5(1)	3171.5(2)	3244.7(3)	3321.2(1)		
4	2997.0(1)	3063.6(3)	3132.4(3)	3203.2(2)			
5	2960.1(2)	3024.8(4)	3091.7(5)	3161.1(1)			
6	2926.6(2)	2989.7(4)[a]	3054.9(3)				

[a] Bands possessing weak secondary heads

Isotope studies of $^{34}S_2$ (70.105).

III. $\underline{C^3\Sigma_u^- \leftarrow X^3\Sigma_g^- \text{ System}}$

Each band possesses from 3 to 6 heads, with a maximum separation between extremes of 1 - 8Å. Isotope effect has been noted for several bands.

Most intense band heads, λ (Intensity) (65.83, 48.55):

v', v''	0	1	2	3	4	5
0	1796.93(9)	1820.46(4)	1844.43(3)	1868.82(1)	1894.50(1)	1919.81(1)
1	1770.75(9)		1816.88(1)	1840.51(1)	1864.65(1)	1889.93(1)
2	1745.57(8)	1768.99(2)				
3	1721.29(5)	1742.89(3)				
4	1697.97(4)	1718.90(2)				
5	1675.39(1)	1695.78(2)	1716.56(1)			
6	1653.60(1)	1673.52(1)	1693.72(1)			

S$_2$

IV.　$C'{}^3\Sigma_u^- \rightarrow X^3\Sigma_g^-$ System

Double-headed bands with separation of ~ 14 cm^{-1} are observed. Most intense band heads, λ (Intensity) (62.71):

(v', v'')	(0, 4)	(0, 3)	(0, 2)	(0, 1)	(0, 0)
λ	1859.49	1835.57	1811.94	1788.84	1766.11
(Intensity)	1	2	2	5	4

V.　$D^3\Pi_u \leftarrow X^3\Sigma_g^-$ System

Each band has 9 heads. Most intense band heads of the a_3, b_3, and c_3 series, λ (Intensity) (65.83, 48.55):

v', v''		0	1	2	3	4
0	a_3	1709.95(10)	1729.18(1)	1750.93(1)		
	b_3	1702.37(8)	1723.44(1)			
	c_3	1694.60(10)	1715.83(1)	1737.02(0)		
1	a_3	1685.32(4)	1705.99(3)	1726.99(0)		
	b_3	1679.88(4)	1700.49(3)			
	c_3	1672.34(6)	1692.75(4)	1714.44(0)		
2	a_3	1663.49(2)	1683.63(2)	1704.08(1)	1724.91(0)	
	b_3	1658.23(2)	1678.25(2)	1698.63(0)		
	c_3	1650.85(2)	1670.87(6)		1711.34(0)	
3	a_3		1662.08(1)	1681.99(1)		
	b_3		1656.85(0)	1676.65(1)		1717.19(0)
	c_3		1649.49(1)	1669.16(1)		1709.36(0)

VI. $\underline{B'^3\Pi_{g,i} \to A^3\Sigma_u^+ \text{ System}}$

Two subsystems – because the $^3\Pi_0$ state is completely predissociated. Only 5 of the 9 possible heads are observed (65.83). Isotope shifts (66.86).

Most intense band heads, λ (66.86, 64.76):

$\quad B'^3\Pi_2 \to A^3\Sigma_u^+; \; (v', v'') \; \lambda \,|\,(0, 1)\; 7785.\,6\,|\,(0, 0)\; 7506.\,8$

$\quad B'^3\Pi_1 \to A^3\Sigma_u^+; \; (v', v'') \; \lambda \,|\,(0, 1)\; 7707.\,4\,|$

VII. $\underline{B'^3\Pi_{g,i} \to A^3\Delta_{u,i} \text{ System}}$

Two subsystems – because the $^3\Pi_0$ state is completely predissociated.

λ (64.76, 62.69):

$B'^3\Pi_2 \to A'^3\Delta_{u,i}; \; (v', v'') \; \lambda \,|\qquad\qquad |\,(0, 2)\; 7583\,|\,(0, 1)\; 7328\,|\,(0, 0)\; 7068$

$B'^3\Pi_1 \to A'^3\Delta_{u,i}; \; (v', v'') \; \lambda \,|\,(0, 3)\; 7759\,|\,(0, 2)\; 7485\,|\,(0, 1)\; 7228\,|\,(0, 0)\; 6984$

VIII. $\underline{f^1\Delta_u \rightleftarrows a^1\Delta_g \text{ System}}$

Single-headed bands. Isotope studies (65.82, 65.80).

Most intense band heads, λ (70.103, 64.77)

v', v''	0	1	2	3	4	5	6
0					2940.49	2999.74	3060.77
1				2847.52	2903.53		
2			2760.14	2813.24			
3		2677.92	2728.33				
4		2648.34	2697.64				
5		2619.78	2668.02				
6	2546.28	2592.52					
7	2520.56	2565.59					
8	2495.77						
9	2471.77						
10	2448.98						

IX. $g\,^1\Delta_u \rightarrow a\,^1\Delta_g$ System

Single-headed bands. Most intense band heads, λ (Intensity) (68.90, 62.71):

v', v''	0	1	2	3	4	5	6	7
0	1914.06 (9)	1939.89 (9)	1966.08 (9)	1992.63 (7)	2019.68 (6)	2047.24 (4)	2075.24 (3)	2103.73 (2)
1	1884.80 (6)		1934.96 (0)	1960.85 (3)	1987.12 (3)	2013.79 (4)	2040.90 (3)	2068.49 (2)
2			1905.09 (2)			1981.57 (2)	2007.91 (2)	2034.59 (1)
3							1976.15 (0)	2002.01 (1)

X. $h\,^1\Sigma_u^+ \rightarrow b\,^1\Sigma_g^+$ System

Most intense band heads, λ (Intensity) (68.90, 67.89, 65.83):

v', v''	0	1	2	3	4	5	6	7
0	1943.25 (4)	1969.75 (5)	1996.80 (5)	2024.18 (5)	2052.04 (3)	2080.47 (2)		
1				1991.44 (9)	2018.27 (4)	2045.87 (4)	2073.77 (3)	
2			1934.20 (5)		1985.95 (1)	2012.44 (3)	2039.66 (3)	2067.19 (2)
3			1905.09 (2)	1929.42 (1)			2006.75 (1)	2033.33 (1)

XI. $i \rightarrow b\,^1\Sigma_g^+$ System

Only a single head is observed. Most intense band heads, λ (Intensity) (68.93, 65.83, 62.71):

(v', v'')	(0, 7)	(1, 8)	(0, 6)	(1, 7)	(0, 5)
λ	1984.52	1979.18	1959.15	1954.07	1934.20
(Intensity)	3	2	0	2	5

(v', v'')	(1, 6)	(0, 4)	(1, 5)	(0, 2)	(1, 1)
λ	1929.44	1909.57	1905.09	1861.73	1811.94
(Intensity)	1	1	2	0	2

SPECTROSCOPIC CONSTANTS

S_2

State	T_o (Observed)	T_o (Calculated)	ω_e	$x_e\omega_e$	B_e	$\alpha_e \times 10^3$	$D_e \times 10^8$	r_e	Remarks	Bibliography
$i?\,^1\Sigma^+_u,\,^1\Delta_u$	$\{55448+b$ $\{55448+a$	~64000 ? ~59900 ?	- -	- -	>0.29 -	- -	- -	<1.9 } -}		(68.93, 65.83, 62.71)
$h\,^1\Sigma^+_u$	51401.3+b	~59900?	819.6	2.70	>0.29	-	~14.52	<1.89	(a)	(65.83, 62.71)
$D\,^3\Pi_u$	58750	58750	793.9	4.0	0.3066	-	~16.293	(1.854)		(69.100, 65.83, 48.55)
$C'?\,^3\Sigma^-_u$	56983.6	56984	-	-	>0.295	-	-	<1.89		(65.83)
$g\,^1\Delta_u$	52244.7+a	~56700	816.4	2.7	0.3217	1.44	20.0	1.811		(68.83)
$C\,^3\Sigma^-_u$	55633.3	55633.3	829.15	3.34	0.32196	1.4	22.0	1.810	(b)	(69.99, 65.83, 48.55)
$f\,^1\Delta_u$	36743.5+a	~41200	438.32	2.70	0.22704	1.78	24.5	2.155	(d)	(70.103, 68.93, 65.83)
$e\,^1\Pi_g$	13451.8+a	~37000	533.7 (c)	-	~0.25	-	-	~2.08		(65.83)
$B'\,^3\Pi_{g,i}$	14144.7+A'	~36000	-	-	0.244	-	-	2.08	(e)	(65.83, 62.69)
$B\,^3\Sigma^-_u$	31689	31689	434	2.75	0.2244	1.8	23.1	2.168		(63.73)
$B''\,^3\Pi_u$	≤31700	≤31700	-	-	>0.2029	-	-	<2.280		(65.83, 63.73)
$A\,^3\Sigma^+_u$	697+A'	~22550	477 (c)	-	-	-	-	-		(65.83, 62.69)
$c\,^1\Sigma^-_u$	c	~23550	533.6 (c)	-	~0.235	-	-	2.122		(62.69)
$A'\,^3\Delta_{u,i}$	A'	~21855	488.6	2.63	0.2284	1.40	19.96	2.148	(f)	(62.69)

501

S_2

SPECTROSCOPIC CONSTANTS

State	T_o (Observed)	T_o (Calculated)	ω_e	$x_e\omega_e$	B_e	$\alpha_e \times 10^3$	$D_e \times 10^8$	r_e	Remarks	Bibliography
$b\,^1\Sigma_g^+$	b	≈ 8500	700.8?	3.4?	-	-	-	-		(65.83)
$a\,^1\Delta_g$	a	≈ 4500	702.35	3.09	0.29262	1.73	20.4	1.8987		(70.103, 68.93)
$X\,^3\Sigma_g^-$ $\begin{cases}^{32}S_2 \\ ^{34}S_2\end{cases}$	0 0	0 0	725.668 704.026	2.844 2.677	0.29541 0.27813	1.58 1.45	21.48 19.59	1.889 1.889	(g) (h)	(n.p. 115) (n.p. 115)

(a) $^3\Pi_2 - {}^3\Pi_1 \approx 462$ cm^{-1}; (b) $\lambda_o = -11.61$ cm^{-1}, $\gamma_o = 0.033$ cm^{-1}; (c) $\Delta G_{1/2}$; (d) $y_e\omega_e = -0.005$ cm^{-1};

(e) $^3\Pi_1 - {}^3\Pi_2 \approx 130$ cm^{-1}; (f) $^3\Delta_2 - {}^3\Delta_1 \approx 303.5$ cm^{-1}; (g) $\lambda_e = 11.82$ cm^{-1}, $\gamma_e = -0.0066$ cm^{-1}; (h) $\lambda_e = 11.73$ cm^{-1},

$\gamma_e = -0.0062$ cm^{-1}

Dissociation energy = 4.4 ± 0.1 eV, 101.5 kcal/mole, 35300 cm^{-1} (71.107).

Perturbations and General Information

Perturbations by a B''$^3\Pi_u$ state are observed for all vibrational levels.   There are three perturbations within each branch.

In emission, the predissociation of the v'' = 0 series stops with the (9, 0) band at 2828Å (31.21).

Higher rotational levels of v' = 17 of the B - X system and all rotational levels of v' ≥ 18 are extremely diffuse.

The B'$^3\Pi_g$ and e$^1\Pi_g$ states are predissociated at v' = 0 (B'$^3\Pi_2$ for J ≥ 34 and B'$^3\Pi_2$ for J ≥ 16) (65.80).

f$^1\Delta_u$ - a'Δ_g systems predissociates for v' ≥ 10 (65.80).

Radiative lifetimes (73.111):

	v'	τ(nsec)
B$^3\Sigma_u^-$ → X$^3\Sigma_g^-$	3	20.7
	4	18.3

Potential energy curves – RKR potential (73.112)

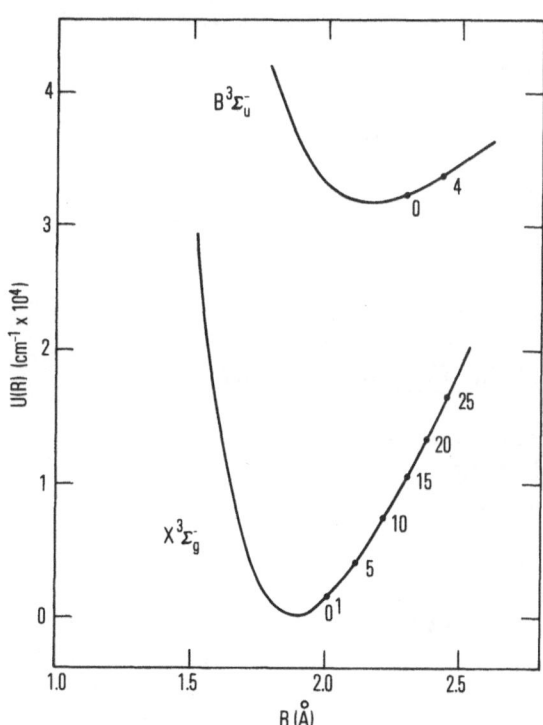

BIBLIOGRAPHY

(27. 1)   Fluorescence, Absorption,
           B. Rosen,
           Z. Physik 43, 69-130

(29. 2)   Resonance Spectra,
           P. Swings,
           Bull. Intern. Acad. Polon. Sci. A 10, 616-20

(29. 3)   Resonance Series,
           P. Swings,
           C. R. Acad. Sci. 189, 982-3

(29. 4)   A. M. Taylor,
           Trans. Faraday Soc. 25, 929-30

(30. 5)   Resonance Series,
           J. Genard,
           Bull. Cl. Sci. Acad. Roy. Belg. 16, 923-30

(30. 6)   Resonance Series,
           J. Genard,
           Bull. Cl. Sci. Acad. Roy. Belg. 16, 1369-77

(30. 7)   Intensity of Resonance Doublet,
           P. Swings,
           C. R. Acad. Sci. 190, 965-7

(30. 8)   Resonance Doublet,
           P. Swings,
           C. R. Acad. Sci. 190, 1010-1

(30. 9)   Resonance Spectra,
           P. Swings,
           C. R. Soc. Polon. Phys. 5, 29-51

(30. 10)   Emission, Predissociation,
           H. H. Van Iddekinge,
           Nature 125, 858

(30. 11)   Flame Spectra,
           V. Kondratjew,
           Z. Physik 63, 322-33

(30. 12) Resonance Spectra,
          P. Swings,
          Z. Physik 61, 681-99

(31. 13) Resonance Series,
          J. Genard,
          Bull. Cl. Sci. Acad. Roy. Belg. 17, 184-90

(31. 14) Resonance Series,
          J. Genard,
          Bull. Cl. Sci. Acad. Roy. Belg. 17, 387-99

(31. 15) Resonance Series,
          J. Genard,
          Bull. Cl. Sci. Acad. Roy. Belg. 17, 583-92

(31. 16) Resonance Series,
          P. Swings,
          Bull. Cl. Sci. Acad. Roy. Belg. 17, 420-4

(31. 17) Resonance Doublet Intensity,
          P. Swings,
          Bull. Cl. Sci. Acad. Roy. Belg. 17, 956-71

(31. 18) Resonance Series,
          P. Swings,
          Bull. Cl. Sci. Acad. Roy. Belg. 17, 972-4

(31. 19) Fine Structure of the Resonance Multiplets,
          P. Swings and A. Legros,
          Bull. Cl. Sci. Acad. Roy. Belg. 17, 808-11

(31. 20) Resonance Spectra,
          J. Fridrichson,
          C. R. Acad. Sci. 192, 737-9

(31. 21) Absorption,
          A. Christy and S. M. Naude,
          Phys. Rev. 37, 903-19

(31. 22) Flame Spectra,
          A. Fowler and W. M. Vaidya,
          Proc. Roy. Soc. A 132, 310-30

(31. 23) W. C. Curtis and S. Tolansky,
          Proc. Univ. Durham Phil. Soc. 8, 323-31

(31. 24) Resonance Spectra,
J. Fridrickson,
Z. Physik 70, 463-7

(31. 25) Predissociation,
V. Henri,
Leipziger Vorträge 130

(34. 26) K. Wieland, M. Wehn, and E. Miescher,
Helv. Phys. Acta 7, 843-9

(35. 27) Ultraviolet Bands,
B. Rosen and M. Desirant,
Bull. Cl. Sci. Acad. Roy. Belg. 21, 723-35

(35. 28) Absorption, Predissociation,
I. I. Agarbiceanu,
C. R. Acad. Sci. 200, 385-6

(35. 29) B - X System, Predissociation,
B. Rosen, M. Desirant, and J. Duchesne,
Phys. Rev. 48, 916

(36. 30) Infrared Bands,
B. Rosen and F. Bouffioux,
Bull. Cl. Sci. Acad. Roy. Belg. 22, 885-93

(36. 31) Dissociation Energy,
P. Goldfinger, W. Jeunehomme, and B. Rosen,
Nature 138, 205-6

(36. 32) Induced Predissociation,
V. Kondratjew and E. Olsson,
Z. Physik 99, 671-6

(36. 33) B - X System, Rotational Analysis, Dissociation Energy,
E. Olsson,
Z. Physik 100, 656-64

(37. 34) B - X System, Rotational Analysis,
E. Olsson,
Arkiv Mat. Astron. Fysik B 25, 1-5

(37. 35) Predissociation,
E. Olsson,
Naturwissenschaften 25, 781-2

(38. 36) B - X System,
E. Olsson,
Arkiv Mat. Astron. Fysik B <u>26</u>, No. 9

(38. 37) Absorption (4370-3600Å) at High Temperature,
B. Rosen and L. Neven,
J. Chim. Phys. <u>35</u>, 58-68

(38. 38) Predissociation,
M. Desirant and B. Rosen,
Physica <u>5</u>, 870-4

(38. 39) Flame Spectra,
E. Schröer,
Z. Phys. Chem. <u>40</u>, 450-4

(38. 40) Predissociation,
W. Lochte-Holtgreven,
Z. Physik <u>109</u>, 147-9

(38. 41) Predissociation,
E. Olsson,
Z. Physik <u>108</u>, 40-4

(39. 42) Absorption,
N. Morguleff,
C. R. Acad. Sci. <u>208</u>, 273-5

(40. 43) Fluorescence, Energy Transfer,
E. Durand,
J. Chem. Phys. <u>8</u>, 46-50

(40. 44) Absorption, Induced Predissociation,
G. Herzberg and L. G. Mundie,
J. Chem. Phys. <u>8</u>, 263-73

(40. 45) Pressure Effect on Predissociation,
B. I. Stepanow,
J. Phys. (USSR) <u>3</u>, 463-6

(40. 46) Flame Spectra,
D. S. Pavlov,
J. Phys. Chem. (USSR) <u>14</u>, 601-4

(40. 47) Transfer From OH*,
M. Miyanisi,
Sci. Papers Inst. Phys. Chem. Res. <u>37</u>, 79-84

(42. 48)   Potential Energy Curves,
           J. W. Linnett,
           Trans. Faraday Soc. 38, 1-9

(42. 49)   H. Zeise,
           Z. Phys. Chem. B 51, 120-5

(42. 50)   S. M. Naude,
           Nature 155, 426-7

(45. 51)   Predissociation,
           B. Rosen,
           Phys. Rev. 68, 124-6

(45. 52)   B - X System, Rotational Analysis,
           S. M. Naude,
           So. African J. Sci. 41, 128-51

(47. 53)   Flame Spectra,
           A. G. Gaydon and G. Whittingham,
           Proc. Roy. Soc. A 189, 313-25

(48. 54)   B - X System, Rotational Analysis,
           S. M. Naude,
           Ann. Phys. 3, 201-22

(48. 55)   Vacuum Ultraviolet Absorption,
           R. Maeder,
           Helv. Phys. Acta 21, 411-28

(48. 56)   Vacuum Ultraviolet Absorption,
           R. Maeder and E. Miescher,
           Nature 161, 393

(49. 57)   Paramagnetism,
           A. B. Scott,
           J. Am. Chem. Soc. 71, 3145-7

(49. 58)   Fluorescence,
           P. Pringham,
           Fluorescence and Phosphorescence
           Interscience Publishers, New York

(50. 59)   B - X System, Perturbations,
           S. M. Naude and H. Verleger,
           Z. Physik 128, 173-9

(52. 60) Dissociation Energy,
H. L. Friedman,
J. Chem. Phys. 20, 1046

(53. 61) B - X System, Rotational Analysis,
K. Ikenoue,
J. Phys. Soc. 8, 646-52

(56. 62) Ionization Potential,
P. Bradt, F. L. Mohler, and V. H. Dibeler,
J. Res. Nat. Bur. Stand. 57, 223-5

(59. 63) Dissociation Energy,
L. Brewer,
J. Chem. Phys. 31, 1143-4

(59. 64) Dissociation Energy,
W. A. Chupka, J. Berkowitz, and C. F. Giese,
J. Chem. Phys. 30, 827-34

(59. 65) Dissociation Energy,
D. G. H. Marsden,
J. Chem. Phys. 31, 1144-5

(60. 66) Dissociation Energy,
P. Colin, P. Goldfinger, and M. Jeunehomme,
Nature 187, 408-9

(60. 67) B - X System, Rotational Analysis,
K. Ikenoue,
Sci. Light 9, 79-97

(62. 68) Ionization Potential,
P. Hagemann,
C. R. Acad. Sci. 255, 1102-3

(62. 69) B' → A, B' - A', B - X, e - c Systems, Rotational Analysis,
J. E. Meakin and R. F. Barrow,
Can. J. Phys. 40, 377-9

(62. 70) Dissociation Energy,
R. Colin and J. Drowart,
J. Chem. Phys. 37, 1120-5

(62. 71) g - a; h, i → b; C' - X Systems, Vibrational Analysis,
Y. Tanaka and M. Ogawa,
J. Chem. Phys. 36, 726-30

S_2

(62. 72)   Dissociation Energy,
           T. M. Sugden and A. Demerdache,
           Nature 195, 596

(63. 73)   B - X System, Triplet Separation,
           R. F. Barrow and J. M. Ketteringham,
           Can. J. Phys. 41, 419-23

(63. 74)   Dissociation Energy,
           L. Herman and P. Felenbock,
           J. Quant. Spectrosc. Radiative Trans. 3, 247-54

(63. 75)   f - a System, vibrational Analysis,
           P. B. V. Haranath,
           Z. Physik 173, 428-31

(64. 76)   B' - A, B' - A', f - a Systems, Rotational Analysis,
           N. A. Narasimham,
           Current Sci. 33, 261-3

(64. 77)   f - a System, Rotational Analysis,
           N. A. Narasimham and J. K. Brody,
           Proc. Indian Acad. Sci. A 59, 345-54

(64. 78)   f - a System, Rotational Analysis,
           J. M. Ketteringham and R. F. Barrow,
           Proc. Phys. Soc. 83, 330-1

(64. 79)   Dissociation Energy,
           P. Colin, P. Goldfinger, and M. Jeunehomme,
           Trans. Faraday Soc. 60, 306-16

(65. 80)   f - a System, Rotational Analysis, Isotope Study,
           N. A. Narasimham and K. S. Gopal,
           Current Sci. 34, 454

(65. 81)   Matrix Isolation,
           L. Brewer, G. D. Brabson, and B. Meyer,
           J. Chem. Phys. 42, 1385-9

(65. 82)   f - a System, Rotational Analysis, Isotope Study,
           N. A. Narasimham and K. M. N. Bhagvat,
           Proc. Indian Acad. Sci. 61, 75-9

(65. 83)   Review of Electronic States,
           R. F. Barrow and R. P. DuParq,
           Elemental Sulfur, B. Meyer, Ed.
           Interscience Publ., New York, 251-63

(66. 84)   Dissociation Energy,
           J. Drowart and P. Goldfinger,
           Quant. Rev. 20, 545-57

(66. 85)   L. Brewer and G. D. Brabson,
           "Ultraviolet Fluorescent and Absorption Spectra of S_2 Isolated in
           Inert-Gas Matrices,"
           J. Chem. Phys. 44, 3274-8

(66. 86)   N. A. Narasimham and K. V. S. R. Apparao,
           "Isotope Shifts in the Near Infra-Red Bands of Diatomic Sulphur,"
           Nature 210, 1034-5

(67. 87)   N. Basco and A. E. Pearson,
           "Reactions of Sulphur Atoms in Presence of Carbon Disulphide,
           Carbonyl Sulphide and Nitric Oxide,"
           Trans. Faraday Soc. 63, 2684-94

(67. 88)   Matrix Isolation,
           B. Petropoulos, O. Dessaux, D. Chaffiol, and P. Goudmand,
           C. R. Acad. Sci. 265, 355-8

(67. 89)   System XIII,
           G. Lakshminarayana and N. A. Narasimham,
           Current Sci. 36, 533

(68. 90)   R. J. Donovan, D. Husain, and P. T. Jackson,
           "Transient Species in the Photolysis of Sulphur Monochloride,
           including $S_2(a^1\Delta_g)$,"
           Trans. Faraday Soc. 64, 1798-805

(68. 91)   Ionization Potential,
           J. Berkowitz and C. Lifshitz,
           J. Chem. Phys. 48, 4346-50

(68. 92)   Dissociation Energy,
           P. Budininkas, R. K. Edwards, and P. G. Wahlbeck,
           J. Chem. Phys. 48, 2859-66

(68. 93)   f, g - a Systems, Rotational Analysis,
           R. F. Barrow and R. P. DuParq,
           J. Phys. B 1, 283-8

(69. 94)   D. J. Meschi and A. W. Searcy,
           "Investigation of the Magnetic Moments of S_2, Se_2, Te_2, Se_6, and
           Se_5 by the Stern-Gerlach Magnetic Deflection Method,"
           J. Chem. Phys. 51, 5134-8

(69.95)   W. H. Smith,
          "Absolute Transition Probabilities for Some Electronic States of
          CS, SO and S_2,"
          J. Quant. Spectrosc. Radiative Trans. 9, 1191-9

(69.96)   M. Elbanowski,
          "Flash Photolysis of Sulphur Vapour,"
          Roc. Chemi. Ann. Soc. Chim. Polon. 43, 1883-91

(69.97)   B - X Systems, Dissociation Energy, Predissociation, Isotope Study,
          M. Ogawa and K. R. Yamawakj,
          Can. J. Phys. 47, 1805-11

(69.98)   Ionization Potential,
          J. Berkowitz and W. A. Chupka,
          J. Chem. Phys. 50, 4245-50

(69.99)   C - X System, Rotational Analysis,
          R. F. Barrow, R. P. DuParq, and J. M. Ricks,
          J. Phys. B 2, 413-8

(69.100) D - X System,
          J. M. Ricks and R. F. Barrow,
          J. Phys. B 2, 906-7

(70.101) Vacuum Ultraviolet Spectra,
          R. J. Donovan, D. Husain, and C. D. Stevenson,
          Trans. Faraday Soc. 66, 1-9

(70.102) J. F. Bott and T. A. Jacobs,
          "Shock-Tube Study of Radiation From S_2,"
          J. Chem. Phys. 52, 3545-50

(70.103) M. Carleer and R. Colin,
          "The $f^1\Delta_u - a^1\Delta_g$ Band System of S_2 in Absorption,"
          J. Phys. B 3, 1715-23

(70.104) A. G. Briggs, R. J. Kemp, L. Batt, and J. H. Holloway,
          "Flash Photolysis of Xenon Difluoride, Carbonyl Sulphide and
          Nitrogen Trifluoride,"
          Spectrochim. Acta 26A, 415-8

(70.105) A. K. Chaudhry, K. N. Upadhya, and K. P. R. Nair,
          "Isotope Shift in the Bands of $B^3\Sigma_u - X^3\Sigma_g$ System of S_2 Molecule,"
          Indian J. Pure Appl. Phys. 8, 52-53

(71.106) A. Tewarson and H. B. Palmer,
          "Origins of Chemiluminescent Emission in Low-Pressure Flames
          of Sulfur-Containing Compounds,"
          Proc. 13th Int. Symp. Combust. 99-107

(71. 107)   B. Rai, J. Singh, and D. K. Rai,
"Dissociation Energies of S_2, SO, Te_2, SeO and P_2 Molecules,"
Israel J. Chem. 9, 563-8

(71. 108)   S. Durmaz and J. N. Murrell,
"The Effect of Rotations on the Predissociation Probabilities of
Diatomic Molecular Spectra,"
Molec. Phys. 21, 209-16

(72. 109)   K. K. Yee, R. F. Barrow, and A. Rogstad,
"Resonance Fluorescence and Raman Spectra of Gaseous Sulphur,"
J. Chem. Soc. Faraday Trans. 68, 1808-11

(72. 110)   V. S. Zuev, S. B. Kormer, L. D. Mikheev, M. V. Sinitsyn,
I. I. Sobel'man, and G. I. Startsev,
"Onset of Inversion in the $^1\Sigma_g^+ \to {}^3\Sigma_g^-$ Transition of Molecular
Sulphur Following the Photodissociation of COS,"
JETP Letters 16, 157-8

(73. 111)   K. A. Meyer and D. R. Crosley,
"Hanle Effect Lifetime Measurements on Selectively Excited
Diatomic Sulfur,"
J. Chem. Phys. 59, 1933-41

(73. 112)   K. A. Meyer and D. R. Crosley,
"Franck-Condon Factors From Selectively Excited Resonance
Fluorescence in the B - X System,"
J. Chem. Phys. 59, 3153-61

(73. 113)   K. A. Meyer and D. R. Crosley,
"Rotational Satellite Intensities and Triplet Splitting in the $B\,^3\Sigma_u^-$
State of S_2,"
Can. J. Phys. 51, 2119-24

(74. 114)   F. D. Wayne, P. B. Davies, and B. A. Thrush,
"The Gas-Phase E. P. R. Spectrum of Diatomic Sulphur Molecules,"
Molec. Phys. 28, 989-96

(n.p. 115)   B - X System, Rotational Analysis, Isotope Study,
R. F. Barrow and J. M. Ricks,
Unpublished

(n.p. 116)   X State, Vibrational Constants,
G. Herzberg,
Unpublished

(75. L117)   V. S. Zuev, L. D. Mikheev, and V. I. Yalovoi
"A Photochemical Laser on the $S_2(^1\Sigma_g^+ - {}^3\Sigma_g^-)$ Electronic Transition,"
Soviet J. Quant. Electronics, 2, 799-806

Sb_2

Methods of Production and Experimental Technique

Absorption at elevated temperatures (800-1600° C).

Thermal emission and microwave discharge.

Fluorescence excited by Hg.

BAND SYSTEMS

System	Transition	Sources	Wavelength Limits	Degrading	Characteristic Bands, λ	Remarks	Bibliography
I	$A \rightleftharpoons X^1\Sigma_g^+$	Absorption	7500-6000	R			(49.6)
II	$B \rightleftharpoons X^1\Sigma_g^+$	Absorption	6000-4500	R			(72.9, 49.6)
III	$D \rightleftharpoons X^1\Sigma_g^+$	Absorption	3400-2830	R	3049.2(6, 2)		(67.8, 35.4)
IV	$F \leftarrow X^1\Sigma_g^+$	Absorption	2340-2150	R	2222.8(2, 1)		(35.4)
V	?	Microwaves	8400-7200	V	8315.5, 7788.1		(67.8)
VI	?	Microwaves	4200-3600	R			(67.8)
VII	?	Microwaves	3000-2900	R		Triplet structure	(67.8)
VIII	?	Absorption	< 2170	R	2138.6		(35.4)

II. $\underline{B \rightleftharpoons X^1\Sigma_g^+ \text{ System}}$

Band heads of ^{121}Sb$_2$, λ (72.9):

(v', v'')	(5, 0)	(4, 0)	(4, 1)	(3, 0)	(3, 1)
λ	5644.6	5562.0	5496.1	5481.4	5417.5

III. $\underline{D \rightleftharpoons X^1\Sigma_g^+ \text{ System}}$

Most intense bands, λ (Intensity):

(v', v'')	(3, 3)	(4, 3)	(7, 4)	(5, 2)	(8, 4)	(6, 2)
λ	3134.7	3114.5	3079.0	3068.9	3059.2	3049.2
(Intensity)	4	4	4	4	4	6

IV. $\underline{F \leftarrow X^1\Sigma_g^+ \text{ System}}$

Most intense band heads, λ (Intensity) (35.4):

(v', v'')	(0, 2)	(2, 3)	(0, 1)	(1, 1)	(2, 1)	(2, 0)
λ	2258.5	2249.7	2244.9	2233.4	2222.8	2209.4
(Intensity)	4	2	5	3	7	5

VIII. $\underline{\text{Band Groups at 2170A}}$

Most intense bands, λ (Intensity) (35.4):

λ	2138.6	2126.8	2115.0	2104.3
(Intensity)	3	2	2	2

Sb$_2$

SPECTROSCOPIC CONSTANTS

State	T_e	ω_e	$x_e\omega_e$	B_e	$\alpha_e \times 10^3$	$D_e \times 10^9$	r_e	Remarks	Bibliography
F	44780	226.0	1.17						(35.4)
D	31605	212	0.2						(35.4)
B	19068.9	218.08	0.537	0.044481 [a]		9.1 [b]			(72.9, 37.5)
A	14991.5	217.0	0.45						(37.5)
$X^1\Sigma_g^+$	0	269.98	0.588	0.050039 [a]		9.4 [b]			(72.9, 37.5)

(a) B_2, (b) D_2

Dissociation energy = 2.37 ± 0.10 eV, 54.7 kcal/mole, 19120 cm^{-1} (73.10).

Perturbations and General Information

D state is vibrationally perturbed (35.4).

D - X system displays predissociation with a peak at 2842Å.   Shorter wavelengths are very diffuse.

BIBLIOGRAPHY

(33. 1)   Fluorescence,
          R. Siksna,
          C. R. Acad. Sci. 196, 1986-7

(33. 2)   Fluorescence,
          J. Genard,
          Phys. Rev. 44, 468-9

(34. 3)   S. M. Naude,
          Phys. Rev. 45, 280

(35. 4)   Ultraviolet Absorption,
          G. Nakamura and T. Shidei,
          Jap. J. Phys. 10, 11-25

(37. 5)   Molecular Constants,
          G. M. Almy and H. A. Schutz,
          Phys. Rev. 51, 62

(49. 6)   L. Gerö and C. Fonô,
          J. Chem. Phys. 17, 345-6

(59. 7)   Dissociation Energy,
          Yu. Ya. Kuzyakov and V. M. Tatevskii,
          Opt. Spectrosc. 7, 467-71

(67. 8)   Microwave Discharge Spectra,
          S. Mrozowski and C. Santaram,
          J. Opt. Soc. Am. 57, 522-30

(72. 9)   J. Sfeila, P. Perdigon, F. Martin, and B. Femelat,
          "The B → X System of Diatomic Antimony,"
          J. Molec. Spectrosc. 42, 239-50

(73. 10) J. Kordis and K. A. Gingerich,
          "Mass Spectroscopic Investigation of the Equilibrium Dissociation of
          Gaseous Sb_2, Sb_3, Sb_4, SbP, SbP_3, and P_2,"
          J. Chem. Phys. 58, 5141-9

Sc$_2$

Spectroscopic Constants

Dissociation energy = 1.12 ± 0.2 eV, 25.9 kcal/mole, 9275 cm^{-1}.

BIBLIOGRAPHY

(69. 1)   K. A. Gingerich,
           "Gaseous Metal Borides. I. On the Dissociation Energy of the
           Molecules ThB, ThP, and Th$_2$, and Predicted Dissociation Energies
           of Selected Diatomic Transition-Metal Borides, "
           High Temp. Sci. 1, 258-67

$$Se_2$$

Methods of Production and Experimental Technique

Absorption at elevated temperatures.

Emission from a microwave discharge in Se vapor.

Laser-induced fluorescence.

BAND SYSTEMS

	System	Transition	Sources	Wavelength Limits	Degrading	Band Head, $v_{0,0}$	Remarks	Bibliography
	I	$B^3\Sigma_u^- \rightleftharpoons X^3\Sigma_g^-$ $\begin{pmatrix} 0_u^+ - 0_g^+ \\ 1_u - 1_g \end{pmatrix}$	Absorption, fluorescence	6700-3250	R			(72.21, 71.19, 66.11)
	II	$C^3\Sigma_u^- \leftarrow X^3\Sigma_g^-$ $\begin{pmatrix} 0_u^+ - 0_g^+ \\ 1_u - 1_g \\ 1_u - 0_g^+ \end{pmatrix}$	Absorption	1960-1868	V			(70.17)
	III	?	Absorption	1856-1843				(72.20)
	IV	$? \leftarrow X^3\Sigma_g^-$ $(1_u - 1_g)$	Absorption	1845-1820				(70.17)
	V	$? \leftarrow X^3\Sigma_g^-$ $\left(0_u^+ - 0_g^+\right)$	Absorption	1826-1812				(70.17)

BAND SYSTEMS

	System	Transition	Sources	Wavelength Limits	Degrading	Band Head, $\nu_{0,0}$	Remarks	Bibliography
	VI	$n \rightarrow a\,^1\Delta_g$ $(1_u \rightarrow 2_g)$	Flourescence					(72.20)

I. $\underline{B^3\Sigma_u^- \rightleftharpoons X^3\Sigma_g^- \left(0_u^+ - 0_g^+,\ 1_u - 1_g\right) \text{Systems}}$

Origins of bands with greatest intensity, λ (66.11):

(v', v'')	(12, 0)	(13, 0)	(14, 0)	(15, 0)	(16, 0)	(17, 0)	(18, 0)
$\lambda\left(^{80}\text{Se}_2\right)$	3483.4	3457.5	3432.1	3407.3	3383.3	3360.0	3337.3
$\lambda\left(^{78}\text{Se}_2\right)$	3479.8	3453.4	3427.8	3402.9	3378.6	3355.1	3332.4

II. $\underline{C^3\Sigma_u^- \leftarrow X^3\Sigma_g^- \text{ Systems}}$

a. $\underline{C\left(0_u^+\right) \leftarrow X\left(0_g^+\right)}$

Strong, diffuse bands with no rotational structure, λ (70.17):

(v', v'')	(0, 2)	(0, 1)	(0, 0)	(1, 0)
λ	1902.04	1888.43	1874.80	1860.36

b. $\underline{C(1_u) \leftarrow X(1_g)}$

Strong bands with sharp rotational structure, λ (70.17):

v', v''	0	1	2	3
0	1896.49	1910.43	1924.50	1938.7
1	1881.29			1922.87
2	1866.45	1879.96	1893.6	
3	1851.97	1865.25		

c. $\underline{C(1_u) \leftarrow X\left(0_g^+\right)}$

Weak bands with sharp structure, λ (70.17):

(v', v'')	(0, 1)	(0, 0)	(1, 0)
λ	1897.18	1883.38	1868.38

Se_2

III.   **? System**

Overlaps a continuum centered at ~ 1845A. Weak bands with sharp structure, λ (70.17):

(v', v'')	(0, 1)	(1, 2)	(0, 0)
λ	1856.53	1855.88	1843.35

IV. $\underline{? \leftarrow X^3\Sigma_g^- (1_u \leftarrow 1_g) \text{ System}}$

Strong bands, λ (70.17):

(v', v'')	(0, 2)	(1, 3)	(0, 1)	(1, 2)	(0, 0)
λ	1846.23	1844.61	1833.26	1831.69	1820.41

V. $\underline{? \leftarrow X^3\Sigma_g^- \left(0_u^+ \leftarrow 0_g^+\right) \text{ System}}$

Band heads, λ (70.17):

(v', v'')	(0, 0)	(1, 1)	(2, 2)	(3, 3)	(1, 0)	(1, 2)
λ	1826.09	1825.47	1824.85	1824.38	1812.81	1812.28

SPECTROSCOPIC CONSTANTS

State	T_e	ω_e	$x_e\omega_e$	B_e	$\alpha_e \times 10^4$	$D_e \times 10^8$	r_e	Remarks	Bibliography
1_u	55276.81	430							(70.17)
0_u^+	54752.48	403.9	1.3	0.0924[a]	3.3		2.133[b]		(70.17)
?	54239.41	404							(70.17)
$C(0_u^+)$	53339[c]								(70.17)
$C(1_u)$	52709.61	428.0	1.22	0.09647[a]	3.33		2.0893		(70.17)
$q(1_u)$[d]	26991	155	2	0.055[e]					(72.20)
$n(1_u)$	~25985.2	183	~0.75						(72.20)
$B(0_u^+)$	25980.36	246.291	1.016	0.07048	3.45	4[f]	2.4464		(66.11)
$B(1_u)$	25912.45	246.42	1.225	0.07086	5.53	2[f]	2.4398		(66.11)
$m(1_u)$	~24000	>154[g]	0.99						(72.20)
$a(2_g)$	~4000	319	0.81						(72.20)
$X(1_g)$	366.7	387.156	0.964	0.09016	2.98	2	2.1630		(71.19, 66.11)

SPECTROSCOPIC CONSTANTS

State	T_e	ω_e	$x_e\omega_e$	B_e	$\alpha_e \times 10^4$	$D_e \times 10^8$	r_e	Remarks	Bibliography
$X(0_g^+)$	0	385.302	0.96363	0.08992	2.88	2.4	2.1659		(71.19, 66.11)

(a) B_o, (b) r_o, (c) T_o, (d) analyzed through perturbation of the B state, (e) B_2, (f) D_o, (g) $\Delta G_{1/2}$

Dissociation energy = 3.164 ± 0.002 eV, 72.9 kcal/mole, 25518 cm^{-1} (72.20).

Perturbations and General Information

B(0_u^+) state is perturbed for all vibrational levels, $v \leq 15$ by m, n, and q
  states. Perturbations for levels of low v are weak (72.20, 63.9).

Both B(0_u^+) and B(1_u) states predissociate (63.9).

Ionization potential (I_p) = 8.88 $\pm$ 0.03 eV (69.15).

BIBLIOGRAPHY

(27. 1)   Absorption and Fluorescence,
           B. Rosen,
           Z. Physik 43, 69-130

(34. 2)   Isotope Study,
           E. Olsson,
           Z. Physik 90, 138-44

(35. 3)   B - X System, Vibrational Analysis,
           T. E. Nevin,
           Philos. Mag. 20, 347-54

(36. 4)   B. Rosen and F. Monfort,
           Bull. Cl. Sci. Acad. Roy. Belg. 22, 215-8

(37. 5)   B - X System, Vibrational Analysis,
           R. K. Asundi and Y. P. Parti,
           Proc. Ind. Acad. Sci. A 6, 207-28

(39. 6)   B - X System, Vibrational Analysis, Perturbations,
           B. Rosen,
           Physica 6, 205-18

(55. 7)   B - X System,
           V. Leelavathi and P. T. Rao,
           Indian J. Phys. 29, 1-10

(63. 8)   Dissociation Energy,
           D. Detry,
           Ind. Chim. Belge 28, 752-3

(63. 9)   Perturbations and Predissociations,
           L. Herman and R. Herman,
           Nature 199, 795

(66. 10)   Dissociation Energy,
           J. Berkowitz and W. A. Chupka,
           J. Chem. Phys. 45, 4289-302

(66. 11)   General Analysis,
           R. F. Barrow, G. G. Chandler, and C. B. Meyer,
           Philos. Trans. Roy. Soc. Lond. A 260, 395-456

(66. 12) J. Drowart and P. Goldfinger,
"The Dissociation Energies of the Group VIA Diatomic Molecules,"
Quant. Rev. 20, 545-7

(68. 13) Dissociation Energy,
P. Budininkas, R. K. Edwards, and P. G. Wahlbeck,
J. Chem. Phys. 48, 2867-9

(68. 14) Dissociation Energy,
R. Colin and J. Drowart,
Trans. Faraday Soc. 64, 2611-21

(69. 15) J. Berkowitz and W. A. Chupka,
"Photoionization of High-Temperature Vapors. VI. S$_2$, Se$_2$ and Te$_2$,"
J. Chem. Phys. 50, 4245-50

(69. 16) D. J. Meschi and A. W. Searcy,
"Investigation of the Magnetic Moments of S$_2$, Se$_2$, Te$_2$ and Se$_5$ by
the Stern-Gerlach Magnetic Deflection Method,"
J. Chem. Phys. 51, 5134-8

(70. 17) R. F. Barrow, W. G. Burton, and J. H. Callomon,
"Absorption Spectrum of Gaseous ^{80}Se$_2$ in the Region 51500-55000
cm^{-1},"
Trans. Faraday Soc. 66, 2685-93

(70. 18) I. R. Beattie, G. A. Ozin, and R. O. Perry,
"The Gas-Phase Raman Spectra of P$_4$, P$_2$, As$_4$ and As$_2$. The
Resonance Fluorescence Spectrum of ^{80}Se$_2$. Resonance Fluorescence-
Raman Effects in the Gas-Phase Spectra of Sulphur and I$_2$. The Effect
of Pressure on the Depolarization Ratios for I$_2$,"
J. Chem. Soc. A 12, 2071-4

(71. 19) R. F. Barrow, I. R. Beattie, W. G. Burton, and T. Gilson,
"Resonance Fluorescence Spectra of ^{80}Se$_2$,"
Trans. Faraday Soc. 67, 583-8

(72. 20) K. K. Yee and R. F. Barrow,
"Absorption and Fluorescence Spectra of Gaseous Se$_2$,"
J. Chem. Soc. Faraday Trans. 68, 1181-8

(72. 21) O. Atabek and R. Lefebvre,
"Evaluation of the Level Shifts Produced on the Discrete Levels of
the BO_u^+ State of the Se$_2$ Molecule by the Interaction With a Repulsive
State,"
Chem. Phys. Letters 17, 167-171

$$Si_2$$

Methods of Production and Experimental Technique

Absorption by flash-photolysis in $C_6H_5SiH_3$ or $BrSiH_3$.

Emission from discharge in SiH_4 and Xe.

BAND SYSTEMS

	System	Transition	Sources	Wavelength Limits	Degrading	Characteristic Bands, λ	Remarks	Bibliography
	I	$H^3\Sigma_u^- - X^3\Sigma_g^-$	Flash-photolysis	4526-3863	R	3979.6(4,1)		(71.6, 63.3, 55.2)
	II	$L^3\Pi_g - D^3\Pi_u$	Discharge and flash-photolysis	3695-3489	R	3568.7(0,1) 3496.0(1,1)		(71.6, 55.2)
	III	$K^3\Sigma_u^- - X^3\Sigma_g^-$	Flash-photolysis	3275-3067	R	3202.0(1,0)		(71.6, 63.3)
	IV	$D^3\Pi_u - X^3\Sigma_g^-$	Flash-photolysis	2900-2700		2882.84 2795.80		(70.5)
	V	$N^3\Sigma_u^- - X^3\Sigma_g^-$	Flash-photolysis	2166-2097	R	2138.35(0,0)		(70.4, 63.3)
	VI	$O^3\Sigma_u^- - X^3\Sigma_g^-$	Flash discharge	2200-1800		1874.28(0,0) 1892.21(0,1)		(70.4)
	VII	$P^3\Pi_g - D^3\Pi_u$	Flash discharge	1870-1900	R	1879.9(0,0) 1898.4(0,1)		(70.4)

I. $\underline{H\,{}^3\Sigma_u^- - X\,{}^3\Sigma_g^-\ \text{System}}$

Band heads, λ (63.3, 55.2):

v', v''	0	1	2	3	4	5	6
0					4427.6	4526.0	
1				4283.1	4375.8	4471.9	
2					4326.0		
3	3942.1						
4	3900.8	3979.6	4060.9				4414.4
5	3863.4						

II. $\underline{L\,{}^3\Pi_g - D\,{}^3\Pi_u\ \text{System}}$

Band heads, λ (71.6, 55.2):

v', v''	0	1	2	3	4
0		3568.7	3634.4	3710.4	3772.3
1		3496.0	3563.1	3632.2	

III. $\underline{K\,{}^3\Sigma_u^- - X\,{}^3\Sigma_g^-\ \text{System}}$

Band heads, λ (63.3):

(v', v'')	(0, 0)	(1, 0)	(2, 0)	(3, 0)	(4, 0)
λ	3248.9	3202.0	3157.8	3115.8	3076.1

IV. $\underline{D\,{}^3\Pi_u - X\,{}^3\Sigma_g^-\ \text{System}}$

Several lines have been observed in absorption but have not been identified (70.5):

$$\lambda\,|\,2882.8\,|\,2838.8\,|\,2795.8\,|\,2758.8$$

V. $N^3\Sigma_u^- - X^3\Sigma_g^-$ System

Band heads, λ (70.4):

v', v''	0	1
0	2138. 35	2161. 78
1	2117. 92	
2	2098. 53	
3	2079. 75	2101. 92
4		2083. 53

VI. $O^3\Sigma_g^- - X^3\Sigma_g^-$ System

Band heads, λ (70.4):

v', v''	0	1	2
0	1874. 28	1892. 21	
1	1860. 53		189. 32
2	1847. 22	1864. 63	

VII. $P^3\Pi_g - D^3\Pi_u$ System

Two red shaded bands have been observed overlapping the O - X
system.   They are tentatively assigned as follows:

1879. 0(0, 0)
1898. 4(0, 1)

SPECTROSCOPIC CONSTANTS

State	T_e	ω_e	$x_e\omega_e$	B_e	$\alpha_e \times 10^3$	$D_e \times 10^6$	r_e	Remarks	Bibliography
$P^3\Pi_g$	88219								(70.4)
$L^3\Pi_g$	63059.1	404.2		0.2370			2.255		(70.4, 55.2)
$O^3\Sigma_u^-$	53341.94		3.0	0.2225	3		2.327		
$N^3\Sigma_u^-$	46762.21	458.6	4.8	0.2193	2.5		2.344		(70.4, 63.3)
$D^3\Pi_u$	~35000	547.94	2.43	0.2596	1.55		2.155		(70.4, 55.2)
$K^3\Sigma_u^-$	30768.77	462.6	5.95	0.2185	3.16		2.349		(70.4, 63.4)
$H^3\Sigma_u^-$	24311.15(a)	275.30	1.99	0.1712			2.6536		(71.6, 70.4, 63.3, 55.2)
$X^3\Sigma_g^-$	0	510.98	2.02	0.2390	1.3		2.246		(70.4, 63.3)

(a)T_o

Dissociation energy = 3.35 ± 0.2 eV, 75 kcal/mole, 26168 cm^{-1}.

Si$_2$

Perturbations and General Information

The bands of the K - X and H - X systems exhibit the presence of perturbations. In the H - X system, the (4, 0) band is sharp, but the (5, 0) band is diffuse and does not appear in emission. All the bands of the K - X system are diffuse.

All the levels above $v' = 0$, $J' = 51$ of the L state are predissociated.

The position of the (2, 0) band in the N - X system is displaced somewhat to the red, indicating a perturbation (70. 4).

BIBLIOGRAPHY

(47. 1)   Preliminary Note,
           A. R. Downie and R. F. Barrow,
           Nature 160, 198

(55. 2)   H - X and L - D Systems,
           A. E. Douglas,
           Can. J. Phys. 33, 801-10

(63. 3)   H - X, K - X, and N - X Systems,
           R. D. Verma and P. A. Warsop,
           Can. J. Phys. 41, 152-60

(70. 4)   A. Lagerqvist and C. Malmberg,
           "New Absorption Systems of the Si$_2$ Molecule in the Vacuum Ultra-
           violet Region, "
           Physica Scripta. 2, 45-9

(70. 5)   D. E. Milligan and M. E. Jacox,
           "Infrared and Ultraviolet Spectra of the Products of the Vacuum-
           Ultraviolet Photolysis of Silane Isolated in an Argon Matrix, "
           J. Chem. Phys. 52, 2594-2608

(71. 6)   I. Dubois and H. Leclercq,
           "Absorption Spectrum of Si$_2$ in the Visible and Near-Ultraviolet
           Region, "
           Can. J. Phys. 49, 3053-4

Sm_2

Sm_2

Spectroscopic Constants

Dissociation energy = 0.52 ± 0.22 eV, 12 kcal/mole, 4200 cm^{-1} (72.1).

BIBLIOGRAPHY

(72. 1)   A. Kant and S. Lin,
           "Dissociation Energies of the Homonuclear Diatomic Rare Earth
           Molecules,"
           Monatshefte für Chemie 103, 757-63

Sn$_2$

Sn$_2$

Band Systems

Bands in the region 4780-4350Å have been attributed to Sn$_2$ but may possibly arise from SnCl$_2$ (62.2).

Spectroscopic Constants

Dissociation energy = 1.99 ±0.18 eV, 45.8 kcal/mole, 16000 cm^{-1} (62.1).

BIBLIOGRAPHY

(62. 1)   Dissociation Energy by Mass Spectra,
          M. Ackerman, J. Drowart, F. E. Stafford, and G. Verhaegen,
          J. Chem. Phys. <u>36</u>, 1557-60

(62. 2)   G. Pannetier and P. Deschamps,
          J. Chim. Phys. <u>59</u>, 517-20

Tb$_2$

Spectroscopic Constants

Dissociation energy = 1.34 ± 0.35 eV, 31 kcal/mole, 11000 cm^{-1} (72.1).

Tb$_2$

BIBLIOGRAPHY

(72. 1)   A. Kant and S. Lin,
           "Dissociation Energies of the Homonuclear Diatomic Rare Earth
           Molecules, "
           Monatshefte für Chemie 103, 757-63

Te$_2$

Methods of Production and Experimental Technique

Absorption.

Emission from microwave discharge.

Fluorescence, laser-induced fluorescence.

BAND SYSTEMS

	System	Transition	Sources	Wavelength Limits	Degrading	Band Head, $\nu_{0,0}$	Remarks	Bibliography
	I	A $0_u^+ \leftarrow$ X 0_g^+	Absorption	5190-4250	R			(69.45, 69.43)
	II	B $0_u^+ \rightleftarrows$ X 0_g^+	Absorption from discharge	6320-3836	R			(69.43, 69.41, 66.36, 42.31, 38.28, 35.16, 27.1)
	III	B $0_u^+ \rightarrow$ X 1_g	Laser fluorescence	5300-6050	R			(72.49)

Te_2

I. $\underline{A\ 0_u^+ \leftarrow X\ 0_g^+\ \text{System}}\ \left(^{130}Te_2\right)$

Band origins, λ (69.43):

v', v''	0	1	2	3	4
0					
...					
6					5190.0
7				5089.7	5153.4
8				5054.8	5117.7
9				5020.7	5082.6
10		4868.5		4987.2	5048.4
11		4837.4			5015.0
12		4806.9	4864.2		
13		4777.1	4833.7		
14			4803.8		
15	4665.2		4774.9		
16	4637.9				
17	4611.3		4664.2		
18	4585.1		4637.5		
19	4559.6		4611.4		
20			4585.8		

II. $\underline{B\ 0_u^+ \rightleftarrows X\ 0_g^+\ \text{System}}\ \left(^{130}Te_2\right)$

Band origins, λ (69.45, 69.41):

v', v''	0	1	2	3
0				
...				
5			4449.1	
6			4418.5	4466.6
7		4341.8	4388.5	4436.0
8		4313.2	4359.3	4406.2
9	4240.5	4285.2	4330.7	
10	4213.7	4257.8	4302.7	
11	4187.5	4231.1		
12	4162.0	4205.0		
13	4137.0	4179.6		
14	4112.6			
15	4088.8			
16	4065.7			
17	4043.1			
18	4021.2			
19	3999.8			
20	3979.1			

B 0_u^+ ⇌ X 0_g^+ System ^{128}Te$_2$

Band origins, λ (69.45, 69.41):

v', v''	0	1	2	...	30	31	32	33
0								
...								
5					6248.7			
6					6188.6	6271.3		
7			4388.8			6210.7	6294.3	
8		4312.9	4359.4				6233.8	6317.7
9		4284.7	4330.2					
10		4257.1						
11	4186.4	4230.3						
12	4160.7	4204.0						
13	4135.6	4178.6						
14	4110.9							
15	4087.2							
16	4064.0							
17	4041.3							
18	4019.2							
19	3997.8							
20	3977.0							

III. ### B 0_u^+ - X 1_g System

Band heads, λ (72.58):

	^{128}Te$_2$	^{130}Te$_2$
v'', v'	0	0
0		
...		
5	5350.0	
6	5421.1	
7	5493.6	5492.7
8	5567.9	5566.1
9	5643.8	5641.6
10	5721.6	5718.8
11	5800.9	5797.8
12	5882.3	5878.7
13	5965.6	5961.1
14	6050.8	6045.5
15	6138.2	
16	6227.7	

Te$_2$

SPECTROSCOPIC CONSTANTS

State	T_e	ω_e	$x_e\omega_e$	$B_e \times 10^2$	$\alpha_e \times 10^4$	$D_e \times 10^9$	r_e	Remarks	Bibliography
					^{130}Te$_2$				
B 0^+_u	22207.4	162.3	0.45	3.254	1.25		2.8244	$y_e\omega_e = -11.09 \times 10^{-3}$	(72.48, 69.45, 69.43)
A 0^+_u	19450.8	143.6	0.45	3.124	1.30		2.8824	$y_e\omega_e = -3.892 \times 10^{-3}$	(72.48, 69.43)
X 1_g	2234	250.00	0.547	3.968(a)	1.06(a)				(72.49, 69.43)
X 0^+_g	0	247.07	0.515	3.968	1.06	4.4	2.5774	$y_e\omega_e = -0.55 \times 10^3$	(72.48, 69.43)
					^{128}Te$_2$				
B 0^+_u	22285.6(b)			3.3121	1.41		2.82442		(72.48, 69.43)
A 0^+_u	19450			3.1740	1.32		2.88226		(72.48, 69.45, 69.43)
X 1_g	2228.5	251.26	0.536	4.0299(a)	1.03(a)				(72.48, 69.45, 69.43)
X 0^+_g	0			4.0299	1.03	4.1	2.55766		(72.48, 69.45, 69.43)

SPECTROSCOPIC CONSTANTS

State	T_e	ω_e	$x_e\omega_e$	$B_e \times 10^2$	$\alpha_e \times 10^4$	$D_e \times 10^9$	r_e	Remarks	Bibliography

(a) It is assumed that the rotational constants for this state are the same as those of the X 0_g^+ state (72.49).

(b) T_o

Dissociation energy = 2.5 ± 0.4 eV, 57.7 kcal/mole, 20200 cm^{-1} (71.47).

Te$_2$

Perturbations and General Information

RKR potential energy curve (n.p. 50) for ^{128}Te$_2$ X 0_g^+ state:

$T_e = 0$ cm^{-1}	v	$T_e + E(v)$ cm^{-1}	r_{min}(Å)	r_{max}(Å)
	0	124.35	2.51335	2.60548
	1	372.26	2.48249	2.64234
	2	619.12	2.46205	2.66878
	3	864.92	2.44591	2.69096
	4	1109.65	2.43229	2.71065
	5	1353.33	2.42037	2.72867
	6	1595.94	2.40971	2.74547
	7	1837.49	2.40000	2.76133
	8	2077.96	2.39108	2.77646
	9	2317.36	2.38286	2.79084
	10	2555.68	2.37513	2.80484
	11	2792.92	2.36782	2.81846
	12	3029.09	2.36095	2.83165
	13	3264.16	2.35441	2.84456
	14	3498.15	2.34819	2.85715
	15	3731.05	2.34227	2.86953

RKR potential energy curve (n.p. 50) for ^{128}Te$_2$ A 0_u^+ state:

$T_e = 19450$ cm^{-1}	v	$T_e + E(v)$ cm^{-1}	r_{min}(Å)	r_{max}(Å)
	0	72.24	2.82474	2.94564
	1	216.01	2.78550	2.99546
	2	358.83	2.75988	3.03171
	3	500.67	2.73987	3.06242
	4	641.50	2.72311	3.08994
	5	781.30	2.70854	3.11533
	6	920.05	2.69557	3.13918
	7	1057.72	2.68383	3.16188
	8	1194.28	2.67307	3.18367
	9	1329.73	2.66317	3.20455
	10	1464.02	2.65389	3.22501
	11	1597.14	2.64511	3.24506
	12	1729.07	2.63686	3.26464
	13	1859.78	2.62900	3.28393
	14	1989.24	2.62153	3.30290
	15	2117.43	2.61435	3.32167

RKR potential energy curve (n.p. 50) for ^{128}Te$_2$ B 0_u^+ state:

T$_e$ = 22285.6 cm^{-1}	v	T$_e$+E(v)cm^{-1}	r$_{min}$(Å)	r$_{max}$(Å)
	0	81.68	2.77021	2.88390
	1	244.31	2.73361	2.93102
	2	405.92	2.70967	2.96521
	3	566.44	2.69099	2.99420
	4	725.80	2.67543	3.02028
	5	883.93	2.66201	3.04448
	6	1040.77	2.65009	3.06729
	7	1196.24	2.63915	3.08889
	8	1350.28	2.62870	3.10929
	9	1502.76	2.61904	3.12893
	10	1653.68	2.60899	3.14728
	11	1803.11	2.60141	3.16717
	12	1950.68	2.59302	3.18565
	13	2096.49	2.58499	3.20395
	14	2240.49	2.57882	3.22345
	15	2382.58	2.57212	3.24213

Franck-Condon factors for ^{128}Te$_2$ $\left(A\ 0_u^+ - X\ 0_g^+ \right)$ (n.p. 50):

	12	13	14	15	16	17	18	19
0	4.985-2	6.959-2	8.846-2	1.032-1	1.110-1	1.102-1	1.014-1	8.679-2
1	7.975-2	7.178-3	5.109-2	2.559-2	5.839-3	2.330-4	1.094-2	3.319-2
2	3.035-2	7.032-3	4.068-4	1.415-2	3.767-2	5.356-2	5.064-2	3.155-2
3	4.838-4	1.598-2	3.871-2	4.614-2	3.100-2	8.307-3	2.854-4	1.440-2
4	2.969-2	4.213-2	2.891-2	6.186-3	1.449-3	1.988-2	3.861-2	3.526-2
5	3.601-2	1.414-2	2.368-6	1.316-2	3.354-2	3.128-2	9.981-3	2.531-4
6	7.044-3	1.933-3	2.217-2	3.304-2	1.643-2	2.123-4	1.101-2	3.076-2
7	3.998-3	2.552-2	2.833-2	7.349-3	1.782-3	2.138-2	2.942-2	1.135-2
8	2.536-2	2.501-2	3.738-3	5.140-3	2.536-2	2.256-2	2.672-3	5.871-3
9	2.428-2	3.104-3	6.396-3	2.556-2	1.740-2	2.713-4	1.184-2	2.638-2
10	4.517-3	5.263-3	2.444-2	1.540-2	2.557-6	1.458-2	2.409-2	6.246-3

Franck-Condon factor followed by factor of ten

Te$_2$

Franck-Condon factors for ^{128}Te$_2$ $\left(\text{B } 0_u^+ - \text{X } 0_g^+\right)$ (n.p. 50):

	9	10	11	12	13	14	15	16
0	8.506-2	1.101-1	1.267-1	1.305-1	1.208-1	1.001-1	7.702-2	5.374-2
1	8.231-2	5.196-2	1.833-2	5.264-4	8.427-3	3.676-2	7.006-2	9.305-2
2	7.989-3	1.866-3	2.622-2	5.616-2	6.270-2	4.059-2	1.115-2	1.710-4
3	1.719-2	4.769-2	5.101-2	2.326-2	6.824-4	1.165-2	4.208-2	5.623-2
4	4.911-2	3.233-2	3.279-3	7.950-3	3.754-2	4.508-2	1.981-2	1.478-4
5	2.209-2	1.318-5	1.961-2	4.190-2	2.498-2	8.072-4	1.316-2	4.003-2
6	1.247-4	2.365-2	3.789-2	1.228-2	1.716-3	2.766-2	3.619-2	1.062-2
7	2.161-2	3.507-2	8.070-3	5.026-3	3.186-2	2.546-2	1.000-3	1.360-2
8	3.429-2	8.942-3	5.096-3	3.117-2	1.933-2	4.950-5	2.135-2	3.063-2
9	1.396-2	2.415-3	2.848-2	1.842-2	2.691-4	2.322-2	2.494-2	1.082-3
10	3.614-5	2.312-2	2.132-2	3.220-6	2.129-2	2.269-2	2.589-4	1.739-2

Franck-Condon factor followed by factor of ten

Perturbations of the v = 0 level of the B 0_u^+ state have been observed.

Ionization cross sections = 17.46 $\pm$ 0.48 $\times$ 10^{-6} cm^2 (66.37).

BIBLIOGRAPHY

(27. 1)   B ⇄ X System in Absorption and Fluorescence,
          B. Rosen,
          Z. Physik 43, 69-130

(29. 2)   Resonance Series,
          W. Kessel,
          C. R. Acad. Sci. 189, 94-6

(29. 3)   Resonance Series,
          W. Kessel,
          C. R. Soc. Polon. Phys. 4, 175-82

(29. 4)   Resonance Series,
          W. Kessel,
          C. R. Soc. Polon. Phys. 4, 183-91

(31. 5)   Fluorescence,
          J. Genard,
          Bull. Cl. Sci. Acad. Roy. Belg. 17, 1241-8

(31. 6)   Resonance Series,
          A. Legros,
          Bull. Cl. Sci. Acad. Roy. Belg. 17, 816-22

(31. 7)   Resonance Series,
          J. Pierard,
          Bull. Cl. Sci. Acad. Roy. Belg. 17, 974-9

(31. 8)   Resonance Series,
          P. Swings and J. Genard,
          Bull. Cl. Sci. Acad. Roy. Belg. 17, 1099-106

(32. 9)   Resonance Series,
          J. Pierard,
          Bull. Cl. Sci. Acad. Roy. Belg. 18, 180-5

(32.10)   Resonance Series,
          J. Pierard and M. Migeotte,
          Bull. Cl. Sci. Acad. Roy. Belg. 18, 246-55

(32. 11) Absorption and Predissociation,
          E. Hirschlaff,
          Z. Physik 75, 315-24

(33. 12) Fluorescence Excited by Magnetic Field,
          R. Smoluchewski,
          Z. Physik 85, 191-200

(34. 13) Resonance Series,
          W. Kessel,
          Acta Phys. Polon. 3, 505-12

(34. 14) Induced Predissociation,
          V. Kondratjew and A. Lauris,
          Z. Physik 92, 741-6

(35. 15) Effect of Magnetic Field,
          I. I. Agarbiceanu,
          C. R. Acad. Sci. 200, 385-6

(35. 16) B→X System, Vibrational Analysis, Isotope Effect,
          E. Olsson,
          Z. Physik 95, 215-20

(36. 17) Fluctuations,
          M. Desirant and A. Minne,
          Bull. Cl. Sci. Acad. Roy. Belg. 22, 646-58

(36. 18) Fluctuations,
          B. Rosen and F. Bouffioux,
          Bull. Cl. Sci. Acad. Roy. Belg. 22, 885-93

(36. 19) Fluctuations,
          M. Desirant and A. Minne,
          C. R. Acad. Sci. 202, 1272-3

(36. 20) Dissociation Energy,
          P. Goldfinger, W. Jeunehomme, and B. Rosen,
          Nature 138, 205-6

(36. 21) Te + Te Recombination Continuum,
          R. Rompe,
          Phys. Z. 37, 807-8

(36. 22) Te + Te Recombination Continuum,
          R. Rompe,
          Z. Phys. 101, 214-33

(37. 23) Resonance Series. Isotope Effect,
          B. Rosen and J. Mat,
          Bull. Cl. Sci. Acad. Roy. Belg. 23, 626-45

(37.24)   Induced Predissociation by Magnetic Field,
          E. Olsson,
          C. R. Acad. Sci. 204, 1182-4

(37.25)   Electron Diffraction,
          L. R. Maxwell and V. M. Mosley,
          Phys. Rev. 51, 648

(38.26)   Fluorescence Polarization,
          S. Mrozowski,
          Acta Phys. Polon. 7, 45-8

(38.27)   Ultraviolet System,
          C. Shin-Piaw,
          Ann. Physique 10, 173-290

(38.28)   B - X System, Vibrational Analysis, Isotope Effect; Effect of Magnetic
          Field on Induced Predissociation,
          E. Olsson,
          Thesis, Stockholm

(40.29)   Electron Diffraction,
          L. R. Maxwell and V. M. Mosley,
          Phys. Rev. 57, 21-3

(40.30)   Te$_2$ $\rightleftarrows$ 2Te; Dissociation Energy,
          H. Zeise,
          Z. Elektrochem. 46, 38-41

(42.31)   Determination of $x_e\omega_e$,
          R. Migeotte,
          Bull. Soc. Roy. Sci. Liege 11, 48-53

(42.32)   Ultraviolet System Between 2450-1950Å,
          R. Migeotte,
          Mem. Soc. Roy. Sci. Liege 5, 549-75

(42.33)   Potential Energy, Internuclear Distance,
          Y. Tanaka and T. Takamine,
          Sci. Papers Inst. Phys. Chem. Res. Japan, 39, 437-46

(44.34)   Predissociation in the B State,
          R. Migeotte and B. Rosen,
          Bull. Soc. Roy. Sci. Liege 13, 248-54

(45.35)   Induced Predissociation,
          B. Rosen,
          Phys. Rev. 68, 124-6

Te$_2$

(66.36) R. P. duParcq and R. F. Barrow,
"The Internuclear Distance in the Te$_2$ Molecule,"
Chem. Com. 9, 270

(66.37) R. F. Pottie,
"Cross Sections for Ionization by Electrons. I. Absolute Ionization
Cross Sections of Zn, Cd, and Te$_2$, II. Comparisons of Theoretical
With Experimental Values for Atoms and Molecules,"
J. Chem. Phys. 44, 916-22

(66.38) J. Drowart and P. Goldfinger,
"The Dissociation Energies of the Group VIA Diatomic Molecules,"
Quant. Rev. 20, 545-57

(67.39) P. Budininkas,
"Dissociation Energies of Gaseous Diatomic Sulfur, Selenium and
Tellurium,"
Thesis, Illinois Institute of Technology

(68.40) P. Budininkas,
"Dissociation Energies of Group VIA Gaseous Homonuclear Diatomic
Molecules. III. Tellurium,"
J. Chem. Phys. 48, 2870-3

(69.41) B - X System of ^{130}Te$_2$, Rotational Analysis,
B. L. Jha and D. R. Rao,
Chem. Phys. Letters 3, 175-6

(69.42) J. Berkowitz and W. A. Chupka,
"Photoionization of High-Temperature Vapors. VI. S$_2$, Se$_2$, and
Te$_2$,"
J. Chem. Phys. 50, 4245-50

(69.43) A, B - X Systems of ^{128}Te$_2$, ^{130}Te$_2$, Rotational Analysis,
R. P. duParcq,
Thesis, Oxford

(69.44) D. J. Meschi and A. W. Searcy,
"Investigation of the Magnetic Moments of S$_2$, Se$_2$, Te$_2$ and Se$_5$ by
the Stern-Gerlach Magnetic Deflection Method,"
J. Chem. Phys. 51, 5134-8

(69.45) B. L. Jha, K. V. Subbaram, and D. R. Rao,
"Electronic Spectra of ^{130}Te$_2$ and ^{128}Te$_2$,"
J. Molec. Spectrosc. 32, 383-97

(71. 46) E. O. Degenkolb, H. Mayfarth, and J. I. Steinfeld,
"Laser-Excited Fluorescence of Tellurium Vapor,"
Chem. Phys. Letters 2, 288-90

(71. 47) B. Rai, J. Singh, and D. K. Rai,
"Dissociation Energies of S_2, SO, Te_2, SeO and P_2,"
Israel J. Chem. 9, 563-8

(72. 48) R. F. Barrow and R. P. duParcq,
"Rotational Analysis of the A 0_u^+, B 0_u^+ - X 0_g^+ Systems of Gaseous Te_2,"
Proc. Roy. Soc. London A 327, 279-287

(72. 49) K. K. Yee and R. F. Barrow,
"Observations on the Absorption and Fluorescence Spectra of Gaseous Te_2,"
J. Chem. Soc. Faraday Trans. II 68, 1397-1403

(n.p. 50) R. F. Barrow,
Private communication (see 72.49)

Th$_2$

Th$_2$

Spectroscopic Constants

Dissociation energy = 2.95 ± 0.35 eV, 68 kcal/mole, 24000 cm^{-1} (69.1).

BIBLIOGRAPHY

(69. 1)   K. A. Gingerich,
"Gaseous Metal Borides. I. On the Dissociation Energy of the
Molecules ThB, ThP, and Th$_2$, and Predicted Dissociation Energies
of Selected Diatomic Transition-Metal Borides,"
High Temp. Sci. 1, 258-267

Ti$_2$

Spectroscopic Constants

Dissociation energy = 1.15 ± 0.17 eV, 28.3 kcal/mole, 9000 cm^{-1} (69.2).

BIBLIOGRAPHY

(64. 1)   A. Kant and B. Strauss,
          "Dissociation Energies of Diatomic Molecules of the Transition
          Elements. II. Titanium, Chromium, Manganese, and Cobalt,"
          J. Chem. Phys. 41, 3806-8

(69. 2)   A. Kant and S. Lin,
          "Dissociation Energies of Ti$_2$ and V$_2$,"
          J. Chem. Phys. 51, 1644-7

(69. 3)   K. A. Gingerich,
          "Gaseous Metal Borides. I. On the Dissociation Energy of the
          Molecules ThB, ThP, and Th$_2$, and Predicted Dissociation Energies
          of Selected Diatomic Transition-Metal Borides,"
          High Temp. Sci. 1, 258-67

Tl$_2$

Tl$_2$

Methods of Production and Experimental Technique

Absorption.

Emission from a hollow chathode and a King furnace.

Band Systems

Five groups of bands have been observed in emission and absorption (65.5, 65.4, 31.2, 31.1).

I.   "Red System" - 6500-4900Å

λ in emission (65.5):

v', v''	0	1	2	3	4
0	6320.3	6375.5	6428.4	6483.0	6537.2
1	6285.4	6339.1	6393.0		
2	6252.0				

The conclusions on the origin of this band system are uncertain. Initial investigation gives $\omega' \sim 88$ cm^{-1} and $\omega'' \approx 136$ cm^{-1} (65.5).

II.   4635-3680Å System

Emission

In emission, the band head appears to be at λ ~ 3770.7Å, with band maxima at:

λ = 4635|4405|4308|4237|4187|4133|4047|4004

diffuse and weak maxima at:

λ = 3923|3857|3800

Absorption

Extensive tables of lines seen in absorption (4400-4200Å) are given in (65.5). There are two tentative assignments given to some of them.

Assignment I:

v', v''	0	1	2	3	4	5	6	7	8
0	4269.9	4287.1	4302.2	4322.2	4340.3				
1		4263.7		4299.1	4360.2	4335.4	4354.2	4372.4	4390.3
2			4251.6	4276.8	4293.9				
3				4255.3	4271.9				
4					4250.8				

Tl$_2$

Assignment II:

v', v''	0	1	2	3	4	5	6	7
0	4400.2	4419.0						
1		4394.3	4412.6	4431.9				
2		4370.2		4406.5	4425.1			
3					4401.9	4420.4		
4						4396.0	4414.2	
5						4372.4	4390.3	4408.7

III. <u>3776-3260Å System</u>

Bands are symmetrical around the lines at 3529 and 3519Å. Maxima at ~ 3600Å.

IV. <u>2850-2740Å System</u>

Bands are asymmetrical around the 2768Å line with an apparent head at 2766.3Å.

V. <u>Visible Continua - 2768Å System</u>

This system arises from the broadening of the lines 3230, 3092, 2922-2919Å. Maxima at λ ~ 3446|3156|3050Å.

<u>Spectroscopic Constants</u>

Dissociation energy = <0.9 eV, <21 kcal/mole, <7300 cm^{-1} (57.3).

BIBLIOGRAPHY

(31. 1) Hollow Cathode,
H. Hamada,
Nature 127, 555

(31. 2) Hollow Cathode,
H. Hamada,
Philos. Mag. 12, 50-67

(57. 3) Dissociation Energy by Mass Spectra,
J. Drowart and R. E. Honig,
J. Chem. Phys. 61, 980-5

(65. 4) Thermal Emission and Absorption Bands,
D. E. S. Ginter, M. L. Ginter, and K. K. Innes,
J. Phys. Chem. 69, 2480-3

(65. 5) D. E. S. Ginter,
"Electronic Spectra of the Homonuclear Molecules of the Group III
Metals,"
Thesis, Vanderbilt University

Tm$_2$

Tm$_2$

Spectroscopic Constants

Dissociation energy = 0.52 ± 0.17 eV, 12 kcal/mole, 4200 cm^{-1} (72. 2).

BIBLIOGRAPHY

(71. 1)   S. Lin and A. Kant,
          "Dissociation Energies of Diatomic Rare Earth Molecules Dy_2, Ho_2,
          Er_2, Tm_2, and Yb_2,"
          Army Materials and Mechanics Research Center, TR No. 71-34

(72. 2)   A. Kant and S. Lin,
          "Dissociation Energies of Homonuclear Diatomic Molecules of the
          Rare Earths,"
          Monatshefte für Chemie 103, 757-63

U$_2$

Spectroscopic Constants

Dissociation energy = 1.73 ± 0.43 eV, 40 kcal/mole, 14000 cm^{-1} (69.1).

U_2

BIBLIOGRAPHY

(69. 1)   K. A. Gingerich,
          "Gaseous Metal Borides. I. On the Dissociation Energy of the
          Molecules ThB, ThP, and Th_2, and Predicted Dissociation Energies
          of Selected Diatomic Transition-Metal Borides, "
          High Temp. Sci. 1, 258-67

V$_2$

Spectroscopic Constants

Dissociation energy = 2.49 ± 0.13 eV, 57.5 kcal/mole, 20100 cm^{-1} (69.1).

V_2

BIBLIOGRAPHY

(69. 1)   A. Kant and S. Lin,
          "Dissociation Energies of Ti_2 and V_2,"
          J. Chem. Phys. 51, 1644-7

Xe$_2$

Methods of Production and Experimental Technique

Absorption.

Emission from electron beam discharge, laser pumping, **α** particles, x rays.

BAND SYSTEMS

	System	Transition	Sources	Wavelength Limits	Degrading	Band Head, $\nu_{0,0}$	Remarks	Bibliography
	I	?	Electron beam, X rays	5000-2600			Continuum	(67.7)
	II	$^{1,3}\Sigma_u^+ \rightleftharpoons X^1\Sigma_g^+$ $\begin{pmatrix} 0_u^+ - 0_g^+ \\ 1_u^- - 0_g^+ \end{pmatrix}$	Electron beam	2250-1470			Continuum	(74.33, 72.14, 65.4, 55.3, 55.2)
	III	$^1\Sigma_u^+ \leftarrow X^1\Sigma_g^+$ $\left(0_u^+ - 0_g^+\right)$	Electron beam	1305-1295				(74.33, 72.14)
	IV	$^1\Sigma_u^+ \leftarrow X^1\Sigma_g^+$ $\left(0_u^+ - 0_g^+\right)$	Electron beam	1207-1192				(74.33, 72.14)
	V	$^3\Sigma_u^+ \leftarrow X^1\Sigma_g^+$ $\left(1_u^- - 0_g^+\right)$	Electron beam	1192-1191				(74.33, 72.14)

Xe_2

II. $\quad \underline{^{1,3}\Sigma_u^+ \rightleftharpoons X^1\Sigma_g^+ \left(0_u^+, \ 1_u - 0_g^+\right) \text{Systems}}$

Upper state correlated to $5p^6 \ ^1S_0 + 6s(3/2)_1^0$ (74.33, 72.14).

III. $\quad \underline{^1\Sigma_u^+ \leftarrow X^1\Sigma_g^+ \left(0_u^+ - 0_g^+\right) \text{System}}$

Upper state correlated to $5p^6 \ ^1S_0 + 6s'(1/2)_1^0$ (74.33, 72.14).

IV. $\quad \underline{^1\Sigma_u^+ \leftarrow X^1\Sigma_g^+ \left(0_u^+ - 0_g^+\right) \text{System}}$

Upper state correlated to $5p^6 \ ^1S_0 + 5d(3/2)_1^0$ (74.33, 72.14).

SPECTROSCOPIC CONSTANTS

State	T_e	ω_e	$x_e \omega_e$	B_e	$\alpha_e \times 10^3$	$D_e \times 10^6$	r_e	Remarks	Bibliography
$X\,^1\Sigma_g^+$ (0_g^+)	0	~ 21.26	~ 0.75	~ 0.013	~ 0.4		~ 4.45	$y_e \omega_e \sim 0.008$	(70.9)

Dissociation energy $\sim 2.4 \times 10^{-2}$ eV, 0.55 kcal/mole, 192.02 cm^{-1} (70.9).

Xe_2

Perturbations and General Information

Quenching of Xe_2 $^{1,3}\Sigma_u^+$ by Xe: $\sigma \approx 10^{-17}$ cm^2 (73.25).

Laser action observed on the $^{1,3}\Sigma_u^+ \rightarrow X^1\Sigma_g^+$ transition at 1720 ± 10Å (74.36, 74.31, 74.30, 73.28, 73.23, 73.22, 73.21, 73.20, 73.19, 73.18).

Radiative lifetime of $^{1,3}\Sigma_u^+ - X^1\Sigma_g^+$

$$\tau = 23 \text{ nsec } (74.32)$$
$$= 130 \text{ nsec } (73.18).$$

Potential energy curves - estimated (70.10):

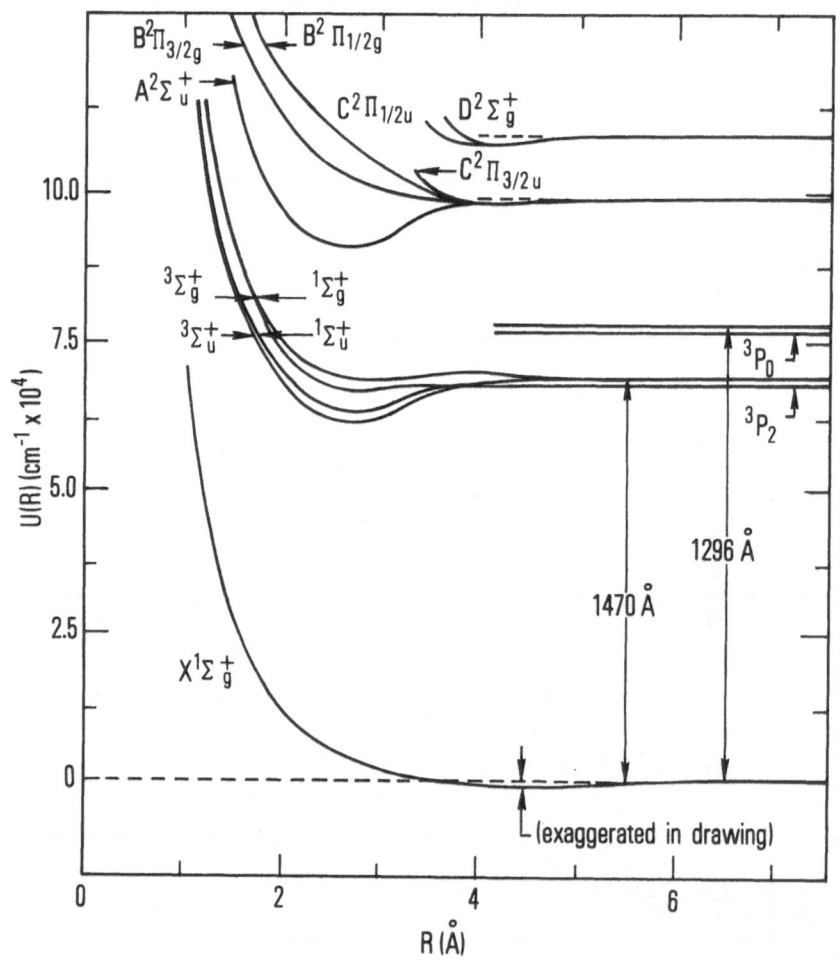

BIBLIOGRAPHY

(54. 1) Y. Tanaka and M. Zelikoff,
"Continuous Emission Spectrum of Xenon in the Vacuum Ultraviolet
Region,"
J. Opt. Soc. Am. 44, 254-5

(55. 2) Condensed Discharge,
Y. Tanaka,
J. Opt. Soc. Am. 45, 710-3

(55. 3) Weak Discharge,
P. G. Wilkinson and Y. Tanaka,
J. Opt. Soc. Am. 45, 344-9

(65. 4) Microwave Discharge,
P. G. Wilkinson and E. T. Byram,
Applied Optics 4, 581-8

(67. 5) J. F. Prince and W. W. Robertson,
"Visible Continua in Xenon, Krypton, and Neon,"
J. Chem. Phys. 46, 3309-13

(67. 6) P. G. Wilkinson,
"The Mechanism of the Argon Emission Continuum in the Vacuum
Ultraviolet. I,"
Can. J. Phys. 45, 1715-27

(67. 7) B. Brocklehurst,
"Luminescence of Gases Excited by High Energy Radiation. Part 2.
Emission Spectra of Molecular Xenon,"
Trans. Faraday Soc. 63, 274-81

(68. 8) L. L. Nichols and W. Vali,
"Pressure Dependence of the Xenon Continuum Radiation Decay Rate,"
J. Chem. Phys. 49, 814-7

(70. 9) J. P. Morucci and A. Lansiart,
"Investigation of Light Emission by Electron Avalanches in Binary
Mixtures Where Xenon is the Parent Constituent. Attempted Interpre-
tation of the Visible Continuum Emitted by Xenon,"
IEEE Trans. Nucl. Sci. 17, 95-106

(70. 10) R. S. Mulliken,
"Potential Curves of Diatomic Rare-Gas Molecules and Their Ions,
With Particular Reference to Xe$_2$,"
J. Chem. Phys. 52, 5170-80

(71.11) I. V. Kozinskaya and L. P. Polozova,
"Molecular Absorption of Xenon in the Vacuum Ultraviolet,"
Optics & Spectroscopy 30, 853-8

(71.12) C. G. Freeman, M. J. McEwan, R. F. C. Claridge, and L. F. Phillips,
"Band Fluorescence of Xenon,"
Chem. Phys. Letters 10, 530-2

(71.13) Xe$_2$ Laser,
N. G. Basov, V. A. Danilychev, and Yu. M. Papov,
Sov. J. Quant. Electron. 1, 18-20

(72.14) M. C. Castex and N. Damany,
"Absorption Spectrum of the Xenon Molecule in the Vacuum Ultra-
violet Region,"
Chem. Phys. Letters 13, 158-161

(72.15) O. Cheshnovsky, B. Raz, and J. Jortner,
"Temperature Dependence of Rare Gas Molecular Emission in the
Vacuum Ultraviolet,"
Chem. Phys. Letters 15, 475-9

(72.16) Xe$_2$ Laser,
H. A. Koehler, L. J. Ferderber, D. L. Redhead, and P. J. Ebert,
Appl. Phys. Letters 21, 198-9

(73.17) A. Gedanken, B. Raz, and J. Jortner,
"Emission Spectra of Homonuclear Diatomic Rare Gas Molecules in
Solid Neon,"
J. Chem. Phys. 59, 1630-3

(73.18) S. C. Wallace, R. T. Hodgson, and R. W. Dreyfus,
"Excitation of Vacuum Ultraviolet Emission From High-Pressure
Xenon by Relativistic Electron Beams,"
Appl. Phys. Letters 23, 22-24

(73.19) S. C. Wallace, R. T. Hodgson, and R. W. Dreyfus,
"Short Pulse Excitation of a Xenon Molecular Dissociation Laser at
172.9 nm by Relativistic Electrons,"
Appl. Phys. Letters 23, 672-4

(73.20) E. R. Ault, M. L. Bhaumik, W. M. Hughes, R. J. Jensen, C. P. Robinson,
A. C. Kolb, and J. Shannon,
"Xenon Molecular Laser in the Vacuum Ultraviolet,"
IEEE J. Quant. Electronics 10, 1031-2

(73.21) J. B. Gerardo and A. W. Johnson,
"1730-Å Radiation Dominated by Stimulated Emission From High-Pressure Xenon,"
J. Appl. Phys. **44**, 4120-4

(73.22) M. Novaro and F. Lagarde,
"Evidence of Stimulated Emission From Xenon Under Pressure,"
C. R. Acad. Sci. B **277**, 671-3

(73.23) P. W. Hoff, J. C. Swingle, and C. K. Rhodes,
"Demonstration of Temporal Coherence, Spatial Coherence, and Threshold Effects in the Molecular Xenon Laser,"
Optics Comm. **8**, 128-131

(73.24) F. H. Mies,
"Stimulated Emission and Population Inversion in Diatomic Bound-Continuum Transitions,"
Molec. Phys. **26**, 1233-46

(73.25) D. J. Bradley, M. H. R. Hutchinson, and H. Koetser,
"Quenching of Vacuum Ultraviolet Fluorescence Emission From Electron Beam Excited Quasi-Molecular Xenon,"
Optics Comm. **7**, 187-190

(73.26) P. W. Hoff, J. C. Swingle, and C. K. Rhodes,
"Observations of Stimulated Emission From High-Pressure Krypton and Argon/Xenon Mixtures,"
Appl. Phys. Letters **23**, 245-6

(73.27) K. K. Docken and T. P. Schafer,
"Spectroscopic Information on Ground-State Ar_2, Kr_2, and Xe_2 From Interatomic Potentials,"
J. Molec. Spectrosc. **46**, 454-9

(73.28) S. E. Harris, A. H. Kung, E. A. Stappaerts, and J. F. Young,
"Stimulated Emission in Multiple-Photon-Pumped Xenon and Argon Excimers,"
Appl. Phys. Letters **23**, 232-4

(73.29) O. Cheshnovsky, B. Raz, and J. Jortner,
"Electronic Energy Transfer in Rare Gas Mixtures,"
J. Chem. Phys. **57**, 3301-7

(74.30) W. M. Hughes, J. Shannon, and R. Hunter,
"Efficient High-Energy-Density Molecular Xenon Laser,"
Appl. Phys. Letters **25**, 85-7

(74.31) A. W. Johnson and J. B. Gerardo,
"Diluent Cooling of a Vacuum-Ultraviolet High-Pressure Xenon Laser,"
J. Appl. Phys. 45, 867-72

(74.32) W. H. Weihofen,
"Spontaneous Decay of the "$^3\Sigma_u^+$" State of Xe$_2$,"
J. Chem. Phys. 60, 445-53

(74.33) M. C. Castex and N. Damany,
"High Resolution Spectrum of Xe$_2$ in the Vacuum Ultraviolet Region.
Molecular Systems Related to the Two Lower Resonance Lines,"
Chem. Phys. Letters 24, 437-40

(74.34) A. G. Molchanov and Yu. M. Popov,
"On the Feasibility of Electroionizational Excitation of the Vacuum
Ultraviolet Generation of Compressed Xenon,"
Kvant. Elekt. 1, 1122-8

(74.35) C. W. Werner, E. V. George, P. W. Hoff, and C. K. Rhodes,
"Dynamic Model of High-Pressure Rare-Gas Excimer Lasers,"
Appl. Phys. Letters 25, 235-8

(74.36) D. J. Bradley, D. R. Hull, M. H. R. Hutchinson, and M. W. McGeoch,
"Megawatt VUV Xenon Laser Employing Coaxial Electron-Beam
Excitation,"
Optics Comm. 11, 335-8

(74.37) S. C. Wallace and R. W. Dreyfus,
"Continuously Tunable Xenon Laser at 1720Å,"
Appl. Phys. Letters 25, 498-500

$$\text{Y}_2$$

Spectroscopic Constants

Dissociation energy = 1.62 ± 0.22 eV, 37.3 kcal/mole, 13050 cm^{-1}.

Y_2

BIBLIOGRAPHY

(69. 1) K. A. Gingerich,
"Gaseous Metal Borides. I. On the Dissociation Energy of the
Molecules ThB, ThP, and Th$_2$, and Predicted Dissociation Energies
of Selected Diatomic Transition-Metal Borides,"
High Temp. Sci. 1, 258-67

Yb$_2$

Methods of Production and Experimental Technique

Knudsen cell effusion.

Spectroscopic Constants

Dissociation energy = 4 ± 4 eV, 92 kcal/mole, 32000 cm^{-1} (72.3).

Yb$_2$

BIBLIOGRAPHY

(71. 1)   S. Lin and A. Kant,
          "Dissociation Energies of Diatomic Rare Earth Molecules Dy$_2$, Ho$_2$,
          Er$_2$, Tm$_2$, and Yb$_2$,"
          Army Materials and Mechanics Research Center, TR No. 71-34

(72. 2)   A. Kant and S. Lin,
          "Dissociation Energies of Homonuclear Diatomic Molecules of the
          Rare Earths,"
          Monatshefte für Chemie 103, 757-63

(72. 3)   M. Guido and G. Balducci,
          "Dissociation Energy of Yb$_2$,"
          J. Chem. Phys. 57, 5611-2

Zn_2

Methods of Production and Experimental Technique

Absorption.

Emission (Tesla coil, hollow cathode).

Fluorescence.

BAND SYSTEMS

	System	Transition	Sources	Wavelength Limits	Degrading	Maximum (λ) in Emission	Remarks	Bibliography
	I		Emission	5350-3890		4450	Continuum	(31.6, 31.5)
	II		Emission	3893-3776		3787	Continuum	(31.6, 31.5)
	III		Emission Absorption	3763-2936		3688	Continuum	(31.6, 31.5, 29.2)
	IV		Absorption Emission Fluorescence	3073-2002		2550	Continuum	(31.6, 31.5, 31.4, 29.2)

Zn_2

III.   3763-2936Å System

Emission

In emission maximum is at λ = 3688Å (31.6, 31.5) and line broadens at 3076A (31.6, 31.5).

Bands superimposed λ |3749|3724|3706|3688|3575|3522|3483|
                                 3454|3431|3411|3052|

Absorption

In absorption bands are without structure and maxima is at ~3050Å (31.6, 29.2).

IV.   3073-2002Å System

Emission (31.6, 31.5)

In emission continuous bands are 2826-2035Å, maximum is at 2550Å, line broadens at 2139Å, and diffuse bands are at λ ~ 2002Å.

Absorption (31.6, 29.2)

In absorption continuous bands are at 2550-2002Å, maxima are at λ = 2139, 2064, and 2002Å, and the line broadens at 2139Å.

Fluorescence (31.3)

Numerous bands in the region 3073-2456Å.

Spectroscopic Constants

Dissociation energy = 0.25 eV(?), 6 kcal/mole(/), 2100 cm^{-1}.

BIBLIOGRAPHY

(29. 1)   Ultraviolet Absorption,
          J. G. Winans,
          Philos. Mag. 7, 555-66

(29. 2)   Ultraviolet Absorption,
          J. M. Walter and S. Barratt,
          Proc. Roy. Soc. A 122, 201-10

(31. 3)   Fluorescence,
          W. Kapuscinski,
          C. R. Soc. Polon. Phys. 5, 401-8

(31. 4)   Emission,
          H. Hamada,
          Nature 127, 555

(31. 5)   Ultraviolet and Visible Emission,
          H. Hamada,
          Philos. Mag. 12, 50-67

(31. 6)   Absorption and Emission, 2139, 2064, 2002Å,
          J. G. Winans,
          Phys. Rev. 37, 902

(33. 7)   R. Siksna,
          Acta Phys. Polon. 2, 253-65

(35. 8)   Van der Waals Molecular Theory,
          W. Finkelnburg,
          Z. Physik 96, 699-713